Ascorbic Acid
in
Aquatic
Organisms

Status and Perspectives

Ascorbic Acid
in
Aquatic Organisms

Status and Perspectives

Edited by
Konrad Dabrowski

CRC Press is an imprint of the
Taylor & Francis Group, an **informa** business

CRC Press
Taylor & Francis Group
6000 Broken Sound Parkway NW, Suite 300
Boca Raton, FL 33487-2742

First issued in paperback 2019

© 2001 by Taylor & Francis Group, LLC
CRC Press is an imprint of Taylor & Francis Group, an Informa business

No claim to original U.S. Government works

ISBN-13: 978-0-8493-9881-0 (hbk)
ISBN-13: 978-0-367-39782-1 (pbk)

Library of Congress Card Number 00-052908

Library of Congress Cataloging-in-Publication Data

Ascorbic acid in aquatic organisms / Edited by Konrad Dabrowski.
      p. cm.
  Includes bibliographical references (p.).
  ISBN 0-8493-9881-9 (alk. paper)
  1. Fishes -- Nutrition. 2. Aquatic organisms--Nutrition. 3. Vitamin C. I. Title

SH156 .D23 2000
639.3--dc21
                                                   00-052908
                                                              CIP

Visit the Taylor & Francis Web site at
http://www.taylorandfrancis.com

and the CRC Press Web site at
http://www.crcpress.com

Preface

This book is written for biologists interested in aquatic ecosystems who would like to include the role of individual nutrients in their comprehensive overview. It is only the beginning in the process of discovery of the importance of the role vitamin C plays in the "health" of natural and man-made ecosystems. I have often been asked by otherwise experienced scientists, who have taught many generations of students about nitrogen and phosphorus cycles, "Where do wild fish obtain their vitamin C?" Synthesis of ascorbic acid in terrestrial plants is well understood, but algae as the only source of ascorbate in the aquatic food pyramid has not been explored. The origin of ascorbate in the animal kingdom has not been studied, and the possible role of the plant/animal symbiosis[1] to provide essential components, to my knowledge, has been largely ignored. King[2] was probably the first who asked an evolutionary question, "Does the loss of one or more genes that permit biological systhesis of the antiscorbutic vitamin.... occur only once, or more often in nature?" There is an expanding field of the culture of aquatic organisms which demand feeds formulated to best meet their requirements for health, growth, and reproduction. With over 100 species now cultured for human consumption, the question arises as to whether we have reliable information for the dietary needs for vitamin C for even a small percentage of these organisms.

Ascorbic Acid in Aquatic Organisms was written as the offshoot of a special session on aquatic organisms and practical aquaculture at the 1998 World Aquaculture Society (WAS) meeting. At that time, the controversies surrounding the definitive scientific information concerning the synthesis, presence, and role of ascorbic acid in fish and crustaceans reached their peak. The intent of the meeting was to present critical "state-of-the-art" knowledge in specific areas of vitamin C related to aquatic organisms. From the onset of the project we were criticized for not including aquatic birds and mammals. I hope that the content of this book will partly answer this criticism. We had a lot of areas to cover, and a more comprehensive approach would probably shift the balance. Therefore, this book is in many ways a "fishing expedition" to the sea of vitamins.

Controversy brings attention, and the scientific community responds by contributing proportionally more to the specific research area. According to the Institute for Scientific Information (ISI)[3] records, the number of research articles published under the key words "vitamin C and fish" averaged 1 to 3 annually in the 1980s, increased to 5 to 15 in the early 90s, and has been steady in the last five years at 25 to 32 research articles per year. Understandably, some contributions add fire to the controversy. By stating that a review takes a critical approach, I expect the author to analyze scientific results and follow a scientific argument to disprove rather than ignore a theory. This is a more difficult task than producing a new hypothesis or new results. It is for the reader to judge whether this has been accomplished.

This book answers whether the need for a critical review of vitamin C in aquaculture. It provides evidence for establishing requirements for particular species and age classes, and questions the validity of some published results because of inadequacy of analytical methods or experimental designs (duration, diet formulations). The book is a combination of good nutrition research and practical guidelines for modern aquaculture practitioners.

We also attempted to challenge researchers and identify new directions in the areas of comparative physiology, subcellular and molecular biology, and possibly aquatic ecology and related global environmental changes. Those who will refer to this book as inspiration would be our dearest friends.

Konrad Dabrowski

1. Dykens, J. A. and Shick, J. M., Oxygen production by endosymbiotic algae controls superoxide dismutase activity in their animal host, *Nature*, 297, 259, 1982.
2. King, C. G., The discovery and chemistry of vitamin C, *Proc. Nutr. Soc.*, 12, 219, 1953.
3. Institute for Scientific Information *(www.isinet.com)*.

Acknowledgments

The session organized during the WAS meeting was possible because of funds made available by a grant from F. Hoffmann La Roche, Switzerland, and particularly the effort of Stewart Anderson. We are grateful to many staff members of the School of Natural Resources (The Ohio State University) for their editorial and secretarial management. Thanks are due to Ms. Pat Polczynski, Ms. Pat Patterson, Ms. Lorrie Schnitzspahn, and Ms. Kellie DiFrischia.

My great thanks are specially due to Mr. John Sulzycki, who "coached" me through the final stages of the book's production process and without whose persuasion, push, and "strong Polish vocabulary" (I guess only your native language makes lasting impressions) this could have taken longer.

Konrad Dabrowski
October 2000
Columbus, OH

Contributors

Nina Åkerblom
Department of Zoophysiology
Goteborg University
Goteborg, Sweden

Sungchul C. Bai
Department of Aquaculture
Pukyong National University
Pusan, Republic of South Korea

Hans Börjeson
Swedish Salmon Research Institute
Sweden

Malcolm R. Brown
CSIRO Division of Marine Research
Tasmania, Australia

Gaetano Caricato
Department of Animal Production
University of Basilicata
Potenza, Italy

Stefano Cecchini
Department of Animal Production
University of Basilicata
Potenza, Italy

Konrad Dabrowski
School of Natural Resources
The Ohio State University
Columbus, OH

Samuel P. Felton
School of Fisheries
University of Washington
Seattle, WA

Lars Förlin
Swedish Salmon Research Institute
Sweden

Jacques Gabaudan
Centre de Recherche en Nutrition
  Animale
Societe Chimique Roche S.A.
Saint-Louis Cedex, France

Marie F. Gouillou-Coustans
Unite Mixte INRA
Fish Nutrition Laboratory
Plouzane, France

Jane Loren Halver
Seattle, WA

John E. Halver
School of Fisheries
University of Washington
Seattle, WA

Kristin Hamre
Directorate of Fisheries
Institute of Nutrition
Bergen, Norway

Zsigmund Jeney
Fishery Research Institute
Szarvas, Hungary

Sadasivam J. Kaushik
Fish Nutrition Laboratory
Unite Mixte INRA
St. Pee-sur Nivelle, France

Philip H. Klesius
Aquatic Animal Health
Research Laboratory
Auburn, AL

Shunsuke Koshio
Faculty of Fisheries
Kagoshima University
Kagoshima, Japan

Patrick Lavens
Laboratory of Aquaculture and
  Artemia Reference Centre
Gent, Belgium

R. T. Lovell
Department of Fisheries
Auburn University
Auburn, AL

Chhorn Lim
Aquatic Animal Health Research
  Laboratory
Auburn, AL

Amund Maage
Directorate of Fisheries
Institute of Nutrition
Bergen, Norway

Michele Maffia
Laboratory of General Physiology
Department of Biology
University of Lecce
Lecce, Italy

Jane A. McLaughlin
Marine Biological Laboratory
Woods Hole, MA

Régis Moreau
Linus Pauling Institute
Oregon State University
Corvallis, OR

Leif Norrgren
Department of Pathology
Swedish University of
  Agricultural Sciences
Uppsala, Sweden

Marco Saroglia
Department of Animal Production
University of Basilicata
Potenza, Italy

Shi-Yen Shiau
Department of Food Science
National Taiwan Ocean University
Keelung, Taiwan

Craig A. Shoemaker
Aquatic Animal Health Research
Laboratory
Auburn, AL

C. Storelli
Laboratory of General Physiology
Department of Biology
University of Lecce
Lecce, Italy

Genciana Terova
Department of Animal Production
University of Basilicata
Potenza, Italy

Viviane Verlhac
Centre de Recherche en Nutrition
  Animale
Societe Chimique Roche S.A.
Saint-Louis Cedex, France

T. Verri
Laboratory of General Physiology
Department of Biology
University of Lecce
Lecce, Italy

Rune Waagbø
Directorate of Fisheries
Institute of Nutrition
Bergen, Norway

Contents

chapter one

Albert Szent-Gyorgyi and the nature of life

Jane A. McLaughlin

Albert Szent-Gyorgyi, a biochemist, followed a lifelong pursuit to understand the nature of life. He received the 1937 Nobel Prize for Physiology and Medicine "for his discoveries in connection with biological combustion processes, with special reference to vitamin C and the catalysis of fumaric acid." His work, carried out in Holland, England, and his native Hungary during the 1920s and 30s, was initiated by his interest in the dispute between Otto Warburg and Heinrick Wieland as to whether hydrogen or oxygen needs to be activated for respiration. He showed that both are needed, that both processes take place. His observations on the oxidative blackening process in a cut piece of potato led him to isolate hexuronic acid, mainly from adrenal glands. After his return to Hungary in 1932, he isolated it from red pepper and paprika, proved it to be Vitamin C, and coined its name, ascorbic acid (AA), for its antiscorbutic action. Rather than continuing with research on Vitamin C, he pursued his interest in biological oxidation. In his studies on C4-dicarboxylic acid, (including fumaric acid) he elucidated the role of H transfer and O_2 utilization involving dehydrogenases and cytochrome oxidase.[1] This discovery later served as the basis for the Hans Krebs citric acid cycle of respiration.

Next, Szent-Gyorgyi turned to muscle to help him understand life. He wrote ". . . it does not matter which material we chose for our study of life, be it grass or muscle, virus or brain. If we only dig deep enough we always arrive at the center, the basic principles on which life was built and due to which it still goes on."[2] During the Second World War, in Hungary, he and his team made important discoveries in muscle biochemistry: the discovery that myosin was actually made up of actin and myosin and that actin is an acti-

0-8493-9881-9/01/$0.00+$.50

vator of the ATP-splitting activity of myosin. In 1949, in the U.S., at the Marine Biological Laboratory in Woods Hole, he permeabilized the muscle membrane with glycerol, added ATP, and watched the muscle contract, thus providing the first proof that the contractile system consists of actin, myosin, and ATP.[2] As early as 1940, he presented preliminary evidence of the presence of myosin in a number of non-muscle cells, and wrote that "a protein fraction, analogous to myosin, should be found in any cell."[3] (This was prior to the discovery that myosin contains a second protein, actin.) Today we know that myosin and actin occur in every type of cell and contribute to a variety of cellular functions. In 1954, he was the recipient of the Albert Lasker Award for his contributions to the understanding of muscle contraction.

He, however, felt that "to understand muscle we have to descend to the electronic level, the rules of which are governed by wave mechanics."[4] But he also recognized that "We will really approach the understanding of life when all structures and functions, all levels, from the electronic to the supramolecular, will merge into one single unit."[4] His observations on the importance of submolecular processes underlying fundamental biochemical and biophysical phenomena thus pioneered the development of submolecular biology. He investigated bioenergetics, triplet states, the role of SH-glutathione, and free radicals.[4] His experimental efforts, however, failed, in the application of these theoretical ideas, to cure cancer. Interestingly, however, there is much current research on the role of free radicals in a number of diseases and the possible use of antioxidants as a preventive measure.

Szent-Gyorgyi was not only a great scientist, he was a warm human being of great moral courage, and good humor. His involvement with all these extraordinarily insightful ideas did not isolate him from the surrounding world. His moral courage was evident in many ways. He objected to experiments being conducted on Italian prisoners of war in World War I. He was outspoken against Hitler and the Communists during their occupation of Hungary. He was instrumental in saving others from oppression and possible death during those regimes.[5] Later, as a U.S. citizen, he expressed concern for mankind in this nuclear age.[6,7,8]

At the end of World War II, officials in Hungary turned over to the U.S., for safe keeping from the Communists, the Crown of St. Stephen, symbol of authority in Hungary. The thousand-years old crown of Hungary's first king was secured at Fort Knox until the late 1970s when a policy decision declared that it was time for it to be returned to Hungary. Szent-Gyorgyi was chosen to escort the crown home to Hungary in 1978. That was his first return to his native land in over 30 years.

His conflict with the Communists had resulted in his departure from Hungary in 1947. When considering where to turn, he remembered a place he had visited during a trip to the U.S. for the International Physiological Congress meeting in Boston in 1929. At the invitation of the Marine Biological Laboratory (MBL) in Woods Hole, an excursion was arranged that brought hundreds of the scientists from that meeting to the renowned MBL. Among other activities of the day, the hosts offered a lobster feast for the visitors. Szent-Gyorgyi remembered the lobsters and the laboratory. In 1947 he returned to the MBL and arranged for a number of his colleagues to join him at what would prove to be his scientific home for almost 40 years.

My acquaintance with Szent-Gyorgyi spanned many years, as his research associate at the MBL. His laboratory was an exciting and stimulating place to work. One worked hard and learned to think about the broader picture while observing experimental details. Laboratory members came together for scientific discussions, at times including leading scientists of the field, visiting at Szent-Gyorgyi's invitation. At other times, they gathered for tea and discussions of general interest, whether scientific, political, or social. Good humor prevailed in the lively group that made up his team.

Szent-Gyorgyi would arrive at the laboratory in the morning eager to share some new idea he had developed overnight. His day was spent at his workbench overlooking the sea, experimenting with his choice of solutions which often resulted in reactions that produced color. He attended to his scientific writings and correspondences at home in the evenings or before breakfast and, in season, before his morning swim.

"Prof," as Szent-Gyorgyi preferred to have us call him ("It takes too much time to say Szent-Gyorgyi."), lived next to the sea in an area where it is possible to see the sun rise and set over water. He swam, fished, and sailed there. His house, The Seven Winds (for it was quite a windy spot), was located near the tip of a peninsula, Penzance Point in Woods Hole, at the southwestern end of Cape Cod. The water passage between a nearby island and the Point is swift and treacherous. Prof, a strong swimmer, gauged the tide flowing through the Woods Hole Passage such that he could swim with the current. He would start from the beach near his house on the one side and swim around the tip of the Point to the other side. He continued to do this until late in his life.

He had a feel for subtle conditions required to catch a fish (Figure 1.1). He could tell, by the combination of the way the currents were flowing and the degree of light at dusk, if he would catch any big fish. He was usually right. Prof said, "I always fish with a big hook; it is more fun not to catch a big fish than not to catch a little fish." This was true of his fishing whether in science or in the sea, and he caught some big fish!

Albert Szent-Gyorgyi said of himself, "I make the wildest theories, connecting up the test tube reaction with the broadest philosophical ideas, but spend most of my time in the laboratory, playing with living matter, keeping my eyes open, observing and pursuing the smallest detail. . . . I must admit that most of the new observations I made were based on wrong theories. My theories collapsed, but something was left afterwards."[5] Prof worked at his bench until shortly before his death in 1986 at age 93. His bibliography[9] covers work spanning three quarters of the 20th century.

Figure 1.1: Dr. Albert Szent-Gyorgi in Woods Hole, around 1959.

References

1. Szent-Gyorgyi, A. In: *Nobel Lectures: Physiology or Medicine, 1922–1941.* Published for the Nobel Foundation by Elsevier, New York. Pages 435–449. 1965.
2. Szent-Gyorgyi, A. *Chemistry of Muscular Contraction.* Academic Press, New York. 1951.
3. Szent-Gyorgyi, A. On protoplasmic structure and function. *Enzymologia Acta Biocatalytica* 9:98–110, 1940.
4. Szent-Gyorgyi, A. *Introduction to a Submolecular Biology.* Academic Press, New York. 1960.
5. Szent-Gyorgyi, A. "Lost in the Twentieth Century." *Annual Review of Biochemistry.* 32:1–14, 1963.
6. Szent-Gyorgyi, A. *Science Ethics and Politics.* Vantage Press, New York. 1963.
7. Szent-Gyorgyi, A. *The Crazy Ape.* Philosophical Library, New York. 1970.
8. Szent-Gyorgyi, A. *What Next?* Philosophical Library, New York. 1971.
9. Publications of Albert Szent-Gyorgyi. *Biological Bulletin* 174:234–240, 1988.

chapter two

Glen King and the Henry Loren family

Jane Loren Halver

Glen King and Henry Loren were students and fraternity brothers at Washington State College in Pullman, WA between 1915 amd 1917. Glen was president of the fraternity and Henry Loren, my father, was studying electrical engineering and playing shortstop for the Cougar Baseball Team.

After graduating, Glen married Hilda and transferred to the University of Pittsburgh for a Ph.D. in Biochemistry. Two years later Henry graduated, married my mother Marieta, and moved to Pittsburgh for two years' resident training at the Westinghouse Training Center for Industrial Electrical Engineering. The Kings and Lorens, friends for several years, lived close together in Pittsburgh. When I was born in 1922, Glen King was my godfather.

Hilda had her son, Bob, at that time, and Marieta and Hilda were close neighbors, visiting often with their newborn children. Glen was busy at the university, isolating and characterizing vitamin C, and at that time urged the Lorens to eat an orange daily to assure vitamin C intake. This habit was adopted by both families for their lifetimes and continues in my home today.

Marieta lived for 101 years, so it must have helped in her health, and Hilda lived over 95 years also on this daily vitamin C intake. In Pittsburgh, I played with Bob and his sister, Dorothy, for two years before moving to Tacoma, Washington, where my father became an industrial engineer for Westinghouse Electric Company. Glen and Hilda visited their relatives in Washington State and also visited us on those occasions.

Dorothy, Bob, and Kendall, the King children, always called my parents "Aunt Marieta and Uncle Henry" during their visits. Sometimes we went to the beaches on Puget Sound to go boating or clam digging. Glen was not a

Figure 2.1: Dr. Charles Glen King during a fishing trip in Maine, Nov. 9, 1948. (Photo courtesy of Dr. Robert King.)

fisherman, but liked to walk in the forest and dig clams on the beach. He was a very quiet, pleasant family man and loyal friend whose company we enjoyed (Figure 2.1).

Later, when I was married to John Emil Halver, Glen (then President of the Nutrition Foundation) was instrumental in the appointment of John as Laboratory Director of the Western Fish Nutrition Laboratory being built by the U.S. Fish and Wildlife Service at Cook, Washington in 1950.

There Glen was most interested in the research on Vitamin C requirements and functions in fish. He and Hilda came to stay with us in our home and review the work at the lab, and Glen was particularly interested in the repair of fish tissues when elevated levels of vitamin C were fed after surgery.

Later, my husband often visited with Glen at meetings and travelled together with him to Atlantic City, New Jersey, where the FASEB meetings were held. Glen and Hilda were also present in Washington, D.C. in 1978 when John signed the membership book of the National Academy of Sciences.

Glen told me that the saddest time of his life was when his son, Kendall, who was following in his footsteps as a biochemist, passed away.

Dorothy and I have corresponded yearly, but have not had the opportunity to meet since childhood.

chapter three

Analytical enigmas in assaying for vitamers C

John E. Halver and Samuel P. Felton

Contents

Abstract

L-ascorbic acid is labile and easily oxidized to dehydroascorbic acid. Early workers carefully extracted tissues using mild treatments in meta-phosphoric acid to protect ascorbic acid. These extraction methods did not extract tissue bound ascorbate. More stringent techniques using stronger acids would extract bound ascorbate esters, but the acidic conditions would hydrolyze the ascorbyl esters yielding variable assay results. A modified technique using 5% TCA for extraction followed by centrifugation, filtration, and injection within one hour resulted in acceptable HPLC assays for several vitamers C. New columns without an ion-binding compound in the mobile phase eluted ascorbic-2-sulfate prior to ascorbic acid, and both compounds were confirmed with electro-ionizing mass spectroscopy.

3.1 Introduction

The vitamers C consist of L-ascorbic acid and derivatives which have ascorbic acid activity in various animals. L-ascorbic acid is labile and oxidized to dehydroascorbic acid (which may be reduced to regenerate active ascorbic acid) and then oxidized further to inactive fragments with no vitamin C activity. Early workers carefully extracted tissues using mild treatments in metaphosphoric acid to stabilize extracted ascorbic acid which was then oxidized to the dehydro form and the diketo form coupled with a chromophore which could be measured spectrophotometrically.[1] When oxidation-stable derivatives of ascorbate were isolated and tested for vitamin C activity, more stringent methods for extraction were necessary to measure tissue concentrations. These included denaturing the protein with trichloracetic acid (TCA) or perchloric acid (PCA) to facilitate extracting the water-soluble components for high-pressure liquid chromatography (HPLC) analysis for the vitamers C and intermediates present.[2] The ascorbyl-2-esters in these extracts were stable for short periods, but were slowly hydrolyzed in the strong acids used. Hot water denaturating the tissues promoted rapid oxidation of ascorbic acid. Microwave tissue denaturation disclosed several vitamers C present, but slow oxidation occurred in the complex mixture of reactants. Homogenizing the tissues in water and adding TCA to make a 5% solution, centrifuging, filtering, and injecting the cold filtrate showed HPLC column separation into L-ascorbic acid, L-ascorbyl-2-sulfate, L-ascorbyl-2-monophosphate, L-ascorbyl-2-diphosphate, and L-ascorbyl-2-triphosphate.[2]

3.2 Vitamers C

L-ascorbic acid is one of the major tissue reducing agents, is very labile, and is easily oxidized to dehydroascorbic acid which can be reduced to regenerate ascorbic acid again for tissue use.[3] Absorption of ascorbate may be by this process with the absorbed dehydroascorbic acid reduced in the circulation system to the active form of vitamin C in fish.[4] More stable derivatives of ascorbic acid have been isolated, and several have anti-scorbutic activity in fish.[3, 5] L-ascorbyl-2-sulfate(C2S) and L-ascorbyl-2-monophosphate(C2MP) have been shown to be active for finfish,[6, 7, 8] and are probably converted to ascorbic acid by hydrolysis. C2S is probably converted to ascorbic acid as needed through circulating feedback control by ascorbyl-2-sulfohydrolase.[9] The L-ascorbyl-2-phosphates(C2MP, C2DP, C2TP) have activity and are easily hydrolyzed to ascorbic acid by tissue phosphatases.[4, 10] The ascorbate-2-glycoside(C2G) has been reported to have vitamin C activity in rats[11] and ascorbate-2-acetate has been identified with a half-life at pH 4.5 of about 24 hours.[12] Other derivatives of ascorbic acid on the carbon 2 and 6 positions may also have activity and need to be included in the list of potential vitamers C.[3, 5]

3.3   Extraction enigmas

Early workers carefully extracted tissues by using mild methods and protected the extracted ascorbic acid with meta-phosphoric acid. The extraction was then deliberately oxidized to form dehydroascorbic acid, and the di-keto structure coupled with a chromophore that could be read spectrophotometrically.[1] Several compounds were used and 2-4-dintrophenylhydrazine (DNPH) has been the method of choice for tissue analysis for vitamin C (reviewed by Dabrowski et al.[4]). Unfortunately, these mild methods of extraction will not detect tissue-bound derivatives of ascorbic acid and more stringent methods must be employed to release the ester derivatives. Earlier techniques used mild extraction at 37 C followed by boiling water to generate "total vitamin C" with the difference estimated as "tissue bound vitamin C."[13, 14] The disadvantage of these indirect techniques was several-fold. Extraction was incomplete and conditions used promoted hydrolysis of extracted esters and interfering compounds that could increase errors in final estimates of either ascorbic acid or of ascorbate derivatives depending upon which fish tissues were assayed. Therefore, efforts were focused upon direct tissue compound analysis using HPLC to identify the results of tissue extraction with combinations of TCA or PCA used for tissue denaturation and ascorbate extraction. TCA at 10% denatured tissue proteins and released the ascorbate esters; however, the sulfate and phosphate esters were rapidly hydrolyzed to ascorbic acid upon standing even at refrigerated temperatures, and HPLC analysis needed to be completed within a few minutes or decreased ester values would be recorded.[2, 15] A compromise technique was developed using ice water homogenization of tissue for 1 minute, followed by dilution with an equal volume of 10% TCA to yield a 5% TCA extraction solution, which was then homogenized for another minute before centrifugation and filtration.[2] The final extract very slowly hydrolyzed the esters, and it was possible to determine the total and ester content of the tissues when assays were conducted within 1 or 2 hours of sample preparation. Microwave denaturation of tissues generated solutions that were prone to oxidation of the ascorbic acid released, and resulted in low ascorbic acid tissue levels. The enigma of what technique to use to extract all the ascorbic acid and active derivatives was not solved, but a partial solution using a combination of water for 1 minute followed by addition of 10% TCA to make a 5% TCA extraction solution did yield the extraction of the esters for HPLC assay within an acceptable assay time limit (see Table 3.1).

3.4   Assay techniques

The DPNH method for ascorbate analysis has been reviewed by Dabrowski et al.[4] and an improved method for minimizing errors from interfering compounds was reported by Dabrowski and Hinterleitner.[16] This method does

Table 3.1 Enigmas in extraction for HPLC assays

Water extracts	Little bound C extracted
Meta phosphoric extracts	Partial total C extracted
Trichloracetic acid (10%)	Hydrolysis of vitamers
Perchloric acid (10%)	Hydrolysis of vitamers
5% TCA extract	Slow hydrolysis
Microwave extract	Oxidation of ascorbic acid
Microwave plus 1% TCA	Best for all vitamers

not measure all ascorbate derivatives and esters that have vitamin C activity, and HPLC methods were used extensively with varying degrees of effectiveness. One improved technique was to use tandem columns to delay peak elution times to enable better resoltuion of the quantitative levels of ester present.[2] Enigmas occurred in the ion binders used for the assays. The elution sequence with one ion-binder was C1–C2S–C2MP–C2DP–C2TP. Interferring compounds often masked the peaks in several tissues and especially in feedstuffs assayed. Column clearing after assay was a long and trying task. The stability of the mobile phase during the extensive times needed for repeated assays was compromised. Column characteristics varied from manufacturers to packing details and to pressures allowed for rapid resolution of compounds desired. Some columns reversed the order of elution of compounds. One Alltec 5μm C18 column had the characteristic of reversing the order of elution when no ion-binding compound was used.[17]

This technique enabled simple quantitative assay for C2S (see Table 3.2) which was authenticated by assaying an aliquot eluted by electrospray ionization mass spectroscopy.[18]

3.5   Conclusions

An improved fish tissue assay technique was developed, using a combination of water homogenization for 1 minute followed by adding an equal volume of 10% TCA to make a 5% TCA solution which was homogenized for an additional minute before immediate centrifugation at $15000 \times g$ for 4 minutes. The supernatant was filtered through a 0.45 μm syringe filter and injected into HPLC equipment containing an isocratic pump with an in-line filter, and an

Table 3.2 Mobile phases for elution

Sodium Acetate 0.1M + 200mg/Liter EDTA
Sodium Acetate 0.1m + 200mg/Liter EDTA + Ion Binder
Ammonium Acetate 0.1 M
K-H-Phosphate 0.25 M
All of above at pH 5.0
Flow rate @ 0.75ml/min at 2200 psi

Table 3.3 Reports of C2S in fish tissues

Halver et al. *Ann. NY Acad. Sci.* 258:81–102 (1975)
Tsujimura et al. *Vitamins* 52:35–44 (1978)
Benitez and Halver. *Proc. Nat. Acad. Sci.* U.S.A. 79:5445–5449 (1982)
Tuckler and Halver. *J. Nutrition* 114:991–1000 (1984)
Tucker and Halver. *Nutrition Rev.* 42:173–179 (1984)
Felton and Halver. *Aquacult & Fish. Mgmt.* 18:387–390 (1987)
Grant et al. *J. World Aquacult. Soc.* 20:143–157 (1989)
Felton et al. *J. Liquid Chromatography* 17:123–1319 (1994)
Felton and Grace. *J. Liquid Chromatography* 18:1563–1581 (1995)
Abdelghany. *J. World Aquaculture* 27:449–455 (1996)

Alltec 5 μm C18 reverse phase column, connected to an UV detector set at 254nm linked to an electrochemical detector set at 0.72 volts.[17] Using C2S as the desired tissue assay compound, the above technique could resolve the presence of this compound in several fish species and concurrently record levels of L-ascorbic acid present. Samples could be stored for several hours before significant loss of extracted C2S was measured (Table 3.3).

References

1. Roe, J.H. and Kuether, C.A. 1943. The determination of ascorbic acid in whole blood and urine through the 2,4-dinophenylhydrazine method. *J. Biol. Chem.* 174:201–208.
2. Felton, S.P. and Halver, J.E. 1987. Vitamin C1 and C2 analysis using double column reverse phase high pressure liquid chromatography. *Aquacult. Fish. Mgmt.* 18:387–390.
3. Seib, P.A. and Tolbert, B.M. (eds) 1982. Ascorbic acid: chemistry, metabolism and uses. *Adv. Chem.*, 200, American Chemical Society, 604 pp.
4. Dabrowski, K., Matusiewicz, and Blom, J.H. 1994. Hydrolysis, absorption and bioavailability of ascorbic acid esters in fish. *Aquaculture* 124:169–192.
5. Seib, P.A. 1985. Oxidation, monosubstitution and industrial synthesis of ascorbic acid. *Int. J. Vit. Nutr. Res.* S27:259–306.
6. Halver, J.E., Smith, R.R., Tolbert, B.M., and Baker, E.M. 1975. Utilization of ascorbic acid in fish. *Ann. N.Y. Acad. Sci.* 258:81–102.
7. Abdelghany, A.E. 1996. Growth response of Nile tilapia to dietary L-ascorbic acid, L-ascorbyl-2-sulfate, and L-ascorbyl-2-polyphosphate. *J. World Aquacult. Soc.* 27:449–455.
8. Grant, B.F., Seib, P.A., Liao, M.L., and Corpron, K. 1989. Polyphosphated L-ascorbic acid: A stable form of vitamin C for aquaculture feeds. *J. World Aquacult. Soc.* 20:143–157.
9. Benitez, L.V. and Halver, J.E. 1982. Ascorbic acid sulfate sulfohydrolase (C2 sulfatase): The modulator of cellular levels of L-ascorbic acid in rainbow trout. *Proc. Natl. Acad. Sci. U.S.A.* 79:5445–5449.
10. Anggawate–Satyabudhly, A.M., Grant, G.F., and Halver, J.E. 1989. Effects of L-ascorbyl phosphates (AsPP) on growth and immunoresistance of rainbow

trout (*Oncorhynchus mykiss*) to infectious hematopoietic necrosis (IPN) virus. *Proc. Third Int. Symp. on Feeding and Nutr. in Fish* Tsoba, Japan, 411–426.

11. Muto, N., Terasawa, K., and Yamamoto, I. 1992. Evaluation of ascorbic acid-2-O-glycoside as vitamin C source: mode of intestinal hydrolysis and absorption following oral administration. *Int. J. Vit. Nutr. Res.* 62:318–323.

12. Paulssen, R.B. 1975. Acylation of ascorbic acid in water. *J. Pharm. Sci.,* 64, 1300–1305.

13. Tucker, B.W. and Halver, J.E. 1984. Ascorbate-2-sulfate metabolism in fish. *Nutr. Rev.* 42:173–179.

14. Tucker, B.W. and Halver, J.E. 1984. Distribution of ascorbate-2-sulfate and distribution, half-life and turnover rates of (1–14C) ascorbic acid in rainbow trout. *J. Nutr.* 114:991–1000.

15. Wang, X.Y. and Seib, P.A. 1990. Liquid chromatography determination of a combined form of L-ascorbic acid (L-ascorbate-2-sulfate) in fish tissue by release of ascorbic acid. *Aquaculture* 87:65–84.

16. Dabrowski, K. and Hinterleitner, S. 1989. Simultaneous analysis of ascorbic acid, dehydroascorbic acid and ascorbic sulfate in biological material. *Analyst* 114:83–87.

17. Felton, S.P., Grace, R., and Halver, J.E. 1994. A non-ion-pairing method for measuring new forms of ascorbate and ascorbic acid. *J. Liquid Chromat.* 17:123–131.

18. Felton, S.P. and Grace R. 1995. Authentication of L-ascorbyl-2-sulfate in salmonid gastric tissue: HPLC/electro-spray ionization mass spectroscopic verification. *J. Liquid Chromat.* 18:1563–1581.

19. Tsujimura, J., Yoshikawa, H., Hasegawa, T., Suzuki, T., Suwa, T., and Kitamura, S. 1978. Studies on the vitamin C activity of ascorbic acid -2-sulfate on the feeding test of new-born rainbow trout. *Vitamins* 52:35–43.

chapter four

Gulonolactone oxidase presence in fishes: activity and significance

Régis Moreau and Konrad Dabrowski

Contents

13

4.1   Introduction

This chapter will concentrate on the distribution of L-ascorbic acid (AA, vitamin C) biosynthesis capacity in vertebrates with an emphasis on fishes. Since a decade ago, research work in fish biology has shed a new light but also brought some controversy on the ability to synthesize AA in the animal kingdom, and the origin and phylogeny of AA. These new data have initiated this review in which earlier and present states of knowledge are presented. The methods frequently employed to measure L-gulono-1,4-lactone oxidase (GLO, EC 1.1.3.8) activity will be critically reviewed. Further research into the mechanism of AA biosynthesis in fish has been initiated and will be compared to the mammalian model. Studying the regulation of AA biosynthesis in fish will have implications in fish nutrition and aquaculture, possibly in determining the vitamin C requirement associated with specific physiological conditions, including reproduction, immune challenge, and stress.

AA is an essential molecule in the overall health of animals, including growth,[1-4] bone formation,[5-7] and reproduction.[8-10] The functions of AA have been frequently reviewed and will not be discussed in detail here.[11] The most widely accepted function of AA is that of a water-soluble antioxidant. AA operates as a chain-breaking antioxidant that scavenges radicals in order to terminate free radical chain reactions. Throughout the body AA (anion at physiological pH) acts as a reducing agent by donating one hydrogen and one electron to a wide array of oxidized biomolecules and trace elements including metal co-enzyme. In turn, AA becomes oxidized. Oxidized AA in the form of ascorbyl radical (one-electron oxidation product) or dehydroascorbic acid (two-electron oxidation product) can be regenerated *in vivo* chemically by reduced glutathione (GSH) or enzymatically by GSH- or NAD(P)H-dependent reductase.[12-15] Ultimately dehydroascorbic acid is metabolized into compounds[16] that lack antioxidant property, such as 2,3-diketogulonic acid, and further hydrolyzed to oxalate and threonate. Endogenous (biosynthesis) and/or exogenous (dietary) supply of AA contribute to maintaining the AA body pool. At first look vertebrates seem to be particularly diversified with respect to AA nourishment. While many vertebrates can synthesize the vitamin, some cannot and thus depend upon a dietary source. To date, no mechanism has been established that could explain such disparity.

4.2   Pathway of L-ascorbic acid synthesis in vertebrates

The pathway of AA synthesis was first studied and elucidated in rats, then generalized to vertebrates. AA is synthesized from either D-glucose or D-galactose as part of the glucuronic acid pathway.[17] This pathway is involved in detoxification mechanisms by glucuronic acid conjugation to foreign compounds.[18] Branching from L-gulonic acid, the biosynthetic pathway of AA comprises three consecutive steps: first, the enzymatic lactonization of L-gulonic acid catalyzed by L-gulonolactone hydrolase (EC 3.1.1.18);[19] second, the oxidation of L-gulonolactone catalyzed by L-gulonolactone oxidase (GLO, EC 1.1.3.8); and third, the spontaneous isomerization of 2-keto-L-gulonolactone leading to AA.[20] The biosynthesis of AA in mammals was reviewed by Sato and Udenfriend.[21] In birds, reptiles, amphibians and fish that synthesize AA, it is accepted that, like in mammals, AA is derived from D-glucose as part of the glucuronic acid pathway. Since GLO is the only enzyme missing in scurvy-prone mammals,[22] it can be argued that GLO is the key enzyme in AA synthesis in animals. Any animal lacking GLO activity is unable to synthesize AA and thus depends upon a dietary source of vitamin C for health, growth, and reproduction.

4.3 Measurement of GLO activity in fish tissues

4.3.1 Tissue preparation

In fish, GLO is found in the kidneys only, but other tissues, such as liver, hepatopancreas, intestine, and muscle, have been analyzed following the same procedure.[23] GLO activity can be assayed in both crude extracts and microsomes prepared from fresh or deep-frozen tissues. In white sturgeon, GLO activity recovery was not significantly different in frozen kidney ($-80°C$) as compared to fresh kidney.[24] If the analysis is to be made on a crude enzyme preparation, a portion of tissue (0.3–0.5 g) is homogenized using a glass hand or spin homogenizer in 2–3 mL of 50 mM phosphate buffer (pH 7.4), containing 1 mM EDTA and 0.2% sodium deoxycholate, and centrifuged at 15,000 g for 15 min at 4°C. To increase the accuracy and the repeatability of the activity assay, microsomal preparation is recommended. To isolate the microsomes, a tissue sample is homogenized in cold 0.25 M sucrose (1 g tissue in 4–6 mL sucrose) and centrifuged at 10,000 g for 15 min at 4°C. The supernatant is spun at 100,000 g for 60 min at 4°C to obtain the microsomal pellet. The pellet is resuspended in 50 mM phosphate buffer (pH 7.4) containing 0.2% sodium deoxycholate to solubilize the membranes and release the enzyme.

4.3.2 Quantitative activity assays

Optimal assay conditions have been previously reported for fish GLO.[23] In short, an aliquot of enzyme preparation is incubated in 50 mM phosphate buffer (pH 7.4) in the presence of 10 mM L-gulonolactone and 2.5 mM reduced glutathione. Sturgeon and sea lamprey GLOs were not significantly affected by assay temperature up to 28°C. The enzyme activity decreased at further elevated temperature. Routinely, fish GLO activity is assayed at 25°C. Q_{10} between 15 and 25°C was found to be 1.73 and 1.65 for sea lamprey and lake sturgeon, respectively.[25, 26]

Table 4.1   Gulonolactone oxidase (GLO) activity assays using
L-gulonolactone as a substrate

End product analysis	Reference
DNPH-derivatives absorbance (524–540 nm)	Roe and Kuether 1943,[27] Ikeda et al. 1963,[28] Ayaz et al. 1976,[29] Terada et al. 1978,[30] Dabrowski and Hinterleitner 1989[31]
2,2'-dipyridyl-derivatives absorbance (525 nm)	Sullivan and Clarke 1955,[32] Zannoni et al. 1974[33]
UV absorbance (265 nm)	Dabrowski 1990[23]
Oxygen consumption	Kiuchi et al. 1982,[34] Dabrowski 1990[23]

GLO activity assays most commonly used with animal tissues rely on the methods listed in Table 4.1.[27-34, 23] These methods can be grouped into four categories: 1) dinitrophenylhydrazine-colorimetric method, 2) 2,2'-dipyridyl-colorimetric method, 3) UV spectrophotometry, and 4) oxygen consumption method. Methods 1, 2, and 3 measure the production of AA. Method 1 quantitates total AA (reduced + oxidized AA), whereas methods 2 and 3 measure reduced AA. Method 4 follows the disappearance of oxygen, one of GLO substrates. Methods 1 and 2 are stopped reactions and require subsamples to be taken at the beginning and at the end of the enzymatic reaction (20–30 min). The enzyme velocity has to be linear throughout the reaction time. GLO activity is expressed as net production of AA. Methods 3 and 4 continuously monitor the course of the reaction. An advantage of these methods over stopped reaction methods is that the operator becomes aware of abnormal enzyme behavior. A disadvantage of Method 4 is that the activities of O_2-requiring enzymes (oxidases, oxygenases) and spontaneous oxidation account for additional O_2 consumption *in vitro* especially when assaying a crude tissue extract. To account for background O_2 uptake, a control without L-gulonolactone or with D-gulonolactone (the inactive isomer) in place of L-gulonolactone should be included.

4.3.3   Qualitative activity assays

A histochemical method for *in situ* identification of GLO was developed by Cohen[35] and modified by Nakajima et al.[36] The method was employed on animal tissue sections[35, 36] including fish.[37] The sequence of events includes the oxidation of L-gulonolactone with transfer of electrons to intermediates and a final electron acceptor, nitroblue tetrazolium, which once reduced forms a dark blue formazan at the site of enzyme activity. Stronger staining was obtained when GLO was activated with menadione,[35] 2,4-dinitrophenyl, and phenazine methosulfate.[36] However, several problems arose. Cytosolic xanthine oxidase[38] and respiratory chain oxidases also react with the couple phenazine methosulfate-nitroblue tetrazolium. Cyanide was used to inhibit cytochrome oxidase. Interfering reduction of nitroblue tetrazolium was reported when D-glucuronate and L-xylulose were produced from L-gulono-lactone by aldonolactonase. In this case ethylenediaminetetraacetic acid (EDTA) which inhibits aldonolactonase activity had to be added.

4.4   Distribution of GLO activity among vertebrates

4.4.1   Distribution of GLO in terrestrial vertebrates

The inability to synthesize AA is rather the exception in vertebrates (Figure 4.1). Many higher vertebrates can synthesize AA due to the presence of a functional GLO associated with the microsomal fraction in either the liver or the kidney or in both organs.[39, 40] Among reptiles and amphibians analyzed to date, all have been shown to possess GLO activity in the kidney.[41–43] In birds, while 16 of the 28 species of passerines examined were reported incapable of producing AA,[44] 12 passerine species and all 11 non-passerines under study had GLO activity in the kidney and/or the liver. A recent phylogenetic reanalysis of Chaudhuri and Chatterjee[44] data showed that the occurrence of the capacity to synthesize AA in passerines was unpredictable and enigmatic.[45] Among mammals, the primates, the guinea pig,[46] and the bats[47] cannot produce AA *de novo*. The enzyme deficiency was the result of missense mutations within the gene for GLO.[48–50] The date of GLO losses was estimated at ~20 and ~70 million years before present in the ancestors of the guinea pigs and humans, respectively.[49, 51] Administering chicken GLO (modified to prevent immune rejection) to the guinea pig induced the production of AA as revealed by elevated plasma levels.[52]

4.4.2   Distribution of GLO in fishes (Table 4.2[23–26, 37, 53–67, 84])

Twenty years ago, it was reported that some fish may biosynthesize AA.[55, 57] Before this work, the understanding was that all fish lack GLO activity, and that the capacity to synthesize AA (by acquisition of GLO) first appeared in amphibians and reptiles.[41, 42] Although these early reports bore a significant importance by questioning the unvarying absence of GLO among living fishes, today some of these data appear suspicious from the analytical point of view, and misleading from the evolutionary perspective. In their studies, Yamamoto et al.[55] and Sato et al.,[56] and later, Soliman et al.[37] and Thomas et al.[58] used the DNPH methods described by Ikeda et al.[28] and Terada et al.[30] to measure GLO activity in teleosts, and claimed to have detected GLO in some teleosts including the common carp (*Cyprinus carpio*) and the goldfish (*Carassius auratus*). As judged by the lack of appropriate correction for interfering DNPH-derivatives, the data reported in the aforementioned studies were artifacts (Figure 4.2). Since the levels of GLO activity found were among the lowest in fishes (≤1 μmol AA/ g tissue/ h), the results by those authors are most likely analytical errors and, in fact, the teleosts examined thus far do

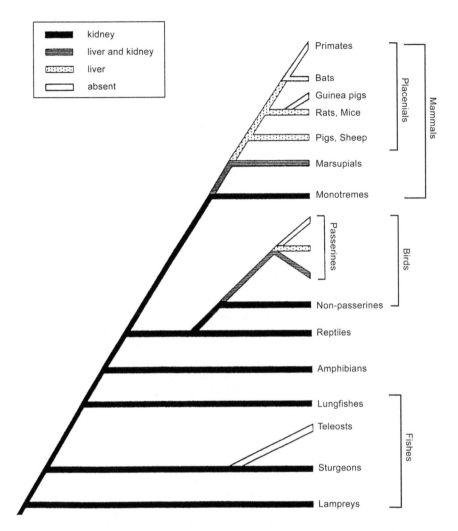

Figure 4.1 Phylogeny and tissue location of L-gulonolactone oxidase (GLO) in verte-brates. The cladogram implies that the most recent common ancestor to all vertebrates expressed GLO in the kidney. In the course of vertebrate evolution, the histological site of expression has shifted to the liver as evidenced by an intermediate state where GLO is expressed in the kidney and liver concomitantly. Ultimately, GLO was lost at more than one occasion, i.e., in teleost fish, passerin birds, guinea pigs, bats, and primates.

not have GLO activity (Table 4.3[23–28, 30–32, 37, 55–61, 63, 67–68]). Repeated analyses of GLO in *C. carpio* in our laboratory using improved methodology, both enzymatic and immunologic, have revealed no detectable activity in the kidney or liver,[67] thus confirming earlier results.[23, 59] On the contrary, substantial evidence suggests that while teleosts cannot synthesize AA, all non-teleost taxa

Table 4.2 Contribution to the distribution of gulonolactone oxidase (GLO) among extant fishes

Reference	Results
Chatterjee 1973b[53]	No GLO activity in *Teleostei:* Anabas testudineus, Apocryptes lanceolatus, Chanos chanos, Catla catla, Cirrhina mrigala, Clarias batrachus, Heteropneustes fossilis, Labeo calbasu, L. rohita, Lates calcarifer, Megalops cyprinoides, Mugil persia, Mystus bleekeri, Notopterus notopterus, Ophiocephalus punctatus, O. striatus, Tilapia mossambica.
Wilson 1973[54]	No GLO activity in *Teleostei:* Ictalurus frucatus, I. punctatus.
Yamamoto et al. 1978[55]	No GLO activity in *Teleostei:* Anguilla japonica, Oncorhynchus masou, Pagrus major, Plecoglossus altivelis, Salmo gairdneri, Seriola quinqueradiata, Stephanolepis cirrhifer, Tilapia nilotica.
	GLO activity reported in *Teleostei:* Carassius carassius, Cyprinus carpio, Parasilurus asotus, Tribolodon hakonensis.
Sato et al. 1978[56]	GLO activity reported in *Teleostei:* Cyprinus carpio.
Dykhuizen et al. 1980[57]	GLO activity reported in *Dipnoi:* Neoceratodus forsteri.
Soliman et al. 1985[37]	No GLO activity in *Teleostei:* Ctenopharyngodon idellus, Oreochromis niloticus, O. macrochir, O. mossanbicus, Salmo gairdneri, S. trutta, Salvelinus fontinatis, Sarotherodon galilaeus, Tilapia buttikoferi, T. zillii.
	GLO activity reproted in *Teleostei:* Cyprinus carpio, Oreochromis aureus, O. spirulus.
Thomas et al. 1985[58]	GLO activity reported in *Teleostei:* Carassius auratus, Fundulus grandis, Mugil cephalus, Tilapia aurea.
Dabrowski 1990[23]	No GLO activity in *Teleostei:* Cyprinus carpio, Leuciscus cephalus.
Dabrowski 1991[59]	No GLO activity in *Teleostei:* Cyprinus carpio.
Dabrowski 1994[60]	GLO activity reported in *Chondrostei:* Acipenser transmontanus, A. fulvescens, Polyodon spathula.
Touhata et al. 1995[61]	No GLO activity in *Teleostei:* Lophiomus setigerus, Paralichthys olivaceus; *Neopterygii:* Lepisosteus oculatus; *Chondrostei:* Polypterus senegalus.
	GLO activity reported in *Dipnoi:* Protopterus aethiopicus; *Elasmobranchii:* Dasyatis akajei, Mustelus manazo; *Agnatha:* Lampetra japonica.
Mishra and Mukhopadhyay 1996[62]	No GLO activity in *Teleostei:* Clarias batrachus.
Moreau et al. 1996[63]	GLO activity reported in *Chondrostei:* Acipenser baeri.
Moreau and Dabrowski 1996[24]	No GLO activity in *Teleostei:* Ictalurus punctatus, Oncorhynchus mykiss.

Table 4.2   Continued

Reference	Results
	GLO activity reported in *Chondrostei:* Acipenser transmontanus.
Young et al. 1997[64]	No GLO activity in *Teleostei:* Morone saxatilis
Mæland and Waagbø 1998[65]	No GLO activity in *Teleostei:* Clupea harengus, Anguilla anguilla, Salmo salar, Gadus morhua, Scomber scombrus, Hippoglossus hippoglossus, Scophthalmus maximus.
	GLO activity reported in *Chondrostei:* Acipenser ruthenus; *Elasmobranchii:* Squalus acanthias.
Moreau and Dabrowski 1998[25]	GLO activity reported in *Neopterygii:* Amia calva; *Agnatha:* Petromyzon marinus.
Mukhopadhyay et al. 1998[66]	No GLO activity in *Teleostei:* Labeo rohita.
Moreau et al. 1999a, b[26, 84]	GLO activity reported in *Chondrostei:* Acipenser fulvescens.
Moreau and Dabrowski 2000[67]	GLO activity reported in *Neopterygii:* Amia calva, Lepisosteus osseus; *Chondrostei:* Polypterus senegalus. No GLO activity in *Teleostei:* C. carpio, C. auratus.

analyzed to date, including the lungfishes,[57, 61] the sharks and rays,[61–65] the sturgeons and paddlefishes,[60, 63, 65] the bowfin (*Amia calva*), the gars (*Lepisosteus osseus*), *Polypterus senegalus*,[67] and the lampreys (*Lampetra japonica* and *Petromyzon marinus*)[61, 25] can produce AA *de novo* in the kidney. Results obtained in fish are in agreement with the view that the kidney is the site of AA biosynthesis in poikilothermic vertebrates, including amphibians and reptiles. The fact that teleosts, which account for ~96% of extant fishes,[69] cannot synthesize AA has overshadowed the remaining 4% for a long time and was responsible for the inaccurate generalization that "fishes are incapable of synthesizing AA."[42, 43] As additional non-teleost species are tested for GLO activity, it becomes clear that a variety of fish taxa possess the ability to synthesize AA (Figure 4.3[69–74]). The survey of modern-day fish is still incomplete, however, as representatives of hagfish, chimaera, and coelacanth taxa are yet to be investigated for the presence of GLO activity.

Based on our data confirming that modern teleosts have lost the ability to synthesize AA, given the monophyly of teleosts, and invoking the principle of parsimony, we argue that the loss is the result of a single event that occurred in the ancestor of modern teleosts. Considering the bowfin (*Amia calva*) as being the living sistergroup to teleost, the time of divergence between the bowfin and teleost lineages should provide a good estimate of the date the trait was lost in teleosts. The genus *Amia* is known as early as the Upper Cretaceous (~140 million years ago) but the earliest known teleosts date back to the late Triassic (~210 million years ago).[74] However, other

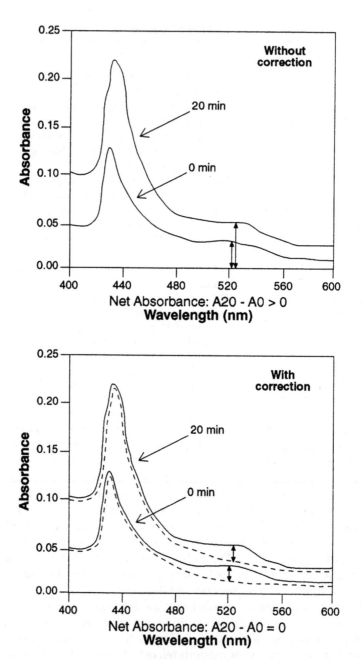

Figure 4.2 Measurement of GLO activity in teleosts using the DNPH method (524 nm) with or without appropriate correction for interfering DNPH-derivatives.

Table 4.3   Gulonolactone oxidase (GLO) actities reported in fish and methods used to determine the production of ascorbic acid *in vitro*

Clade/species	GLO activity (μmol AA/g tissue/h)	Method	Correction for DNPH Interference
Teleostei			
Cyprinus carpio	0.92[a], 0.95[b], 1.01[d]	DNPH[4]	No
	0[f,g]	DNPH[6], UV[7]	Yes
	0[n]	UV[7]	n.a.
Carassius carassius	0.75[a]	DNPH[4]	No
C. auratus	0.77[e]	DNPH[5]	No
	0[n]	UV[7]	n.a.
Tribolodon hakonensis	0.17[a]	DNPH[4]	No
Parasilurus asotus	0.24[a]	DNPH[4]	No
Oreochromis spirulus	1.0[d]	DNPH[4]	No
O. aureus	0.80[d], 0.07[e]	DNPH[4,5]	No
Mugil cephalus	0.03[e]	DNPH[5]	No
	0[f]	DNPH[6], UV[7]	Yes
Fundulus grandis	0.11[e]	DNPH[5]	No
Amiiformes			
Amia calva	2.5–4.6[n]	UV[7]	n.a.
Semionotiformes			
Lepisosteus osseus	3–10[n]	UV[7]	n.a.
Acipenseriformes			
Acipenser transmontanus	1.4[j], 5.6[h]	DNPH[6], UV[7]	Yes
A. fulvescens	2.6[m], 2.8[h]	DNPH[6], UV[7]	Yes
A. baeri	0.39[k]	DNPH[6]	Yes
Polyodon spathula	1.6[h]	DNPH[6]	Yes
Polypteriformes			
Polypterus senegalus	2.3[n]	UV[7]	n.a.
Dipnoi			
Neoceratodus forsteri	0.05[c]	DNPH[1], DCIP[3]	?
Protopterus aethiopicus	0.37[i]	Dipyridyl[2]	n.a.

Table 4.3   Continued

Clade/species	GLO activity (μmol AA/g tissue/h)	Method	Correction for DNPH Interference
Elasmobranchii			
Mustelus manazo	0.46[i]	Dipyridyl[2]	n.a.
Dasyatis akajei	0.14[i]	Dipyridyl[2]	n.a.
Petromyzontiformes			
Lampetra japonica	0.12[i]	Dipyridyl[2]	n.a.
Petromyzon marinus	4.6–5.3[l]	UV[7]	n.a.

[a]Yamamoto et al. 1978[55], [b]Sato et al. 1978[56], [c]Dykhuizen et al. 1980[57], [d]Soliman et al. 1985[37], [e]Thomas et al. 1985[58], [f]Dabrowski 1990[23], [g]Dabrowski 1991[59], [h]Dabrowski 1994[60], [i]Touhata et al. 1995[61], [j]Moreau and Dabrowski 1996[24], [k]Moreau et al. 1996[63], [l]Moreau and Dabrowski 1998[25], [m]Moreau et al. 1999a[26], [n]Moreau and Dabrowski 2000[67].

[1]Roe and Kuether 1943[27], [2]Sullivan and Clarke 1955[32], [3]Chatterjee et al. 1958[68], [4]Ikeda et al. 1963[28], [5]Terada et al. 1978[30], [6]Dabrowski and Hinterleitner 1989[31], [7]Dabrowski 1990[23].

n.a. = not applicable.

members of the Amiidae family, today extinct, are known from the Upper Jurassic (~200 million years ago). Therefore, as a first approximation, the loss of the AA synthesis pathway in teleosts may have occurred 200–210 million years before present.

The presence of GLO activity in the Australian lungfish (*Neoceratodus forsteri*) and African lungfish (*Protopterus aethiopicus*) bears a considerable importance in linking terrestrial and aquatic vertebrates with respect to the evolution of AA biosynthesis. There is no doubt that land vertebrates have evolved from lobe-finned fish (Sarcopterygii, represented today by the lung-fishes and the coelacanth) early in the Devonian period (~400 million years ago). It can be argued, therefore, that AA biosynthesis is homologous among the vertebrates, aquatic and terrestrial, and that the origin of the trait dates back to the Cambrian (500–550 million years ago) when the ancestor of present-day lampreys presumably existed. Achieving the purification of GLO from sturgeon kidney using a methodology originally designed for chicken GLO[34] implies that the fish and bird enzymes have structural simi-larities likely passed on by common ancestry, and supports the character homology.[67]

4.5   Rate of AA synthesis in vivo

All animals require vitamin C whether or not they can synthesize AA. Many fish species raised in aquaculture, such as the salmonids, ictalurids, cichlids, and cyprinids, are teleosts and thus require a dietary source of vitamin C. Their vitamin C requirements vary with species, strain,[75] age, environmental conditions,[76] and induced stressors.[77] All aforesaid factors could significantly

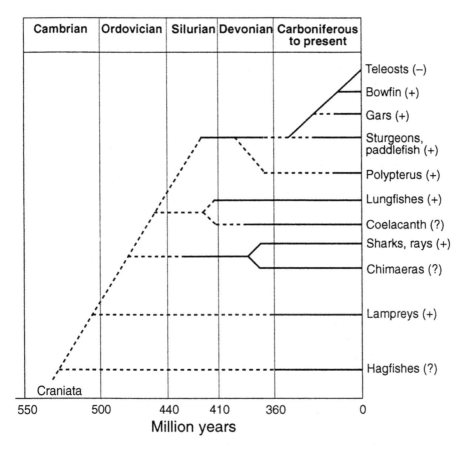

Figure 4.3   Distribution of GLO activity among extant fishes through time (adapted from Nelson 1994).[69] The continuous lines represent the fossil records and the dotted lines represent the inferred phylogeny (Greenwood 1984[70], Gardiner 1984[71], Schultze and Wiley 1984[72], Wiley and Schultze 1984[73], Carroll 1988[74], Nelson 1994[69]).

affect GLO activity and/or expression. Studying the rate of AA production in a fish like sturgeon under controlled environmental and physiological conditions should allow nutritionists to identify factors that affect AA requirements in teleosts as well. In order to meet their vitamin C requirement, teleost species cultured for economic purposes typically are fed compound diets supplemented with 50–100 mg AA/kg diet.[78] Daily, a salmonid should ingest 1 mg AA/kg body weight to meet its vitamin C requirement. The recent finding that some fish synthesize AA poses the question of whether they are able to produce enough vitamin C to meet their requirement.

    Among the fish species able to synthesize AA, a few are easily available for experimentation. Since several species of sturgeon have been domesticated and their progeny raised in tanks and fed commercial or semipurified diets, sturgeon stand out as the most available model today. The lengthy juvenile stage in sturgeon would allow a researcher to distinguish between life stages prior to sexual maturity. Alternatively, lamprey could become a valuable model at the time of reproduction as we found that AA status among upstream migrants was consistent over two successive years. Using *in vitro* kinetics data, we have applied to adult sea lamprey[25] and juvenile lake sturgeon[26] the mathematical procedure used by others[79] to quantify the rate of AA synthesis in mammals. The rate of AA synthesis per day in fish was calculated at 15°C based on GLO maximum velocity (V_{max}) measured at 25°C. The *in vitro* Michaelis constant (K_m) of white sturgeon GLO for L-gulonolactone (0.92 mM[26]) was used for calculation. Projections show that a 1-kg sea lamprey synthesizes 4 mg AA daily, and a 1-kg sturgeon, 3 mg at 15°C. These estimates compare favorably with the daily requirement of a 1-kg rainbow trout (1 mg AA[80]). Therefore, it appears that large juvenile sturgeon produce adequate amounts of AA to meet their needs. A comparison with a mammal able to synthesize AA, such as the rat, reveals, however, that the rate of synthesis in fish is approximately 7 times slower than in the rat per unit of weight. Indeed, a 0.1-kg rat was shown to produce 2.6 mg AA per day.[81] Another major difference between rodent and fish AA metabolism is that AA turnover is faster in these mammals than in fish as judged by half-lives of 3 days in the guinea pig[82] vs. 40–70 days in the rainbow trout.[75, 83] In short, rodent AA metabolism is characterized by fast rate of synthesis and short half-life (or fast turnover), whereas in sturgeon both the rate of synthesis and the turnover are slower. Overall the rate of AA synthesis in vertebrates may be positively correlated with the basal metabolic rate.

4.6   Factors affecting GLO activity in fish

Little is known about the factors influencing GLO activity and regulation in fish. Dietary AA and fish sex have been studied as two potential factors affecting GLO activity. Research in white and lake sturgeon[60, 26, 84] has shown that increasing dietary AA did not exert a negative feedback on GLO activity, unlike in rabbits[85] and mice.[86] On the contrary, GLO activity seemed to be slightly stimulated by dietary AA in sturgeon. Gender effect on GLO was evaluated in adult sea lamprey (*P. marinus*) while migrating to the spawning grounds.[25] The study did not show any significant difference of GLO activity between sexes.

4.7   Importance of synthesizing AA for a fish

Throughout the animal kingdom, natural diets of AA-synthesizing species do not differ dramatically from those of scurvy-prone species with respect to AA intake. Since vitamin C occurs in significant amounts (up to 3 g/100 g in the Australian fruit *Terminalia fernandiana*[87]) in food sources (animal organs, fruits, and vegetables, except in grains and nuts), feral species are rarely exposed to long periods of AA deprivation, and being incapable of synthesizing AA does not undermine fitness. However, in situations where access to food is restricted as a result of human activities, such as for domesticated animals, scurvy-prone species may suffer the consequences of vitamin C deficiency (impaired bone formation, retarded wound healing), whereas species synthesizing AA will not be affected. But such situations are exceptional, anthropogenically driven, and most importantly unrelated to the evolutionary process that led to the loss of AA biosynthesis long before the origin of humans. We suggested that ~500 million years ago, the ancestor of present-day vertebrates was able to synthesize AA, and that the trait was lost secondarily on at least five separate occasions, in the teleosts, the passerine birds, the guinea pigs, the bats, and the primates. Despite the character loss, the teleosts have been and continue to be among the most successful taxa to radiate and colonize the earth, as evidenced by the explosive speciation in African cichlids.[88–89]

4.7.1   Is the ability to synthesize AA adaptive?

The pathway for AA synthesis may likely have arisen in fish when adequate supplies were not available in the food (500–550 million years ago). But in the course of evolution AA may have become more available in the food chain or vertebrates may have modified their diets to include food items rich in AA to a point where retaining the pathway did not benefit to its bearer in the sense of promoting fitness. As a consequence, secondary reversal could be functionally neutral.[90] If having GLO activity has become superfluous, natural selection is no longer capable of correcting point mutations, and they will tend to accumulate as in the guinea pigs[49] and the primates.[51] Modern fishes may also provide a convincing illustration that retaining the pathway is a vestigial character. Indeed, while fish species that have ancient lineage have retained the AA synthesis pathway (lampreys, sharks, and sturgeons), recently evolved teleosts have lost the function. In spite of the loss, supposedly attributable to the inevitable degeneration by mutation of unneeded function, teleosts account for ~96% of today's fish biodiversity. Furthermore, in the sturgeon, a species that synthesizes AA, the absence of regulation of the key enzyme, GLO, by dietary AA may demonstrate the functional neutrality of the pathway that was abandoned.

4.7.2   Is the loss of AA synthesis adaptive or non-adaptive?

The reason(s) why the teleosts have lost the ability to synthesize AA are unknown. So are the explanatory mechanism for its disappearance in some passerines and mammals. Was it the consequence of selection or genetic drift? It is expensive for an organism to maintain its genes in good working condition. Retaining the ability to synthesize AA while dietary AA is abundant for many generations does not seem most cost-effective. Adaptive change, wherein the new allele becomes fixed because the carrier of the new form (in the present case a variant unable to carry out the synthesis of AA) is more fit than the homozygote of the original form, can be inferred to have occurred at the molecular level. A mutation shutting down the AA biosynthesis could be beneficial,[91] thus selected for. Although instances of rat[92] and swine mutants[93] unable to synthesize AA have been reported, no benefit has been associated with the reversal. On the contrary, homozygote osteogenic-disorder rats (*od/od*) were found to be sterile. Consequently, the adaptive hypothesis[94] for the loss of AA biosynthesis is not well documented, perhaps because fitness is so difficult to measure.

Genetic drift, however, can be an important evolutionary force. Its effects within population are the random fixation and loss of alleles. Because genetic drift is a sampling process, its effects are most pronounced in small populations. Most species are thought to have originated with low population sizes. Although the population may subsequently grow in size and later consist of a large number of individuals, like for the teleosts, the gene pool of subsequent generations derive from the genes present in the original founders. A mutation that occurred in the founding population of early teleosts, bats, or primates may have become fixed and was passed on to the subsequent generations. Jukes and King[95] suggested that the loss of AA biosynthesis in primates and guinea pigs was non-adaptive; in the feral state these vertebrates consume large quantities of AA. However, even beneficial functions may be lost by genetic drift provided that the selective advantage and/or the population size becomes small.

Acknowledgments

Thanks to Pat Patterson for artwork.

References

1. McLaren, B. A., Keller, E., O'Donnell, D. J., and Elvehjem, C. A., The nutrition of rainbow trout. I. Studies of vitamin requirements, *Arch. Biochem. Biophys.*, 15, 169, 1947.
2. Halver, J. E., Ashley, L. M., and Smith, R. R., Ascorbic acid requirement of coho salmon and rainbow trout, *Trans. Am. Fish Soc.*, 4, 762, 1969.
3. Wilson, R. P. and Poe, W. E., Impaired collagen formation in the scorbutic channel catfish, *J. Nutr.*, 103, 1359, 1973.
4. Lim, C. and Lovell, R. T., Pathology of the vitamin C deficiency syndrome in channel catfish (*Ictalurus punctatus*), *J. Nutr.*, 108, 1137, 1978.
5. Peterkofsky, B., Ascorbate requirement for hydroxylation and secretion of procollagen: relationship to inhibition of collagen synthesis in scurvy, *Am. J. Clin. Nutr.*, 54, 1135S, 1991.
6. Phillips, C. L., Tajima, S., and Pinnell, S. R., Ascorbic acid and transforming growth factor-β1 increase collagen biosynthesis via different mechanisms: Coordinate regulation of Proα1(I) and Proα1(III) collagens, *Arch. Biochem. Biophys.*, 295, 397, 1992.

7. Gosiewska, A., Wilson, S., Kwon, D., and Peterkofsky, B., Evidence for an *in vitro* role of insulin-like growth factor-binding protein-1 and -2 as inhibitors of collagen gene expression in vitamin C-deficient and fasted guinea pigs, *Endocrinology*, 134, 1329, 1994.

8. Blom, J. H. and Dabrowski, K., Reproductive success of female rainbow trout (*Oncorhynchus mykiss*) in response to graded dietary ascorbyl monophosphate levels, *Biol. Reprod.*, 52, 1073, 1995.

9. Ciereszko, A. and Dabrowski, K., Sperm quality and ascorbic acid concentration in rainbow trout semen are affected by dietary vitamin C: an across-season study, *Biol. Reprod.*, 52, 982, 1995.

10. Luck, M. R., Jeyaseelan, I., and Scholes, R. A., Ascorbic acid and fertility, *Biol. Reprod.*, 52, 262, 1995.

11. Carr, A. C. and Frei, B., Towards a new recommended dietary allowance for vitamin C based on antioxidant and health effects in humans, *Am. J. Clin. Nutr.*, 69, 1086, 1999.

12. Rose, R. C., Renal metabolism of the oxidized form of ascorbic acid (dehydro-L-ascorbic acid), *Am. J. Physiol.*, 256, F52, 1989.

13. Choi, J. L. and Rose, R. C., Regeneration of ascorbic acid by rat colon, *Proc. Soc. Exp. Biol. Med.*, 190, 369, 1989.

14. Meister, A., Glutathione-ascorbic acid antioxidant system in animals, *J. Biol. Chem.*, 269, 9397, 1994.

15. Winkler, B. S., Orselli, S. M., and Rex, T. S., The redox couple between glutathione and ascorbic acid: a chemical and physiological perspective, *Free Radical Biol. Med.*, 17, 333, 1994.

16. Niemelä, K., Oxidative and non-oxidative alkali-catalysed degradation of L-ascorbic acid, *J. Chromatogr.*, 399, 235, 1987.

17. Hassan, M. and Lehninger, A. L., Enzymatic formation of ascorbic acid in rat liver extracts, *J. Biol. Chem.*, 223, 123, 1956.

18. Smith, R. L. and Williams, R. T., Implication of the conjugation of drugs and other exogenous compounds, in *Glucuronic Acid Free and Combined*, Dutton, G. J., Ed., Academic Press Inc., 457, 1966.

19. Stubbs, D. W. and Haufrect, D. B., Effects of actinomycin D and puromycin on induction of gulonolactone hydrolase by somatotrophic hormone, *Arch. Biochem. Biophys.*, 124, 365, 1968.

20. Chatterjee, I. B., Chatterjee, G. C., Ghosh, N. C., Ghosh, J. J., and Guha, B. C., Biological synthesis of L-ascorbic acid in animal tissues: Conversion of L-gulonolactone into L-ascorbic acid, *Biochem. J.*, 74, 193, 1960.

21. Sato, P. and Udenfriend, S., Studies on ascorbic acid related to the genetic basis of scurvy, *Vit. Horm.*, 36, 33, 1978.

22. Sato, P., Nishikimi, M., and Udenfriend, S., Is L-gulonolactone-oxidase the only enzyme missing in animals subject to scurvy?, *Biochem. Biophys. Res. Com.*, 71, 293, 1976.

23. Dabrowski, K., Gulonolactone oxidase is missing in teleost fish—The direct spectrophotometric assay, *Biol. Chem. Hoppe-Seyler*, 371, 207, 1990.

24. Moreau, R. and Dabrowski, K., The primary localization of ascorbate and its synthesis in the kidneys of acipenserid (*Chondrostei*) and teleost (*Teleostei*) fishes, *J. Comp. Physiol. B.*, 166, 178, 1996.

25. Moreau, R. and Dabrowski, K., Body pool and synthesis of ascorbic acid in adult sea lamprey (*Petromyzon marinus*): An agnathan fish with gulonolactone oxidase activity, *Proc. Natl. Acad. Sci. U.S.A.*, 95, 10279, 1998.

26. Moreau, R., Dabrowski, K., and Sato, P. H., Renal L-gulono-1,4-lactone oxidase as affected by dietary ascorbic acid in lake sturgeon (*Acipenser fulvescens*), *Aquaculture*, 180, 359, 1999a.
27. Roe, J. H. and Kuether, C. A., The determination of ascorbic acid in whole blood and urine through the 2,4-dinitrophenylhydrazine derivative of dehydroascorbic acid, *J. Biol. Chem.*, 147, 399, 1943.
28. Ikeda, S., Sato, M., and Kimura, R., *Bull. Japan. Soc. Sci. Fish.*, 29, 757, 1963.
29. Ayaz, K. M., Jenness, R., and Birney, E. C., An improved assay for L-gulonolactone oxidase, *Anal. Biochem.*, 72, 161, 1976.
30. Terada, M., Watanabe, Y., Kunitomo, M., and Hayashi, E., Differential rapid analysis of ascorbic acid and ascorbic acid 2-sulfate by dinitrophenylhydrazine method, *Anal. Biochem.*, 84, 604, 1978.
31. Dabrowski, K. and Hinterleitner, S., Applications of a simultaneous assay of ascorbic acid, dehydroascorbic acid and ascorbic sulphate in biological materials, *Analyst*, 114, 83, 1989.
32. Sullivan, M. X. and Clarke, H. C. N., A highly specific procedure for ascorbic acid, *J. Assoc. Offic. Agr. Chemists*, 38, 514, 1955.
33. Zannoni, V., Lynch, M., Goldstein, S., and Sato, P., A rapid micromethod for the determination of ascorbic acid in plasma and tissues, *Biochem. Med.*, 11, 41, 1974.
34. Kiuchi, K., Nishikimi, M., and Yagi, K., Purification and characterization of L-gulonolactone oxidase from chicken kidney microsomes, *Biochemistry*, 21, 5076, 1982.
35. Cohen, R. B., Histochemical localization of L-gulonolactone oxidase activity in tissues of several species, *Proc. Soc. Exp. Biol. Med.*, 106, 309, 1961.
36. Nakajima, Y., Shantha, T. R., and Bourne, G. H., Histochemical detection of L-gulonolactone:phenazine methosulfate oxidoreductase activity in several mammals with special reference to synthesis of vitamin C in primates, *Histochemie*, 18, 293, 1969.
37. Soliman, A. K., Jauncey, K., and Roberts, R. J., Qualitative and quantitative identification of L-gulonolactone oxidase activity in some teleosts, *Aquacult. Fish Manag.*, 16, 249, 1985.
38. Nandi, A. and Chatterjje, I. B., Interrelation of xanthine oxidase and dehydrogenase and L-gulonolactone oxidase in animal tissues, *Indian J. Exp. Biol.*, 29, 574, 1991.
39. Birney, E. C., Jenness, R., and Hume, I. D., Ascorbic acid biosynthesis in the mammalian kidney, *Experientia*, 35, 1425, 1979.
40. Birney, E. C., Jenness, R., and Hume, I. D., Evolution of an enzyme system: Ascorbic acid biosynthesis in monotremes and marsupials, *Evolution*, 34, 230, 1980.
41. Roy, R. N. and Guha, B. C., Species difference in regard to the biosynthesis of ascorbic acid, *Nature*, 182, 319, 1958.
42. Chatterjee, I. B., Evolution and the biosynthesis of ascorbic acid, *Science*, 182, 1271, 1973a.
43. Nandi, A., Mukhopadhyay, C. K., Ghosh, M. K., Chattopadhyay, D. J., and Chatterjee, I. B., Evolutionary significance of vitamin C biosynthesis in terrestrial vertebrates, *Free Radic. Biol. Med.*, 22, 1047, 1997.
44. Chaudhuri, C. R. and Chatterjee, I. B., L-Ascorbic acid synthesis in birds: phylogenetic trend, *Science*, 164, 435, 1969.
45. del Rio, C. M., Can passerines synthesize vitamin C? *The Auk*, 114:513–516.
46. Burns, J. J., Missing step in man, monkey and guinea pig required for the biosynthesis of L-ascorbic acid, *Nature*, 180, 553, 1957.

47. Birney, E. C., Jenness, R., and Ayaz, K. M., Inability of bats to synthesize L-ascorbic acid, *Nature,* 260, 626, 1976.
48. Kawai, T., Nishikimi, M., Ozawa, T., and Yagi, K., A missense mutation of L-gulono-γ-lactone oxidase causes the inability of scurvy-prone osteogenic disorder rats to synthesize L-ascorbic acid, *J. Biol. Chem.,* 267, 21973, 1992.
49. Nishikimi, M., Kawai, T., and Yagi, K., Guinea pigs possess a highly mutated gene for L-gulono-γ-lactone oxidase, the key enzyme for L-ascorbic acid biosynthesis missing in this species, *J. Biol. Chem.,* 267, 21967, 1992.
50. Nishikimi, M., Fukuyama, R., Minoshima, S., Shimizu, N., and Yagi, K., Cloning and chromosomal mapping of the human nonfuctional gene for L-gulono-γ-lactone oxidase, the enzyme for L-ascorcbic acid biosynthesis missing in man, *J. Biol. Chem.,* 269, 13685, 1994.
51. Nishikimi, M. and Yagi, K., Molecular basis for the deficiency in humans of gulonolactone oxidase, a key enzyme for ascorbic acid biosynthesis, *Am. J. Clin. Nutr.,* 54, 1203S, 1991.
52. Sato, P. H. and Grahn, I. V., Administration of isolated chicken L-gulonolactone oxidase to guinea pigs evokes ascorbic acid synthetic capacity, *Arch. Biochem. Biophys.,* 210, 609, 1981.
53. Chatterjee, I. B., Vitamin C synthesis in animals: evolutionary trend, *Science and Culture,* 39, 210, 1973b.
54. Wilson, R. P., Absence of ascorbic acid synthesis in channel catfish, *Ictalurus punctatus* and blue catfish, *Ictalurus fructatus, Comp. Biochem. Physiol.,* 46B, 635, 1973.
55. Yamamoto, Y., Sato, M., and Ikeda, S., Existence of L-gulonolactone oxidase in some teleosts, *Bull. Jap. Soc. Sci. Fish.,* 44, 775, 1978.
56. Sato, M., Yoshinaka, R., and Yamamoto, Y., Nonessentiality of ascorbic acid in the diet of carp, *Bull. Jap. Soc. Sci. Fish,* 44, 1151, 1978.
57. Dykhuizen, D. E., Harrison, K. M., and Richardson, B. J., Evolutionay implications of ascorbic acid production in the Australian lungfish, *Experientia,* 36, 945, 1980.
58. Thomas, P., Bally, M. B., and Neff, J. M., Influence of some environmental variables on the ascorbic acid status of mullet, *Mugil cephalus* L., tissues. II. Seasonal fluctuations and biosynthesis ability, *J. Fish. Biol.,* 27, 47, 1985.
59. Dabrowski, K., Administration of gulonolactone does not evoke ascorbic acid synthesis in teleost fish, *Fish. Physiol. Biochem.,* 9, 215, 1991.
60. Dabrowski, K., Primitive Actinopterigian fishes can synthesize ascorbic acid, *Experientia,* 50, 745, 1994.
61. Touhata, K., Toyohara, H., Mitani, T., Kinoshita, M., Satou, M., and Sakaguchi, M., Distribution of L-gulono-1,4-lactone oxidase among fishes, *Fish. Sci.,* 61, 729, 1995.
62. Mishra, S. and Mukhopadhyay, P. K., Ascorbic acid requirement of catfish fry *Clarias batrachus* (Linn.), *Indian J. Fish.,* 43, 157, 1996.
63. Moreau, R., Kaushik, S. J., and Dabrowski, K., Ascorbic acid status as affected by dietary treatment in the Siberian sturgeon (*Acipenser baeri* Brandt): tissue concentration, mobilisation and L-gulonolactone oxidase activity, *Fish Physiol. Biochem.,* 15, 431, 1996.
64. Young, K. R., Brougher, D. C., and Soares, J. H., Ascorbic acid requirements, and L-gulonolactone oxidase enzyme activity in striped bass (*Morone saxatilis*), Book of Abstracts, *World Aquaculture '97,* Feb. 19–23, 1997, Seattle, WA, p. 511.

65. Mæland, A. and Waagbø, R., Examination of the qualitative ability of some cold water marine teleosts to synthesise ascorbic acid, *Comp. Biochem. Physiol.*, 121A, 249, 1998.

66. Mukhopadhyay, P. K., Nandi, S., Hassan, M. A., Dey, A, and Sarkar, S., Effect of dietary deficiency and supplementation of ascorbic acid on performance, vertebral collagen content, tissue vitamin and enzyme status in *Labeo rohita*, in *Technological Advancements in Fisheries*, Publ. #1-School Indl. Fish, Hameed, M.S. and Kurup, B.M., Eds., Cochin Univ. Sci. Tech., Cochin, 1998, 101–107.

67. Moreau, R. and Dabrowski, K., Biosynthesis of ascorbic acid by extant Actinopterygians, *J. Fish Biol.*, 57, 733, 2000.

68. Chatterjee, I. B., Ghosh, J. J., Ghosh, N. C., and Guha, B. C. Effect of cyanide on the biosynthesis of ascorbic acid by an enzyme preparation from goat-liver tissue, *Biochem. J.*, 70, 509, 1958.

69. Nelson, J. S., *Fishes of the World*, John Wiley & Sons, New York, 1994, 600 p.

70. Greenwood, P. H., *Polypterus* and *Erpetoichthys*: Anachronistic Osteichthyans. In *Living Fossils*, Eldredge, N., and Stanley, S. M., Eds., Springer–Verlag, New York, 1984, 143

71. Gardiner, B. G., Sturgeons as living fossils. In *Living Fossils*, Eldredge, N., and Stanley, S. M., Eds., Springer–Verlag, New York, 1984, 148.

72. Schultze, H.-P. and Wiley, E. O., The Neopterygian *Amia* as a living fossil. In *Living Fossils*, Eldredge, N., and Stanley, S. M., Eds., Springer–Verlag, New York, 1984, 153.

73. Wiley, E. O. and Schultze, H.-P. Family Lepisosteida (gars) as living fossils. In *Living Fossils*, Eldredge, N., and Stanley, S. M., Eds., Springer–Verlag, New York, 1984, 160.

74. Carroll, R. L., *Vertebrate paleontology and evolution*, W. H. Freeman. New York, 1988, 698 p.

75. Matusiewicz, M., Dabrowski, K., Völker, L., and Matusiewicz, K., Regulation of saturation and depletion of ascorbic acid in rainbow trout, *J. Nutr. Biochem.*, 5, 204, 1994.

76. Wahli, T., Frischknecht, R., Schmitt, M., Gabaudan, J., Verlhac, V., and Meier, W., A comparison of the effect of silicone coated ascorbic acid and ascorbyl phosphate on the course of ichthyophthiriosis in rainbow trout, *Oncorhynchus mykiss* (Walbaum), *J. Fish. Dis.*, 18, 347, 1995.

77. Wise, D. J., Tomasso, J. R., and Brand, T. M., Ascorbic acid inhibition of nitrite-induced methemoglobinemia in channel catfish, *Prog. Fish. Cult.*, 50, 77, 1988.

78. National Research Council, *Nutrient Requirements of Fish*, National Academy Press, Washington, D.C., 1993.

79. Rucker, R. B., Dubick, M. A., and Mouritsen, J., Hypothetical calculations of ascorbic acid synthesis based on estimates *in vitro*, *Am. J. Clin. Nutr.*, 33, 961, 1980.

80. Hilton, J. W., Cho, C. Y., and Slinger, S. J., Effect of graded levels of supplemental ascorbic acid in practical diets fed to rainbow trout (*Salmo gairdneri*), *J. Fish. Res. Bd. Can.*, 35, 431, 1978.

81. Burns, J. J., Mosbach, E. H., and Schulenberg, S., Ascorbic acid synthesis in normal and drug-treated rats, studied with L-ascorbic-1-C^{14} acid, *J. Biol. Chem.*, 207, 679, 1954.

82. Bannay, M. and Dimant, E., On the metabolism of L-ascorbic acid in the scorbutic guinea pig, *Biochim. Biophys. Acta.*, 59, 313, 1962.

83. Blom, J. H. and Dabrowski, K., Ascorbic acid metabolism in fish: Is there a maternal effect on the progeny?, *Aquaculture,* 147, 215, 1996.

84. Moreau, R., Dabrowski, K., Czesny, S., and Chila, F., Vitamin C-vitamin E interaction in juvenile lake sturgeon (*Acipenser fulvescens* R.), a fish able to synthesize ascorbic acid, *J. Appl. Ichthyol.,* 15, 250, 1999b.

85. Jenness, R., Birney, E. C., and Ayaz, K. L., Ascorbic acid and L-gulonolactone oxidase in lagomorphs, *Comp. Biochem. Physiol.,* 61B, 395, 1978.

86. Tsao, C. S. and Young, M., Effect of exogenous ascorbic acid intake on biosynthesis of ascorbic acid in mice, *Life Sci.,* 45, 1553, 1989.

87. Brand, J. C., Cherikoff, V., Lee, A., and Truswell, A. S., An outstanding food source of vitamin C, *Lancet,* 2, 873, 1982.

88. Stiassny, M. L. J. and Meyer, A., Cichlids of the rift Lakes, *Sci. Am.,* 280, 64, 1999.

89. Seehausen, O. and van Alphen, J. M., Can sympatric speciation by disruptive sexual selection explain rapid evolution of cichlid diversity in Lake Victoria?, *Ecol. Lett.,* 2, 262, 1999.

90. King, J. L. and Jukes, T. H., Non-Darwinian evolution: Most evolutionary change in proteins may be due to neutral mutations and genetic drift, *Science,* 164, 788, 1969 .

91. Bánhegyi, G., Csala, M., Braun, L., Garzó, T., and Mandl, J., Ascorbate synthesis dependent glutathione consumption in mouse liver, *FEBS Lett.,* 381, 39, 1996.

92. Mizushima, Y., Harauchi, T., Yoshizaki, T., and Makino, S., A rat mutant unable to synthesize vitamin C, *Experientia,* 40, 359, 1984.

93. Vögeli, P., Congenital ascorbic acid (vitamin C) deficiency in the pig, 1997, http://www.zb.inw.agrl.ethz.ch/wel_en.htm.

94. Pauling, L., Evolution and the need for ascorbic acid, *Proc. Natl. Acad. Sci. U.S.A.,* 67, 1643, 1970.

95. Jukes, T. H. and King, J. L., Evolutionary loss of ascorbic acid synthesizing ability, *J. Human Evol.,* 4, 85, 1975.

chapter five

Critical review of the requirements of ascorbic acid in cold and cool water fishes (salmonids, percids, plecoglossids, and flatfishes)

Jacques Gabaudan and Viviane Verlhac

Contents

0-8493-9881-9/01/$0.00+$.50
© 2001 by CRC Press LLC

5.1   Introduction

Vitamin C (ascorbic acid, AA) is essential for aquatic animals, as they cannot synthesize this nutrient and therefore depend fully on an exogenous supply. Vitamin C acts as a cofactor in hydroxylation reactions and as a strong biological reducing agent in tissues and cells. Substantial experimental evidence indicates that it is involved in several physiological processes, including growth, reproduction, response to stressors, wound healing, and immune response. Knowledge of the quantitative AA requirements for cold and cool water fish, accumulated in the past years, has resulted in the disappearance of clinical signs of AA deficiency in fish farming. Nevertheless, there are discrepancies among reports in the literature as to the quantitative AA requirements and what practical supplementation levels should be. A difficulty associated with studies on vitamin C has been the highly labile nature of this compound, causing ingested amounts to be often less than what was thought to be contained in the experimental diets. The development of chemically stabilized ascorbate forms with full AA bioavailability for fish, such as ascorbate-phosphate forms, has made possible the accurate determination of the minimum requirement based essentially on two criteria: growth and absence of clinical signs of deficiency.

Nutrient Requirements of Fish (NRC, 1993) is the standard reference stating nutritional requirements for maximum growth of major cultured species under laboratory conditions. It is intended to serve as a basis for setting minimum nutrient levels. However, for practical purposes vitamin C requirements should be considered as variable quantities depending on the criteria selected for their determination. Indeed, dietary vitamin C levels necessary to support and possibly enhance functions other than growth are also being investigated. For example, research with rainbow trout (*Oncorhynchus mykiss*) indicates that optimal reproduction performance, health status, and response to stressors require higher AA intakes than the amount needed for growth (Halver, 1995). Recent studies show that cellular immune parameters positively respond to enhanced cellular concentrations in vitamin C (Verlhac et al., 1995). This chapter reviews the established vitamin C requirements of salmonids, percids, plecoglocids, and flatfishes, and highlights the biological and environmental factors that may affect the dietary requirements of this micronutrient.

5.2 The NRC requirements for vitamin C

For the fish species of interest in this review, the *Nutrient Requirements of Fish* handbook of the NRC (1993) lists vitamin C requirements for Atlantic salmon (*Salmo salar*), Pacific salmon (*Oncorhynchus* spp), and rainbow trout (Table 5.1). A requirement of 50 mg AA/kg diet is attributed to Atlantic salmon on the basis of growth and absence of deficiency signs, and Pacific salmon on the basis of maximum kidney storage. In rainbow trout, 40 mg AA/kg diet are needed for growth and absence of deficiency signs, while maximum kidney storage requires 100 mg AA/kg diet. The NRC emphasizes that these values represent the minimum allowance for maximum growth under experimental conditions. The criteria used in the studies reported by the NRC are weight gain, absence of deficiency signs, and/or maximum kidney storage, and do not take into consideration factors such as size, age, growth rate, nutrient interrelationships, energy density of the feed, processing, storage losses, and environmental conditions. These factors are known to affect the vitamin C requirements (Halver, 1995) and, as it will be proposed in this chapter, must be taken into account when feeds are formulated for commercial production.

The NRC handbook does not list requirements for percids, plecoglocids, and flatfishes because of insufficient data, and the focus is on cultured species of importance for North America. Lovell (1994) suggested that when allowances are not available, analogies should be made within coldwater or warmwater, freshwater or saltwater, and carnivorous or omnivorous species. Data are available (Table 5.2) which support the extrapolations that can be made with the requirements of more widely studied species. Few nutritional requirements have been determined for percids, and therefore recommendations are made on the basis of trout and salmon requirements (Brown et al., 1996). The essentiality of AA has been demonstrated for Ayu *Plecoglossus altivelis* (Yamamoto, 1982). The effect of AA on schooling behaviors of ayu revealed that the optimal values for these parameters were obtained for dietary AA contents of 150 and 541 ppm, compared to 0 and 18 ppm (Koshio et al., 1997). However, the highest liver and brain vitamin C concentrations were obtained in the fish fed the diet containing 541 ppm.

Table 5.1 NRC vitamin C requirements: salmonids

Fish	Requirement (mg/kg diet)	Criteria	Reference
Atlantic salmon	50	Weight gain Absence of deficiency signs	Lall et al., 1990
Pacific salmon	50	Maximum kidney storage	Halver et al., 1969
Rainbow trout	100	Maximum kidney storage	Halver et al., 1969
	40	Weight gain Absence of deficiency signs	Hilton et al., 1978

Table 5.2   Vitamin C requirements of percids, plecoglossids, and flatfishes

Fish	Requirement (mg/kg diet)	Criteria	Reference
Yellow perch	required	Weight gain Absence of deficiency signs	Brown et al., 1996
Ayu	200	Weight gain	Yamamoto, 1982
	150	Distance to nearest neighbour Spontaneous activity	Koshio et al., 1997
Flounder	28–47	Weight gain	Teshima et al., 1991
Plaice	200	Weight gain	Rosenlund et al., 1990
Turbot	required	Renal granulomatous disease	Coustans et al., 1990
Turbot larva	1500–2500 µg *Artemia* DM	Constant tissue concentration	Merchie et al., 1996a
Turbot nursery	20	Weight gain and survival	Merchie et al., 1996b
	200	Hepatocyte ultrastructure	

Behavioral pattern was chosen as the response criteria to dietary vitamin C supplementation levels because it is related to the quality of fish raised for releasing projects. Quantitative vitamin C requirements for flatfishes have also not been systematically studied although its essentiality has been demonstrated for Japanese flounder (*Paralichthys olivaceus*), plaice (*Pleuronectes platessa*), and turbot (*Scophthalmus maximus*). In Japanese flounder, best weight gain and feed conversion ratio were obtained with diets containing 28 to 47 mg AA equivalent/kg diet, the source of vitamin C being the stable L-ascorbyl-2-phosphate-Mg (Teshima et al., 1991). A specific AA requirement has also been demonstrated for young plaice (Rosenlund et al., 1990) on the basis of growth, mortality, hepatosomatic index, AA liver concentrations, and vertebral hydroxyproline concentrations. The authors concluded that 200 mg AA/kg diet supplied as coated AA covers the requirement. Messager et al. (1986) demonstrated that turbot fed a diet deficient in vitamin C exhibited hypertyrosinemia, tyrosine crystal deposits in several organs, and cornea opacity. Later, Coustans et al. (1990) showed that three months after complete AA tissue depletion, turbot do not exhibit signs of scurvy as recorded in other fish species but a severe impairment of tyrosine metabolism leading to renal granulomatous disease which can cause high mortalities. The quantitative requirement has unfortunately not been determined for juvenile turbot. An attempt to determine the role of AA in the early development of turbot larvae was made by enriching live prey with two levels of ascorbyl palmitate (Merchie et al., 1996a). The enrichment of *Artemia* at levels of 1500 to 2500 µg AA/g dry weight did not lead to better growth and survival compared to the control containing 500 µg AA/g dry weight but did improve the pigmentation of the larvae, their resistance to experimental infection with *Vibrio anguillarum* and to a salinity stress test. Nursery stages of turbot raised with diets containing 0, 20, 200, and 2000 mg AA/kg exhibited no differences in production characteristics between treatments but the group fed the 200 mg/kg diet had the best hepatocyte ultrastructure with high storage of glycogen and lipid (Merchie at al., 1996b).

5.3 Ascorbic acid concentrations in tissues and organs

5.3.1 Ascorbic acid distribution

Vitamin C is concentrated in many vital organs with an active metabolism. The concentration of vitamin C in various tissues is related to the dietary intake of the vitamin. Moreover, some tissues such as brain and thymus, as well as leukocytes, accumulate in high concentrations. In these tissues, AA levels seem to be retained longer in the case of dietary vitamin C depletion compared to storage organs such as liver. Figure 5.1 presents the results of an experiment carried out in our laboratory, showing the distribution of AA in various tissues of rainbow trout fed 200 or 1000 mg AA equivalents per kg of feed, for one month. The very high levels found in thymus and brain tissues confirm the hypothesis of the importance of AA in preserving vital tissues from oxidation processes.

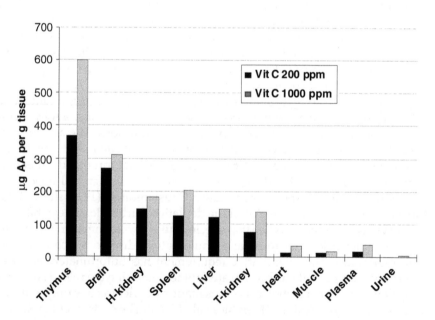

Figure 5.1 Tissue AA distribution in rainbow trout fed vitamin C as ascorbate phosphate at 200 and 1000 mg AA equivalents per kg of feed for one month. (H-kidney: headkidney, T-kidney: trunk kidney).

Liver and head kidney are important storage organs for vitamin C in fish. The high level found in the head kidney is likely to be related to the presence of lymphopoietic tissues. Trunk kidney and spleen are also able to store a large amount of vitamin C. The trunk kidney is the site of the chromaffin cells which are responsible for catecholamine biosynthesis. AA is concentrated at the site of catecholamine biosynthesis and is released with newly synthesized corticosteroids in response to stressors.

5.3.2   Leukocyte content

Leukocytes accumulate high concentrations of AA. Results presented in Table 5.3 show the increase in intracellular AA concentrations in relation to dietary intake of vitamin C in rainbow trout. It seems that the capacity of leukocytes to store vitamin C plateaued between 40 and 45 nmoles $AA/10^8$ cells, corresponding to 7.04 to 7.92 μg $AA/10^8$ cells (Verlhac et al., 1995). This has been observed in several experiments, even when rainbow trout were fed a 1000 ppm dose for two weeks (Verlhac et al., 1998). Although it is not possible to strictly compare the amount stored by the liver expressed in μg per g with the leukocyte content determined by number of cells, 7 μg of AA stored by 10^8 cells represents a very high amount suggesting that vitamin C plays an important role in protecting immune cells from oxidative damage.

5.3.3   Effect of environmental factors on vitamin C tissue concentrations

Environmental factors that can elicit a physiological response may affect the ascorbate requirement. Thomas (1990) showed that various environmental factors cause changes in the AA status of striped mullet (Table 5.4). Changes in AA concentrations of gills, kidney, liver, and brain had been studied after fish were subjected to different changes in environmental factors such as salinity, trauma, temperature, cadmium, and fuel oil. All the factors investi-

Table 5.3   Leukocytes and liver ascorbic acid content in rainbow trout fed the experimental diets containing different doses of vitamin C for three weeks

Dietary treatment	Vit C at 20 mg AA equiv./kg feed	Vit C at 200 mg AA equiv./kg feed	Vit C at 2000 mg AA equiv./kg feed	Vit C at 4000 mg AA equiv./kg feed
Liver AA in μg/g	22.7 ± 0.5	117.0 ± 0.1	161.0 ± 11.1	178.0 ± 3.8
Leukocyte AA in nmole/10^8 cells	11.2 ± 0.3	26.6 ± 3.1	40.7 ± 2.6	39.41 ± 4.2
in μg/10^8 cells	1.97 ± 0.05	4.68 ± 0.54	7.16 ± 0.46	6.93 ± 0.74

Note: Results are expressed as the mean $\pm$ SD of 2 replicate tanks for each treatment (Verlhac et al., 1995).

Table 5.4 Summary of environmental effects on tissue AA levels in striped mullet

Environmental factor	Interval (day)	Changes in tissue AA				Reference
		Gill	Kidney	Liver	Brain	
Salinity	7	⇑	⇓	—	⇑	Thomas, 1984
Trauma	4	⇓	⇓	—	—	"
Temperature	14	—	—	—	⇓	"
Cadmium	42	⇓	⇓	⇓	⇓	Thomas et al., 1982
Fuel oil	7	⇓	⇓	⇓	—	Thomas, 1987

Note: adapted from Thomas, 1990.

gated induced changes in AA stores, but variations were tissue specific. These observations led to the conclusion that AA depletion in the kidney can be regarded as a nonspecific response to environmental stressors.

5.4   Criteria for the determination of ascorbic acid requirement

The establishment of vitamin C requirements heavily depends upon the choice of the criteria used. Historically, the criteria commonly used were the prevention of deficiency symptoms, maximum growth, and survival rates. In order to determine that such minimum requirements do not lead to inadequate intakes, vitamin C requirements should also be based on vitamin C functions, that is, on achieving a vitamin C concentration in the organs and cells which give additional benefits. The requirement must also take into account the variability among individuals in a population and the variation between strains. The vitamin C requirement should therefore be assessed for specific purposes and culture conditions. The requirement has been shown to be affected by size, age, feeding rate, metabolic rate, physiological state, nutrient interrelationships, health status, and environmental factors (Lovell, 1994; Halver, 1995).

5.4.1   Age and metabolic rate

Most studies on vitamin C nutrition in fish have been carried out using young fish (Sandnes, 1991), but there is evidence that the need for vitamin C changes as the fish grow. On the basis of an experiment which lasted over 300 days, Hilton et al. (1978) suggested that the vitamin C requirement for rainbow trout for growth decreases with time. It also appeared to be related to the stage of development. Larval stages of turbot have a higher requirement in vitamin C than nursery stages (Merchie et al., 1996b). An experiment with two strains of rainbow trout exhibiting growth rate differences revealed that a higher growth rate induced a higher demand for vitamin C (Matusiewcz et al., 1994). The authors further suggested that mechanisms regulating AA turnover changed as the fish grew but when and why such changes occurred requires further elucidation.

5.4.2   Reproduction

The role of AA on the reproductive system of male fish has not been extensively studied although beneficial effects of vitamin C on human sperm quality have been demonstrated. In rainbow trout, sperm concentrations and
motility were found to be correlated to seminal plasma concentrations of
vitamin C (Ciereszko and Dabrowski, 1995). The dietary AA levels needed to
raise seminal concentrations of this vitamin resulting in beneficial effects
were 130 to 270 mg AA/kg which is well above the NRC (1993) recommendations for optimal growth (40 mg/kg). In a subsequent experiment, a correlation between hatching rate and the seminal plasma concentrations of AA
showed that survival of embryos from rainbow trout males with less than 7.3
μg AA/ml of seminal plasma was reduced (Dabrowski and Ciereszko, 1996).
Fecundity and embryo survival also increased in female rainbow trout with
increased dietary vitamin C levels (Blom and Dabrowski, 1995) and hatching
performance was affected by dietary vitamin C (Sandnes et al., 1984). Studies
by Waagbo et al. (1989) demonstrated a relationship between AA intake and
circulating levels of oestradiol-17 β and vitellogenin, confirming the role of
AA in hormone synthesis in endocrine tissues. Figure 5.2 shows a significant
increase in total AA concentration in the ovaries with increasing levels of
dietary AA after ten months of experimental feeding. Concurrently, a significant increase in egg mass produced (Figure 5.3), fecundity and embryo sur-

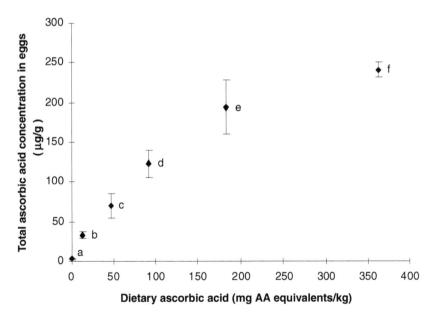

Figure 5.2   Concentration of total AA in eggs of rainbow trout at the time of ovulation after 10 months of feeding with graded levels of dietary ascorbate monophosphate (from Blom and Dabrowski, 1995).

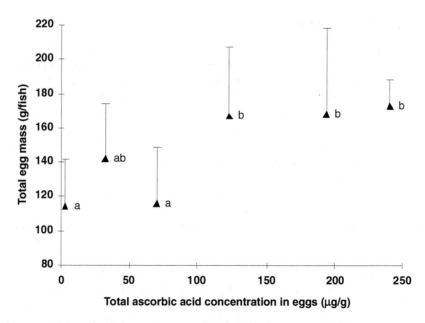

Figure 5.3   Influence of the total AA concentration in eggs of rainbow trout on the total egg mass produced (from Blom and Dabrowski, 1995).

vival were observed with increasing ascorbate intake and ovary ascorbate concentrations (Blom and Dabrowski, 1995). A dietary level of 350 to 400 mg AA/kg diet is therefore needed to saturate the ovaries and optimize reproductive performance in rainbow trout.

5.4.3   Nutrient interrelationships

AA interacts with nutritionally important metallic elements. In monogastric animals, vitamin C is known to stimulate the absorption of dietary nonheme iron and to reduce that of copper. In rainbow trout, the toxicity of waterborn copper leading to anemia is decreased with elevated levels of dietary AA (Yamamoto et al., 1981). AA appeared to have little effect on the absorption of dietary copper (Lanno et al., 1985). The reasons for this difference and the influence of vitamin C on the metabolism and excretion of copper in trout are not well understood. A deficiency in AA caused a reduction in circulating iron levels and a redistribution of iron tissue concentrations in rainbow trout (Hilton, 1989). AA enhanced the intestinal absorption of nonheme iron by reducing the ferric form of the element into the ferrous form and then forming a soluble chelate which is well absorbed. Although there is evidence that AA also influences iron metabolism at tissue levels in mammals, such effects were not yet elucidated in fish.

Vitamin C and vitamin E are *in vivo* chain-breaking antioxidants present in the aqueous phase of cytoplasm and at the cell membrane level, respectively. Studies in higher vertebrates indicated a cooperative inhibition of oxidative processes by both vitamins and a possible regeneration of vitamin E by vitamin C. Such vitamin E and C interactions have been observed both in homogeneous solutions and in liposomal membrane systems (Niki, 1987) thus suggesting that vitamin C is linked to the protection of membranes against free-radical damage. Vitamin C interrelationships with other nutrients require further research in aquatic animals in order to clarify their impact on practical dietary ascorbate supplement recommendations.

5.4.4   Health

Vitamin C plays an important role in the health status of fish, not only because it is essential but also because it acts as an antioxidant and therefore protects important functions of the immune system from oxidative damage. Maintaining an efficient system of defense is of major importance for growth as well as for improved resistance to environmental stressors and pathogens. Indeed, growing healthy fish requires them to be able to develop strong defense mechanisms against pathogen invasion. Improving the immune functions also leads to a better vaccine efficiency. Intensively reared fish may also be exposed to stressful conditions which frequently result in a depressed immune status. Consequently, fish with enhanced immune functions will be better prepared to combat both stress and pathogen invasion.

5.4.4.1   Immune response

As an antioxidant, vitamin C plays an important role in protecting cells from oxidative damage. Leukocytes are able to store a large amount of AA (Verlhac et al., 1995). Figure 5.4A shows how increased dietary doses of vitamin C correlated with higher intracellular levels of AA in rainbow trout leukocytes. Figure 5.4B presents the relationship between the intracellular AA concentrations of rainbow trout leukocytes and their oxidative burst response. The activation of the oxidative burst by a pathogen induces the production by the cell of very high amounts of reactive oxygen species that will attack the bacterial cell wall. These reactive oxygen species are released intra- and extracellularly and are also toxic to the cell itself. Both intra-cellular and extra-cellular AA protects the cell from auto-oxidation and prolongs their activity although they are committed to oxidation.

Studies on the efficiency of elevated doses of dietary vitamin C as an immunomodulator have been reviewed by Verlhac and Gabaudan (1997). Although some studies did not show a positive effect of vitamin C, a majority of studies demonstrate modulating properties of vitamin C on macrophage activities such as oxidative burst, phagocytic activity, and pinocytosis. Concerning lysozyme, some studies have shown an impact of vitamin C on lysozyme (Dabrowski et al., 1996) while others have not (Verlhac et al., 1996). Complement activity activated through the alternative pathway did not seem to be influenced by vitamin C. Those parameters characterize the nonspecific immune defense mechanisms.

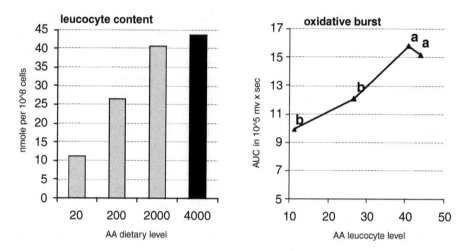

Figure 5.4 A. Leukocyte AA content in relation to dietary vitamin C intake in rainbow trout fed 20, 200, 2000, and 4000 mg AA equivalents / kg feed for 3 weeks. B. Oxidative burst of rainbow trout macrophages in relation to leukocyte AA concentration (from Verlhac et al., 1995).

Very few studies have demonstrated an enhancement of antibody response specific to an antigen in relation to increased dietary intake in vitamin C, while proliferation of lymphocytes, a cellular mechanism of specific immunity, clearly seems to be influenced by vitamin C. This might be because of the fact that these cells are also able to store a large amount of vitamin C to protect themselves from oxidative damage and better perform their activity. A complement activated through the classical pathway related to specific immunity is generally not affected by vitamin C (Verlhac and Gabaudan, 1997).

When considering the influence of elevated doses of dietary vitamin C, it is necessary to take into account the feeding duration. Vitamin C was first demonstrated to act after long-term feeding periods (Verlhac and Gabaudan, 1994). However, considering the use of vitamin C as a prophylactic feed additive to boost the immune system of fish and improve their health status before a predictable stress situation, it was demonstrated that a diet containing 1000 mg/kg vitamin C fed for two weeks gave similar results in terms of enhancing nonspecific cellular functions (Verlhac et al., 1998). In fact, it seemed that intracellular concentrations of rainbow trout leukocytes reached a plateau at a level that can be reached after a two-week feeding period at 1000 mg/kg. In this respect, most studies lack the determination of AA storage in target organs or tissues, which makes it difficult to compare studies. Experimental conditions and growth performance of the experimental fish also represent a determinant factor in comparing the observed effects in terms of vitamin C modulation.

5.4.4.2 *Disease resistance*

The difficulty to control challenge experiments has to be considered when evaluating the effect of feeding an elevated dose of vitamin C on resistance to infectious diseases. Among the studies that have been performed to evaluate the effect of vitamin C on disease resistance, the variety of the fish species, the pathogens, and the modes of infection tested make the results difficult to interpret even when the feeding duration is ignored (Verlhac and Gabaudan, 1997). Since vitamin C is not a therapeutic agent, no beneficial effect can reasonably be expected when 80% of mortality is reached within a few days after infection. Challenge experiments should be designed in order to reach a moderate level of mortality. From an economic point of view, even a 5% increase in survival could be extremely valuable. Navarre and Halver (1989) have clearly demonstrated the impact of the level of infection administered to rainbow trout fed graded levels of vitamin C on their survival (Figure 5.5). A significant effect of vitamin C could only be observed when the dose of infection was reduced.

5.4.4.3 *Wound healing*

AA plays a role in the collagen synthesis by acting as a coenzyme in the hydroxylation of proline into hydroxyproline. The capacity of vitamin C to improve wound repair has been evaluated in few studies. However, a significant

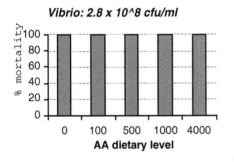
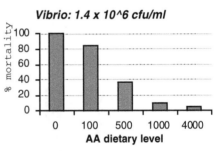
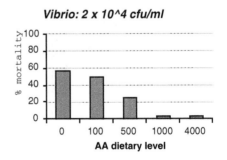

Figure 5.5 Effect of graded dietary doses of vitamin C on mortality of rainbow trout after challenge with vibrio at different concentrations (adapted from Navarre and Halver, 1989).

improvement has been demonstrated by Halver (1972) in rainbow trout and coho salmon fed graded doses of vitamin C as crystalline AA. Despite the difficulty to run such studies, it would be of great interest to reevaluate the effect of vitamin C on maintaining the integrity of natural epithelial barriers which prevent the pathogens from gaining access to the body tissues and circulation.

5.4.4.4 Resistance to stressors

All physiological stressors influence the health status of the fish and have an effect on diverting body reserves of micronutrients away from essential physiological functions to those necessary for immediate survival. Immunodepression is known to be a major secondary effect in the response of an organism to stress. Many situations such as transport, crowding, handling, and deteriorated water quality can cause a stress response in fish. The fish will react by secreting high levels of stress hormones (corticosteroids and catecholamines) which are known to be immunodepressive. Therefore, any measures alleviating stress would allow nutrients to be utilized for growth and improvement of health status.

The role of feeding a high dose of vitamin C in reducing stress effect could be explained by the fact that the release of stress hormones by interrenal and chromaffin cells is accompanied by a high consumption of AA. Therefore, by feeding an elevated level of vitamin C, the tissue stores of AA can be kept at a high level sufficient to satisfy hormonal changes, maintain immunocompetence, and allow fish to grow under normal conditions. Ishibashi et al. (1992) (Figure 5.6) have demonstrated that vitamin C fed at a

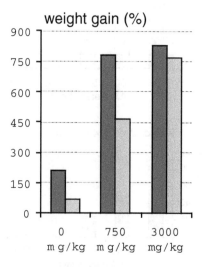
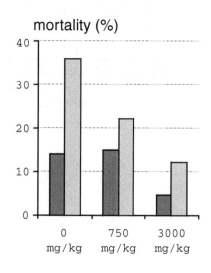

■Control □stressed

Figure 5.6 Effect of dietary vitamin C on growth and mortality of parrot fish submitted to an intermittent hypoxic stress (adapted from Ishibashi et al., 1992).

high level to fish subjected to an intermittent hypoxic stress did sustain a normal growth compared to that of control fish. Furthermore, mortality resulting from the effect of stress was decreased as the dietary vitamin C dose increased.

5.5 Conclusion

The vitamin C requirements stated in the NRC (1993) handbook represent the minimum allowance needed to sustain good growth and survival of salmonids under controlled laboratory conditions. Nevertheless, more information is needed on the vitamin C requirement in relation to specific criteria. It is likely that the requirement during winter feeding should be adapted to a reduced feed consumption in order to maintain appropriate AA tissue levels as feed intake is related to water temperature. Growth rates, particularly in salmonids fed high energy diets, have improved considerably in recent years and in some instances are much higher than those observed in most laboratory studies. The impact of fast growth of salmonids on the vitamin C needs of the fish should be evaluated. The synergistic effect of vitamin C and vitamin E should also be reconsidered in relation to high-energy diets. Environmental and health conditions and husbandry techniques are likely to affect the vitamin C requirement and such an impact should be taken into account. Ascorbate status based on tissue concentrations must be defined in relation to specific physiological conditions. Further, it would be relevant to evaluate the role of AA at the cellular level and to take into consideration inter-individual variability. This additional knowledge will be helpful to guarantee vitamin C concentrations in organs and cells amenable to each cultured fish species under its particular farming conditions.

References

Blom, J. H., Dabrowski, K., 1995. Reproductive success of female rainbow trout (*Oncorhynchus mykiss*) in response to graded dietary ascorbyl monophosphate levels. *Biol. Reprod.* 52, 1073–1080.

Brown, P. B., Dabrowski, K., Garling, D. L., 1996. Nutrition and feeding of yellow perch (*Perca flavescens*). *J. Appl. Ichthyol.* 12, 171–174.

Ciereszko, A., Dabrowski, K., 1995. Sperm quality and AA concentration in rainbow trout semen are affected by dietary vitamin C : an across-season study. *Biol. Reprod.* 52, 982–988.

Coustans, M.-F., Guillaume, J., Metailler, R., Dugornay, O., Messager, J.-L., 1990. Effect of an AA deficiency on tyrosinemia and renal granulomatous disease in turbot (*Scophtalmus maximus*) interaction with a slight polyhypovitaminosis. *Comp. Biochem. Physiol.* 97a, 145–152.

Dabrowski K., Ciereszko, A., 1996. AA protects against male infertility in a teleost fish. *Experientia* 52, 97–100.

Dabrowski, K., Matusiewicz, K., Matusiewicz, M., Hoppe, P. P., Ebeling, J., 1996. Bioavailability of vitamin C from two ascorbyl monophosphate esters in rainbow trout, *Oncorhynchus mykiss* (Walbum). *Aquacul. Nutr.* 2, 3–10.

Halver, J. E., 1972. The role of AA in fish disease and tissue repair. *Bull. Jap. Soc. Sci. Fish.* 38, 79–92.

Halver, J. E., 1995. Vitamin requirement study techniques. *J. Appl. Ichthyol.* 11, 215–224.

Halver, J. E., Ashley, L. M., Smith, R. R., 1969. AA requirements of coho salmon and rainbow trout. *Trans. Am. Fish Soc.* 90, 762–771.

Hilton, J. W., 1989. The interaction of vitamins, minerals and diet composition in the diet of fish. *Aquaculture* 79, 223–244.

Hilton, J. W., Cho, C. Y., Slinger, S. J., 1978. Effect of graded levels of supplemental AA in practical diets fed to rainbow trout (*Salmo gairdneri*). *J. Fish. Res. Board Can.* 35, 431–436.

Ishibashi, Y., Kato, K., Ikeda, S., Murata, O., Nasu, T., Kumai, H., 1992. Effects of dietary AA on tolerance to intermittent hypoxic stress in Japanese parrot fish. *Nippon Suisan Gakkaishi* 58, 2147–2152.

Koshio, S., Sakakura, Y., Lida, Y., Tsukamoto, K., Kida, T., Dabrowski, K., 1997. The effect of vitamin C intake on schooling behavior of amphidromous fish, ayu *Plecoglossus altiveli. Fish. Sci.* 63, 619–624.

Lall, S. P., Olivier, G., Weerakoon, D. E. M., Hines, J. A., 1990. The effect of vitamin C deficiency and excess on immune response in Atlantic salmon (*Salmo salar* L.). In: M. Takeda and T. Watanabe (Editors), *Proceedings of the Fish Nutrition Meeting*, Tokyo, pp. 427–441.

Lanno, R. P., Slinger, S. J., Hilton, J. W., 1985. Effect of AA on dietary copper toxicity in rainbow trout (*Salmo gairdneri* Richardson). *Aquaculture* 49, 269–287.

Lovell, R. T., 1994. Dietary nutrient allowances for fish. *Feedstuffs* July 20, 91–97.

Matusiewicz, M., Dabrowski, K., Völker, L., Matusiewicz, K., 1994. Regulation of saturation and depletion of AA in rainbow trout. *J. Nutr. Biochem.* 5, 204–212.

Merchie, G., Lavens, P., Dhert, Ph., Ulloa Gomez, M. G., Nelis, H., De Leenheer, A., Sorgeloos, P., 1996a. Dietary AA requirements during the hatchery production of turbot larvae. *J. Fish Biol.* 49, 573–583.

Merchie, G., Lavens, P., Storch, V., Übel, U., Nelis, H., De Leenheer, A., Sorgeloos, P., 1996b. Influence of dietary vitamin C dosage on turbot (*Scophtalmus maximus*) and european sea bass (*Dicentrarchus labrax*) nursery stages. *Comp. Biochem. Physiol.* 114a, 123–133.

Messager, J.-L., Ansquer, D., Metailler, R., Person-Le Ruyet, J., 1986. Induction expérimentale de l'hypertyrosinémie granulomateuse chez le turbot d'élevage (*Scophtalmus maximus*) par une alimentation carencée en acide ascorbique. *Ichtyophysiologica Acta* 10, 201–214.

NRC (National Research Council), 1993. *Nutrient Requirements of Fish.* Washington, D.C., National Academy Press.

Navarre, O., Halver, J. E., 1989. Disease resistance and humoral antibody production in rainbow trout fed high levels of vitamin C. *Aquaculture* 79, 207–221.

Niki, E., 1987. Interaction of ascorbate and α-tocopherol. *Ann. N.Y. Acad. Sci.* 498, 186–199.

Rosenlund, G., Jorgensen, L., Waagbo, R., Sandnes, K., 1990. Effects of different dietary levels of AA in plaice (*Pleuronectes platessa* L.). *Comp. Biochem. Physiol.* 96a, 395–398.

Sandnes, K., 1991. Vitamin C in fish nutrition—a review. *Fisk. Dir. Ser. Ernaering* 4, 3–32.

Sandnes, K., Ulgenes, Y., Braekkan, O. R., Utne, F., 1984. The effect of AA supplementation in broodstock feed on reproduction of rainbow trout (*Salmo gairdneri*). *Aquaculture* 43, 167–177.

Teshima, S. I., Kanazawa, A., Koshio, S., Itoh, S., 1991. L-ascorbyl-2-phosphate-MG as vitamin C source for the japanese flounder (*Paralichthys olivaceus*). In: S. J. Kaushik and P. Luquet (Editors), *Fish Nutrition in Practice*. Coll. les Colloq. , No. 61. INRA, Paris, pp. 157–166.

Thomas, P., 1984. Influence on some environmental variables on the ascorbic acid status of mullet, *Mugil cephalus* L., tissues. I. Effect of salinity capture—stress and temperature. *J. Fish Biol.* 25, 711–721.

Thomas, P., 1987. Influence on some environmental variables on the ascorbic acid status of striped mullet, *Mugil cephalus* Linn., tissues. III. Effects of exposure to oil. *J. Fish Biol.* 30, 485–494.

Thomas, P., 1990. Molecular and biochemical responses of fish to stressors and their potential use in environmental monitoring. *American Fisheries Symposium* 8, 9–28.

Thomas, P., Bally, M., Neff, J. M., 1982. AA status of mullet, *Mugil cephalus* Linn., exposed to cadmium. *J. Fish Biol.* 20, 183–196.

Verlhac, V., Gabaudan, J., 1994. Influence of vitamin C on the immune system of salmonids. *Aquaculture* 25, 21–36.

Verlhac, V., Gabaudan, J., 1997. The effect of vitamin C on fish health. Brochure No. 51002. F. Hoffmann-La Roche AG 4070 Basel, Switzerland.

Verlhac, V., Gabaudan, J., Schüep, W., 1995. Immunomodulation in fish : II. Effect of dietary vitamin C. In: K. Kurmaly (Editor), Proceedings of the 2nd Roche Aquaculture Center Conference on Nutrition and Disease. Bangkok, Thailand.

Verlhac, V., Gabaudan, J., Obach, A., Schüep, W., Hole, R., 1996. Influence of dietary glucan and vitamin C on non-specific and specific immune response of rainbow trout (*Oncorhynchus mykiss*). *Aquaculture* 143, 123–133.

Verlhac, V., Obach, A., Gabaudan, J., Schüep, W., Hole, R., 1998. Immunomodulation by dietary vitamin C and glucan in rainbow trout (*Oncorhynchus mykiss*). *Fish & Shellfish Immunol.* 8, 409–424.

Waagbo, R., Thorsen, T., Sandnes, K., 1989. Role of dietary AA in vitellogenesis in rainbow trout (*Salmo gairdneri*). *Aquaculture* 80, 301–314.

Yamamoto, S., 1982. The rearing of larval ayu (*Plecoglossus altivelis*) using a test diet of compound food supplemented with vitamin C and phosphatic salts. *Can. Transl. Fish. Aquat. Sci.* No. 4851, 2 pp.

Yamamoto, Y. K., Hayama, K., Ikeda, S., 1981. Effect of dietary ascorbic acid on copper poisoning in rainbow trout. *Bull. Jp. Soc. Sci. Fish.* 47, 1085–1089.

chapter six

Ascorbic acid requirement in freshwater and marine fish: is there a difference?

Marie F. Gouillou-Coustans and Sadasivam J. Kaushik

Contents

6.1   Introduction

An overview of vitamin requirements of cultured young fish was published recently.[1] As indicated by the author, the difference of vitamin requirement values through the two National Research Council (NRC) bulletins[2, 3] may reflect a difference in philosophy as to the purpose of the Nutrient Requirement handbooks of the NRC (Table 6.1). Thus, the NRC data[3] list the minimum level of vitamins necessary to prevent signs of deficiency for warm-water fish, especially channel catfish, *Ictalurus punctatus*, which remains the more extensively studied species, compared to common carp, *Cyprinus carpio*, or red sea bream, *Chrysophrys major*, while the NRC data[2] enlist recommended

0-8493-9881-9/01/$0.00+$.50

Table 6.1 Recommended levels and data on minimal vitamin requirements for absence of deficiency signs and maximal weight gain of different species

Vitamins	Recommended level[2]	Minimum level to prevent sign of deficiency[3]	Minimal requirement for maximal weight gain[4-6]			
	Coldwaterfish	Channel catfish	Rainbow trout	Channel catfish	Chick	Pig
A, IU	2500	1000–2000	2500	1000–2000	1500	2200
D_3, IU	2400	500–1000	2400	500	200	220
E, IU	30	30	50	50	10	16
K, mg	10	R	R	R	0.5	0.5
Ascorbic acid, mg	100	60	50	25–50	NR	NR
Thiamin, mg	10	1	1	1	1.8	1.5
Riboflavin, mg	20	9	4	9	3.6	4
Pyridoxine, mg	10	3	3	3	3	2
Pantothenic acid, mg	40	10–20	20	15	10	12
Niacin, mg	150	14	10	14	27	20
Biotin, mg	1	R	0.15	R	0.15	0.08
Folic acid, mg	5	NR	1	1.5	0.55	0.3
B_{12}, mg	0.02	R	0.01	R	0.009	0.02
Choline, mg	3000	R	1000	400	1300	600
Myo-Inositol, mg	400	NR	300	NR	NR	NR

Note: Level expressed as IU or mg.kg^{-1} dry diet

NR: not required

R: required

levels for coldwater fish. The latter values are two-fold (vitamin D_3, riboflavin, ascorbic acid [AA] and pantothenic acid), three-fold (pyridoxine), or ten-fold (thiamin, niacin) higher than the former. The NRC subcommittee on the Nutrient Requirements of Fish felt that combining the two previous bulletins was warranted because the current evidence of nutrient requirements does not vary greatly among fish species, with perhaps a few exceptions. It appears that the preferred source of certain nutrients may vary more widely than the actual requirement for specific nutrients for various fish species.[7] Thus, the latest NRC edition combines all species of finfish having commercial significance.[4] So, the values in the nutrient requirement table represent minimum requirements for maximum performance of fish under experimental conditions. For rainbow trout, *Oncorhynchus mykiss*, and channel catfish, a similitude of requirements can be noted for several vitamins such as vitamin A, vitamin E, thiamin, riboflavin, pyridoxine, pantothenic acid, niacin, folic acid, and AA. There are differences for vitamin D_3, choline and myo-inositol. For channel catfish, biotin and vitamin B_{12} are required in their diet but quantities are not yet determined. Similarly, the quantitative requirements for most of the vitamins have been established for Pacific salmon, *O. kisutch*, while only some of the needs are known for common carp and blue tilapia, *Oreochromis aureus*.[4] Finally, a comparison between phylogenetically distant species such as rainbow trout, channel catfish, chick, and pig demonstrate that the dietary requirements for most water-soluble vitamins are very similar (Table 6.1). Since vitamin requirements are comparable between these diverse species, we can hypothesize that differences in vitamin requirements will be negligible between freshwater and marine species.

In this general context, an attempt was made to verify whether vitamin supplementation at the minimal requirement level as defined by the NRC[4] is sufficient for two freshwater species such as rainbow trout and chinook salmon, *O. tschawytscha*, and two marine species, European sea bass, *Dicentrarchus labrax*, or red drum, *Sciaenops ocellatus*, fed practical diets.[8,9] The results demonstrate that supply of all vitamins at their minimal requirement levels is sufficient to support rapid and maximal weight gain for diverse species of freshwater or marine finfish, namely trout and chinook salmon juveniles, European seabass or red drum were fed diets with practical ingredients (Table 6.2).

Given this general background on overall vitamin requirements, it seems interesting to compare freshwater fish and marine fish requirements for a given vitamin. The literature describes a wide range of data on AA requirements within the same species, as well as among freshwater or marine species. Are these differences apparent or true?

6.2   AA requirement in fish

For all vitamins, data provided by the NRC[4] are currently the best estimate of requirements. Where several values appear in the literature, the NRC

Table 6.2   Growth, feed utilisation and daily nitrogen gain of freshwater and
marine fish fed practical diets containing graded levels of vitamin mix

	Vitamin mix[4]	**FBW (g)**	**FE**	**N gain (mg)**	**Ref**
R6	Excess	41.4	1.40	520	8
R7	1 ×	41.4	1.39	532	8
C8	1.25 ×	24.7	0.68	ND	8
C7	1 ×	21.9	0.57	ND	8
S6	2 ×	7.7	0.91	507	8
S5	1 ×	7.8	0.91	516	8
RD4	4 ×	121	0.98	699	9
RD3	3 ×	124	0.98	728	9
RD2	2 ×	122	1.00	737	9
RD1	1 ×	124	0.97	687	9

R6, R7, Rainbow trout (IBW = 2.8 g), grown at 15°C, over 12 weeks

C8, C7, Chinook salmon (IBW = 9.8 gr), grown at 12°C, over 16 weeks

S6, S5, Seabass (IBW = 0.8 gr), grown at 20°C, over 12 weeks

RD4 to RD1, Red Drum (IBW = 4.5 gr), grown at 28–31°C, over 7 weeks

IBW: initial body weight

FBW: final body weight

FE: (feed efficiency): wet weight gain/dry feed intake

ND: not determined

subcommittee appears to have retained the most reasonable estimate. In the case of rainbow trout, Woodward[1] listed different studies estimating the minimal dietary requirement for all vitamins and for the vitamin-like nutrients. For six vitamins (vitamin D_3, thiamin, riboflavin, pyridoxine, pantothenic acid, niacin), the NRC values correspond to the estimate of minimal dietary requirements. For the following vitamins (vitamin E, folic acid, biotin, and vitamin B_{12}), the values reported by the NRC[4] are two-fold higher than the estimates (Table 6.3). However, AA remains an exception for which the value of a 50 mg AA equivalent (eq).kg^{-1} diet is proposed with a safety margin corresponding to five-fold the minimum requirement level for young rainbow trout. The level of 10 mg AA eq.kg^{-1} diet was estimated with ascorbate monophosphate, a stable and biologically available form, using survival and weight gain as end point criteria.[10]

A great number of studies to establish dietary requirements for vitamin C in fish have used free AA (L-isomer form of AA) as the dietary source (unprotected or coated form). The lack of stability of this compound in fish feed and the high water solubility resulting in leaching have complicated accurate studies to evaluate the requirement. Consequently, quantification of vitamin C requirement by supplying unprotected AA was not possible. A precise estimate of minimal requirement requires a stable, biologically available form. From the extensive literature, it appears that ascorbate phosphates can serve as the stable and bioavailable form of vitamin C allowing nutri-

Table 6.3 Minimum vitamin requirements data based on best available estimated values obtained under experimental conditions for maximum performance of rainbow trout

Vitamin	Minimal requirement[4]	Estimate of minimal dietary requirement[1]
A, IU	2500	2500–5000
D_3, IU	2400	2400
E, IU	50	27.5
K, mg	R	0.45
Ascorbic acid, mg	50	10
Thiamin, mg	1.0	1.0
Riboflavin, mg	4.0	3.6
Pyridoxine, mg	3.0	2.0
Pantothenic acid, mg	20	19.1
Niacin, mg	10	10
Biotin, mg	0.15	0.08
Folic acid, mg	1.0	0.6
B_{12}, mg	0.01	0.007
Choline, mg	1000	430–4000
Myo-inositol, mg	300	250

Note: Level expressed as IU or $mg.kg^{-1}$ dry diet

R: Required

tionists to determine minimal dietary requirements much more readily and accurately than is possible with crystalline AA.[11–13] The AA supplied as ascorbyl monophoshate (AMP) or ascorbyl polyphoshate (APP) has been tested for diverse freshwater and marine species (Tables 6.4 and 6.5).

It follows from the above that care should be taken when comparing studies using different dietary sources of vitamin C and different experimental methods. In Tables 6.4 and 6.5 a selected number of studies are listed which are following the most desired experimental conditions:

a) Semi-purified diets (SPD) or practical diets (PD) tested under controlled environmental conditions on young rapidly growing animals (in most cases, the initial body weight [IBW] less than 10g);

b) Supplementation of AA in different forms, stable or not, as L-AA, ascorbyl palmitate (AP), AMP, or APP (studies using other AA forms were not included);

c) Graded levels of dietary AA with at least three values (including the zero level);

d) AA requirement values determined in relation to one or several response criteria, such as absence of deficiency symptoms, weight gain, feed efficiency, collagen or hydroxyproline concentration, and maximum AA storage in liver, in kidney, or in whole body (only studies analysing tissue AA concentrations reaching a plateau were included).

Table 6.4 Dietary AA requirement for growing freshwater fish determined under laboratory conditions

Species	IBW g	Diet	AA form	AA dietary graded levels mg.kg⁻¹	Requirement mg.kg⁻¹	Response criteria	Reference
Oncorhynchus mykiss	0.3	SPD	L-AA	0, 50, 100, 200, 400, 1000	100	WG, ADS	14
	6.7	PD	L-AA	0, 20, 40, 80, 160, 320, 640, 1280	40	WG, ADS	15
	0.2	SPD	L-AA	0, 20, 50, 100, 200, 500, 2000	20 50–100 500	WG, ADS C/H MLS	16
	NS	SPD	APP (AA eq)	0, 20, 40, 80	20	WG, ADS	11
	*	PD	APP (AA eq)	0, 20, 80, 320	20 320	WG IM/DR	17
	0.07	PD	AMP (AA eq)	0, 5, 10, 20, 40, 80, 160, 320	5	WG, ADS, FE, C/H	18
	*						
Oncorhynchus kisutch	2.9	PD	AMP (AA eq)	0, 10, 20, 40	10	WG, ADS, FE	10
	0.4	SPD	L-AA	0, 50, 100, 200, 400, 1000	50–100	WG, ADS	14
Salmo salar	3.3	PD	L-AA	0, 50, 100, 200, 500, 1000, 2000	50 1000	WG, ADS MKS	19
	0.2	PD	AMP (AA eq)	0, 10, 20, 40, 80, 160	10 20	WG, ADS C/H	20
	*						
Ictalurus punctatus	2.3	SPD	L-AA	0, 25, 50, 100, 200, 400	50	WG, FE	21
	2.3	SPD	L-AA	0, 30, 60, 90, 120, 180, 240	60	WG, ADS, C/H	22
	7.9	SPD	L-AA	0, 25, 50, 75, 100, 200	25 50	ADS WG, FE	23

Table 6.4 Continued

Species	IBW g	Diet	AA form	AA dietary graded levels mg.kg^{-1}	Requirement mg.kg^{-1}	Response criteria	Reference
	4.0	SPD	L-AA	0, 30, 60, 150	30 150	WG, ADS, C/H IM/DR	24
	13	SPD	AMP (AA eq)	0, 11, 22, 44, 132	11	WG, ADS, C/H	25
	6.5	PD	APP (AA eq)	0, 50, 150, 250	50 150	WG, FE, IM/DR MLS	26
Oreochromis aureus	1.5	SPD	L-AA	0, 25, 50, 65, 80, 95, 110	50	WG, ADS, FE	27
O. niloticus × O. aureus	1.1	SPD	L-AA	0, 40, 60, 80, 100, 125, 150, 200	79	WG, FE, MLS	30
Oreochromis niloticus	1.5	SPD	AMP (AA eq)	0, 30, 50, 70, 90, 120	37–42(17–20)	WG, C/H	28
	1.0	PD	L-AA	0, 500, 750, 1000, 1250, 3000, 4000	1250 (420)	WG, FE	31
Cirrhina mrigala	*	SPD	L-AA	0, 60, 300, 600, 900, 1200	650–750	WG, ADS	32
Morone chrysops × M. saxatilis	0.55	SPD	APP (AA eq)	0, 10, 20, 30, 45, 60, 75, 150	22 45	WG, ADS MLS	33
Cichlasoma urophthalmus	0.2	SPD	L-AA	0, 40, 80, 160, 320, 640, 1280, 2560, 5120, 10240	40	WG	34
Piaractus mesopotamicus	8.7	PD	AP	0, 50, 100, 200	110 139	ADS WG, ADS	35
Clarias gariepinus	20	PD	L-AA	0, 30, 60, 90, 120, 240	60 (46)	ADS, C/H	36

Table 6.4 Continued

Species	IBW g	Diet	AA form	AA dietary graded levels mg.kg^{-1}	Requirement mg.kg^{-1}	Response criteria	Reference
Clarias batrachus	1.5	SPD	L-AA	0, 50, 100, 150, 200, 250	200 (69)	WG	37
Cyprinus carpio	*	SPD	APP (AA eq)	0, 10, 30, 90 270, 810	45 354	WG MBS	38

IBW: initial body weight

PD: practical diet

SPD: semipurified diet

L-AA: L-ascorbic acid

AP: ascorbyl palmitate

AMP: ascorbyl monophosphate

APP: ascorbyl polyphosphate

WG: weight gain

FE: feed efficiency (wet weight gain/dry feed intake)

ADS: absence deficiency symptoms

MKS: maximum kidney storage

MLS: maximum liver storage

MBS: maximum body storage

IM: Immunoresistance

DR: disease resistance

C/H: collagen or hydroxyproline concentration

NS: not specified

* First feeding larvae

Table 6.5 Dietary AA requirement for growing marine fish determined under laboratory conditions

Species	IBW g	Diet	AA form	AA dietary graded levels mg.kg^{-1}	Requirement mg.kg^{-1}	Response criteria	Reference
Oplegnathus fascietus	NS	SPD	L-AA	0, 250, 500, 750, 1000, 3000	250 / 500	WG, ADS / MLS, MKS	42
Pleuronectes platessa	2.9	PD	L-AA	0, 200, 2000	200	WG, ADS, C/H	43
Sparus aurata	0.5	SPD	L-AA	0, 50, 100, 300, 600, 1200, 3200	100 (63)	ADS	44
Sebastes schlegeli	7	SPD	L-AA	0, 25, 50, 75, 150, 1500	144	WG	45
Seriola quinqueradiata	2.7	PD	AMP (AA eq)	0, 14, 28, 47	14–28	WG, ADS, C/H	46
Paralichthys olivaceus	43	SPD	AMP (AA eq)	0, 5, 14, 28, 47	28–47	WG, ADS, FE	47
Scophthalmus maximus		SPD	APP (AA eq)	0, 20, 200, 2000	20	WG	48
Scianops ocellatus	43	SPD	APP (AA eq)	0, 10, 20, 30, 45, 60, 75, 150	15	WG, FE, ADS	49
Lates calcarifer	0.9	PD	L-AA	0, 500, 700, 900, 1100, 2500, 5000	500–700 / 1100 / 2500	WG, ADS, FE / MLS / MKS	50
	1.9	PD	AMP (AA eq)	0, 30, 60, 100	30 (12.6)	WG, ADS, FE, C/H	51
	5.8	PD	AMP	0, 30, 60, 100	30	WG, ADS, FE	52

Table 6.5 Continued

Species	IBW g	Diet	AA form	AA dietary graded levels mg.kg^{-1}	Requirement mg.kg^{-1}	Response criteria	Reference
Dicentrarchus labrax	3.3	PD	L-AA	0, 60, 200, 420	200	WG, ADS, MLS	53
		SPD	APP (AA eq)	0, 20, 200, 2000	20	WG	48
	0.7	PD	APP (AA eq)	0, 5, 10, 20, 40, 80, 160, 320	5	WG, ADS, FE,	54
					5–31	C/H	
					121	MLS	

IBW: initial body weight

PD: practical diet

SPD: semipurified diet

L-AA: L-ascorbic acid

AP: ascorbyl palmitate

AMP: ascorbyl monophosphate

APP: ascorbyl polyphosphate

WG: weight gain

FE: feed efficiency (wet weight gain/dry feed intake)

ADS: absence deficiency symptoms

MKS: maximum kidney storage

MLS: maximum liver storage

MBS: maximum body storage

IM: immunoresistance

DR: disease resistance

C/H: collagen or hydroxyproline concentration

NS: not specified

* First feeding larvae

6.2.1   AA requirement in freshwater fish

Data from studies undertaken under laboratory conditions to evaluate AA needs in 14 freshwater species are summarized in Table 6.4.

In *O. mykiss* and *O. kisutch*, the minimum AA requirement based on weight gain (WG) and absence of deficiency signs (ADS) is about 20–100 mg.kg^{-1} diet while the minimum requirement to maintain a normal collagen formation (C/H) in the tissues was estimated to be 50–100 mg.kg^{-1}.[14–16] These estimations were obtained by adding L-AA into practical or semi-purified diets. Only one experiment shows maximum AA liver storage (MLS) for a dietary supplementation of 500 mg.kg^{-1} corresponding to 25-fold the allowance for good growth.[16] Since trials were compared,[10, 11, 17, 18] the AA requirement values obtained under similar conditions (APP or AMP supplementation, fry from first-feeding or young fish, tight AA gradient) corresponded to 5 to 20 mg AA eq.kg^{-1} diet. These data are much lower than those obtained using L-AA. It is noteworthy that these minimum requirements vary four-fold (5 to 20 mg AA eq.kg^{-1} diet). However, are these differences due to diet formulations, the choice of the lowest AA supplementation level or differences in initial body weights? Finally, results suggested that even if 20 mg AA eq.kg^{-1} allows fish good growth, 320 mg AA eq.kg^{-1} is needed to increase immunoresistance (IM) of fish exposed to infectious hematopoietic necrosis (IHN) virus.[17]

In Atlantic Salmon, *Salmo salar,* the supplemental AA level of 50 mg.kg^{-1} is sufficient for normal growth and prevention of deficiency signs and 1000 mg.kg^{-1} is required for maximum kidney storage (MKS).[19] Using AMP as AA source, Sandnes et al.[20] concluded that the minimum dietary requirement for optimal growth and normal development was in the range of 10–20 mg AA eq.kg^{-1} during the period studied.

Studies on the AA requirements of channel catfish conducted with semi-purified diets supplemented with L-AA indicated that 25–60 mg.kg^{-1} was required for maximal growth, feed efficiency (FE), absence of deficiency symptoms and collagen or hydroxyproline concentration (C/H).[21–24] Increased resistance against infection (DR) was provided when the level of supplemental vitamin C is increased to the highest dietary level (150 mg.kg^{-1}) at a water temperature of 23°C.[24] Thus, these estimations obtained with L-AA are very close to those observed in salmonids. El-Naggar and Lovell demonstrated that diets supplemented with the lowest concentration of AA in the form of L-AA or AMP provided for normal growth, bone collagen content, and hematocrits, with no overt signs of scurvy, in channel catfish. The concentrations (11 mg AA eq.kg^{-1} for AMP and 8.8 mg.kg^{-1} for L-AA, when fed) are undoubtedly near the minimum requirements for young channel catfish.[25] Results of the study by Li et al,[26] indicated that channel catfish require no more than 50 mg AA eq.kg^{-1} for normal growth, stress response, and disease resistance. It is worth noting that this value was the lowest level fed. As dietary vitamin C increases, ascorbate concentration in liver increases, reaching plateau values at 220 µg.g^{-1}. Thus, these estimations of AA requirement of channel catfish obtained with phosphate derivatives of AA are very close to those observed in salmonids.

In *O. aureus*, growth and feed conversion did not improve at dietary AA levels above 50 mg.kg^{-1}.[27] In hybrid tilapia, *O. niloticus* × *O. aureus*, weight gain and vertebral collagen concentration analysed by broken-line regression indicated that the adequate dietary AA was about 37–42 mg of AMP supplied as L-ascorbyl-2-monophosphate-Mg (17–20 mg AA eq.kg^{-1} diet).[28] Recently, these authors compared the biopotency of two ascorbate sources such as L-ascorbyl-2-monophosphate-Mg and L-ascorbyl-2-monophosphate-Na.[29] Based on growth, the results indicated that AMP-Na was more effective in meeting the vitamin C requirement for tilapia than AMP-Mg (15.98 and 18.82 mg AA eq.kg^{-1} diet, respectively). In a previous study using L-AA, the AA requirement was estimated to be 79 mg.kg^{-1}.[30] These estimations corresponded to requirement values obtained in salmonids and catfish. It is thus surprising that in nile tilapia, *O. niloticus*, based on the nutritional and pathological parameters investigated, the recommended dietary inclusion level was 1250 mg.kg^{-1} dry diet, equivalent to a net requirement (after processing and storage) of 420 mg.kg^{-1} dry diet.[31] Similar results have been reported in an Indian major carp, *Cirrhina mrigala*.[32] Statistical analysis of growth, deformity, and survival rates at the different AA dietary levels suggested an optimum requirement of 650–750 mg.kg^{-1} during early life of this species. The authors concluded that besides the high environmental temperature 25–35°C, which seemed the major factor for this high AA requirement, the larvae were given the diets as soon as they started feeding. The possible loss of the highly water-soluble AA into water in case of microparticulate diets was definitely involved.

For hybrid striped bass, *Morone chrysops* × *M. saxatilis*,[33] the first quantification of vitamin C needs showed that based on non-linear least squares regression analysis of weight gain, the minimum dietary requirement was 22 mg of active vitamin C.kg^{-1}. AA analysis in liver indicated that liver ascorbate did not continue to increase in fish fed diets containing greater than 45 mg active vitamin C.kg^{-1}.

For each of the following species, only one experiment has also been undertaken, and moreover with the use of unstable AA forms.[34–37] Thus, in *Cichlasoma urophthalmus*[34] the minimum level of vitamin C required for normal growth was 40 mg.kg^{-1} diet and the minimum level required to ensure health of fish was 110 mg.kg^{-1}. A study using ascorbyl palmitate suggested that 50 mg.kg^{-1} was sufficient to improve development of *Piaractus mesopotamicus* fingerlings but the optimum level under aquarium conditions, determined by regression analysis was 139 mg.kg^{-1} dry ration.[35] In African catfish, *Clarias gariepinus*, a concentration of 46 mg.kg^{-1} (inclusion level in the diet was 60 mg.kg^{-1}) prevented "broken-skull disease" and allowed optimum synthesis of vertebral collagen.[36] In *Clarias batrachus*,[37] best growth performance was obtained with 69 mg.kg^{-1} (inclusion level in the diet was 200 mg.kg^{-1}). The vitamin C needs for the purpose of weight gain and normal development were evaluated in the range of 40–139 mg.kg^{-1} for these species.

Our own results in the first-feeding common carp larvae showed that the requirement can be much lower.[38] Based on growth performance of the first-feeding larvae and maximal whole body AA concentrations, a dietary need of about 45 mg eq AA.kg^{-1} and 354 mg eq AA.kg^{-1} has been calculated, respectively, in common carp. For the purpose of growth performance, the dietary vitamin C needs of common carp are similar to those determined for salmonids, catfish, and tilapia. These results confirm the dietary essentiality of vitamin C for common carp, as already demonstrated by Dabrowski et al.[39] However, Reddy and Ramesh report that there is no need to supplement AA in the diet of this species as it may be synthesized under normal conditions.[40] In Siberian sturgeon, *Acipenser baeri*, a non-teleost fish, the presence of L-gulonolactone oxidase activity in kidney was combined with the absence of gross scorbutic signs in AA-free diet fed fish. It was found in this study that all the groups expressed very good growth rates and suggested that there was no need of dietary AA.[41]

6.2.2 AA requirement in marine fish

Data from studies undertaken under laboratory conditions to evaluate the AA requirement in 10 marine species are summarized in Table 6.5. The first point to note is that studies on vitamin C needs in marine species are more recent than those undertaken in freshwater species. Only tropical seabass, *Lates calcarifer* and European seabass are the subject of several studies.

The sufficient supplementary AA level to maintain normal growth and health of Japanese parrot fish *Oplegnatus fascietus* was estimated to be 250 mg.kg^{-1} and 500 mg.kg^{-1} in order to saturate hepatic and renal tissue AA concentrations.[42] Rosenlund et al.[43] suggested that a supplementation of 200 mg.kg^{-1} covers the nutritional requirement for normal growth and development in young plaices *Pleuronectes platessa*; but it is worth noting that in this study this was the lowest level tested. A study on gilthead bream, *Sparus aurata*, was made by Alexis et al.[44] to determine the pathological signs developed as a result of ascorbate deficiency. Fish fed semi-purified diets containing more than 100 mg.kg^{-1} (63 mg.kg^{-1} L-AA retained) showed normal appearance and behaviour during the course of the experiment. In juvenile Korean rockfish, *Sebastes schlegeli*, results showed that 144 mg AA.kg^{-1} was required for maximum growth.[45] In these forementioned experiments, L-AA was used as the vitamin C source and the estimated requirements were as high as 63–250 mg.kg^{-1} to support normal growth and to prevent pathological signs.

A level of about 14–28 mg eq AA.kg^{-1} diet was sufficient to support good growth and to prevent gross vitamin C deficiency symptoms in yellowtail, *Seriola quinqueradiata*. The supplementation of AMP elevated the AA concentrations of serum and liver and the ratio of hydroxyproline/proline in the bone.[46] Teshima et al.[47] found that 28–47 mg eq AA.kg^{-1} diet was required to support good growth of the Japanese flounder *Paralichthys olivaceus*. Approximately 37-day old post-larvae of turbot, *Scophthalmus maximus*, previously fed live prey were subsequently weaned to a semi-purified diet containing graded APP levels.[48] Reduced growth performance was only observed in the AA-free treatment, indicating that a marginal level of 20 mg eq AA.kg^{-1} diet was sufficient during post-larval stages. Finally, the regression analysis of weight gain data using the broken-line model resulted in a minimum vitamin C requirement of 15 mg active vitamin C.kg^{-1} for red drum, *Sciaenops ocellatus*,[49] reared in brackish water (6 g L^{-1}). Although studies referred to above were undertaken with different species, the comparison of these 2 groups of experiments strongly suggests that the use of AMP or APP leads to a significantly reduced level of "minimum requirement" (63–250 mg.kg^{-1} to 14–47 mg eq AA.kg^{-1}).

Such a re-evaluation of the minimum vitamin C need can be also seen in the different experiments undertaken with AA supplied as L-AA or AMP or APP forms in tropical seabass and European seabass. In a first experiment, a dietary AA level of 700 mg.kg^{-1} diet was found to be required for a maximum growth and prevention of external deficiency signs for tropical seabass fingerlings. A nearly 1100 mg.kg^{-1} diet was required to maintain an optimum liver AA concentration and it was suggested as the preferable level for safety reasons.[50] Crystalline AA was used as the vitamin C source and losses during feed processing, storage, and leaching made it difficult to estimate the exact amount of AA ingested by the fish. In a subsequent trial with Asian seabass fed experimental diets containing different levels of AMP, the same authors found that supplementing 30 mg (equivalent to 12.6 mg AA) per kilogram of diet was sufficient for normal growth, prevention of external deficiency signs and normal level of body hydroxyproline concentration.[51] These results have been confirmed recently.[52] In European seabass,[53] growth does not seem to be influenced by the vitamin C supply during the first 24 weeks, but after that and up to week 42, reduced growth becomes more and more evident in groups fed with zero and 60 mg.kg^{-1} in comparison with the groups fed 200 and 420 mg.kg^{-1}. Maximum liver storage was observed at a dietary AA level of 200 mg.kg^{-1}. The 41-day old post-larval of seabass, previously fed live prey, were subsequently weaned to a semi-purified diet containing graded APP levels. Reduced production results were only obtained in the AA-free treatment, indicating that a level of 20 mg AA eq.kg^{-1} diet was sufficient for post-larval of European seabass.[48] In another study, seabass juveniles fed AA free diet showed deficiency symptoms and poor growth, compared with fish fed AA supplemented diets (5 to 320 mg AA eq.kg^{-1}); a dose-dependent response of hepatic AA concentration and skin or whole-body hydroxyproline concentrations to dietary vitamin C levels was also recorded. The requirements were estimated to be 5, 31, and 121 mg AA eq.kg^{-1} for optimal growth or skin hydroxyproline concentrations, whole body hydroxyproline concentrations, and maximum liver storage, respectively.[54] So based on the weight gain, absence of deficiency signs, and tissue hydroxyproline concentrations, the minimum vitamin C requirements for Asian and European seabass were in the range of 5–31 mg AA eq.kg^{-1}.

Very little data is available on the quantitative vitamin C needs of marine flatfishes. As mentioned earlier, data of Rosenlund et al.[43] in juvenile plaice are higher than what has recently been reported for post-larval turbot.[48] It is worth noting that turbot is the only species where the most striking signs of AA deficiency is hypertyrosinemia, leading to a renal granulomatous disease. In deficient turbot, no external signs of scurvy were found, though proline hydroxylation was depressed by AA deficiency in skin and vertebrae.[55] However, hypertyrosinemia followed by tyrosine crystals deposition in renal tissues was identified and reinforced when the AA deficient diet was also low in several other vitamins (Table 6.6). For this temperate marine species, tyrosinemia remains the best biochemical indicator for AA deficiency. Alterations in kidney were observed in a vitamin C deficient euryhaline cichlid infected with *Mycobacterium* spp.[56] and in Asian seabass fed a diet deficient in AA.[52] The vitamin C deficiency also produces the same kidney pathology in gilthead bream.[44] Although the etiology of granulomatosis observed in these three species may be in relation to AA deficiency, more investigation with tyrosinemia measurement is necessary.

6.3   Is there a difference in requirement due to salinity?

The list of the different works in freshwater and marine fish indicated that in regard to vitamin C requirement, the response criteria such as weight gain

Table 6.6   Plasma tyrosine levels and hypertyrosinemia % in turbot fed diets supplemented or not with ascorbic acid (AA) and vitamin mixture (VM)

Parameters	Day	AA$^+$* VM$^+$	AA$^+$* VM$^-$	AA$^-$*VM$^+$	AA$^-$*VM$^-$
Tyrosine	38	66 ± 9	68 ± 4	81 ± 9	69 ± 5
hypertyrosinemia %		0	0	0	0
Tyrosine	65	68 ± 5	66 ± 7	71 ± 4	85 ± 8
hypertyrosinemia %		0	0	0	0
Tyrosine	94	55 ± 3	54 ± 2	240 ± 138	1173 ± 311
hypertyrosinemia %		0	0	7	33
Tyrosine	108	49 ± 2	46 ± 2	655 ± 228	1707 ± 334
hypertyrosinemia %		0	0	23	50
Tyrosine	142	71 ± 4	75 ± 3	1157 ± 292	3466 ± 349
hypertyrosinemia %		0	0	40	80
Tyrosine	159	68 ± 4	67 ± 4	1666 ± 477	3344 ± 355
hypertyrosinemia %		0	0	47	100

Header spanning: Diets[55]

Note: Plasma tyrosine level expressed as µMoles.L^{-1}.

(WG), absence of deficiency symptoms (ADS) and collagen or hydroxyproline (C/H) concentration were the most studied. The absolute values in relation to these criteria were in the range of 5–45 mg AA eq.kg^{-1} and 5–47 mg AA eq.kg^{-1} in freshwater and in marine species, respectively. Based on the above, it can be concluded that AA requirement was very similar in freshwater and marine species as indicated by optimal weight gain and normal development. Moreover, it is worth pointing out that in many experiments, the lowest dietary AA level used corresponded to the minimum requirement based on the weight gain, absence of deficiency symptoms, and collagen or hydroxyproline concentrations.

The accurate quantification of the requirement may depend upon various factors, such as fish size, age, growth rate, stage of sexual maturity, smoltification, type of diet, processing, and storage time of the diet as well as environmental stressors, such as disease, water temperature, and water quality and levels of environmental toxicants. The variability due to dietary factors can be reduced by using stable forms of vitamin C. The minor differences observed in experiments using the same species and AA-phosphate forms were probably due to the variability in the protocols. Consequently, a standardisation of the experimental protocols could facilitate comparisons. These minimum conditions are suggested: very small fish, if possible first-feeding larvae, to provide maximum growth response; a semi-purified diet, well accepted by fish; phosphate derivatives of AA in order to secure stability and bioavalability of vitamin C derivatives; tight gradient of AA supplementation with a minimum of 5–6 levels; and a minimum of response criteria such as growth performance and feed utilisation, absence of deficiency symptoms, tissue (liver, kidney) vitamin C concentrations and tissue (vertebrae, skin, muscle, whole body) collagen or hydroxyproline concentrations.

Special attention must be given to vitamin C storage. Tissue levels of a given vitamin may be useful as an index of nutritional status of the animal with respect to that vitamin. In that sense, they are complementary to growth studies in assessing the adequacy of the dietary concentration. With increasing dietary concentrations of the water-soluble vitamin, tissue level of the vitamin would be expected to reach a maximum or saturation level. Above this level they would not increase along with a further increase in dietary concentration—the excess intake being either excreted or metabolized.[10] Generally, requirement values based on maximum liver or kidney storage were much higher than the requirement values for the maximum weight gain and absence of deficiency symptoms. The actual requirement can vary considerably depending upon which criterion was used.

Vitamin C requirement levels determined under controlled laboratory cultures do not generally include surpluses. In practice, however, a margin of safety is commonly added to compensate for processing and storage losses, variation in composition and bioavailability of nutrients in feed ingredients, and variation in requirements caused by environmental effects. The requirements for vitamins are determined with diets containing purified and chemically defined ingredients that are highly digestible by fish. This fact should be considered when formulating diets from natural feedstuffs in which the bioavailability of the nutrients is markedly less than that in the laboratory diets.[4] In the case of AA, the NRC recommendations (50 mg.kg^{-1}) are already five-fold higher than the data on requirements of salmonids (10 mg.kg^{-1}). The recent evaluations by Kaushik et al.[8] and Gouillou-Coustans et al.[18] suggested that reduction of this level to 25 or 5 mg.kg^{-1} supplied as AMP in practical diet had no adverse effects on growth and health of juvenile rainbow trout. Whether this is applicable to stressful, high density aquaculture conditions needs to be verified.

6.4   Conclusion

A survey of literature on vitamin C requirements, expectedly, presented a picture of high requirements and high variability. A great part of the variability was due to the use of unstable forms of dietary vitamin C. Closer analysis showed that the quantitative requirement for AA was very similar between freshwater and marine species as measured by optimal weight gain and normal development. As already stated by Woodward,[1] differences among species in dietary vitamin requirement levels were probably the exception rather than the rule and large apparent differences in vitamin requirements can be attributed to methodological artifacts. This principle seemed to be verified in the case of vitamin C requirement of finfish and such comparative studies must be undertaken for the other fat and water-soluble vitamins.

References

1. Woodward, B., Dietary vitamin requirements of cultured young fish, with emphasis on quantitative estimates for salmonids, *Aquaculture*, 124, 133, 1994.
2. National Research Council, *Nutrient Requirements of Coldwater Fishes*, National Academy Press, Washington, D.C., 1981.
3. National Research Council, *Nutrient Requirements of Warmwater Fishes and Shellfishes*, National Academy Press, Washington, D.C., 1983.
4. National Research Council, *Nutrient Requirements of Fish*, National Academy Press, Washington, D.C., 1993.
5. National Research Council, *Nutrient Requirements of Poultry*, National Academy Press, Washington, D.C., 1984.
6. National Research Council, *Nutrient Requirements of Swine*, National Academy Press, Washington, D.C., 1988.
7. Lovell, R. T., and Wilson, R. P., Nutrient requirements of fish: revised NRC bulletin, in *Fish Nutrition in Practice*, INRA, Eds., Paris, 1993, 839.
8. Kaushik, S. J., Gouillou-Coustans, M. F., and Cho, C. Y., Application of the recommendations on vitamin requirements of finfish by NRC (1993) to salmonids and seabass using practical and purified diets, *Aquaculture*, 161, 463, 1998.
9. Gouillou-Coustans, M. F., Falguière, J. C., Philippe, R., and Kaushik, S. J., Response of red drum *Sciaenops ocellatus* to different dietary levels of vitamin

premix under tropical culture conditions, in *Island Aquaculture and Tropical Aquaculture,* Martinique 97, European Aquaculture Society, Ostende, Belgium, March 1997, 145.

10. Cho, C. Y. and Cowey, C. B., Utilization of monophosphate esters of ascorbic acid by rainbow trout (*Oncorhynchus mykiss*), in *Fish Nutrition in Practice,* INRA, Eds., Paris, 1993, 149.

11. Grant, B. F., Seib, P. A., Liao, M.-L., and Corpron, K. E., Polyphosphorylated L-ascorbic acid: a stable form of vitamin C for aquaculture feeds, *J. World Maricult. Soc.,* 20, 143, 1989.

12. Sandnes, K., Vitamin C in fish nutrition. A review, *Fisk. Dir. Skr. Ser, Ernoering,* 4, 3, 1991.

13. Dabrowski, K., Matusiewicz, M., and Blom, J. H., Hydrolysis, absorption and bioavailability of ascorbic acid esters in fish, *Aquaculture,* 124, 169, 1994.

14. Halver, J. E., Ashley, L. M., and Smith R. R., Ascorbic acid requirements of coho salmon and rainbow trout, *Trans. Amer. Fish. Soc.,* 98, 762, 1969.

15. Hilton, J. W., Cho, C. Y., and Slinger, S. J., Effect of graded levels of supplemental ascorbic acid in practical diet fed to rainbow trout (*Salmo gairdneri*), *J. Fish. Res. Board Can.,* 35, 431, 1978.

16. Sato, M., Kondo, T., Yoshinaka, R., and Ikeda, S., Effect of dietary ascorbic acid levels on collagen formation in rainbow trout, *Bull. Jpn. Soc. Sci. Fish.,* 48, 553, 1982.

17. Anggawati-Satyabudhy, A. M., Grant, B. F., and Halver, J. E., Effects of L-ascorbyl phosphates (AsPP) on growth and immunoresistance of rainbow trout (*Oncorhynchus mykiss*) to infectious hematopoietic necrosis (IHN) virus, in *The Current Status of Fish Nutrition in Aquaculture,* Takeda, M. and Watanabe, T., Eds., Tokyo University of Fisheries, Tokyo, 1989, 411.

18. Gouillou-Coustans, M. F., Métailler, R., Lebrun, L., Huelvan, C., Desbruyères, E., Moriceau, J., and Kaushik, S., Dietary ascorbic acid requirement of rainbow trout (*Oncorhynchus mykiss*) fry from first-feeding, presented at the IX International Symposium on Nutrition and Feeding in Fish. Miyazaki, May 22 to 24, 2000.

19. Lall, S. P., Olivier, G., Weerakoon, D. E. M., and Hines, J. A., The effect of vitamin C deficiency and excess on immune response in atlantic salmon (*Salmo salar* L.), in *The Current Status of Fish Nutrition in Aquaculture,* Takeda, M. and Watanabe, T., Eds., Tokyo University of Fisheries, Tokyo, 1989, 427.

20. Sandnes, K., Torrissen, O., and Waagbo, R., The minimum dietary requirement of vitamin C in Atlantic salmon (*Salmo salar*) fry using Ca ascorbate-2-monophosphate as dietary source, *Fish Physiol. and Biochem.,* 10, 315, 1992.

21. Andrews, J. W. and Murai, T., Studies of the vitamin C requirements of the channel catfish (*Ictalurus punctatus*), *J. Nutr.,* 105, 557, 1975.

22. Lim, C. and Lovell, R. T., Pathology of the vitamin C deficiency syndrome in channel catfish (*Ictalurus punctatus*), *J. Nutr.,* 108, 1137, 1978.

23. Murai, T., Andrews, J. W., and Bauernfeind, J. C., Use of L-ascorbyl acid, ethocel coated ascorbic acid and ascorbate 2-sulfate in diets for channel catfish (*Ictalurus punctatus*), *J. Nutr.,* 108, 1761, 1978.

24. Durve, V. S. and Lovell, R. T., Vitamin C and disease resistance in channel catfish (*Ictalurus punctatus*), *Can. J. Fish., Aquat. Sci.,* 39, 948, 1982.

25. El Naggar, G. O. and Lovell, R. T., L-ascorbyl-2-monophosphate has equal antiscorbutic activity as L-ascorbic acid but L-ascorbyl-2-sulfate is inferior to L-ascorbic acid for channel catfish, *J. Nutr.,* 121, 1622, 1991.

26. Li, M. H., Wise, D. J., and Robinson, E. H., Effect of dietary vitamin C on weight gain, tissue ascorbate concentration, stress response, and disease resistance of channel catfish (*Ictalurus punctatus*), *J. World Aquacult. Soc.*, 29, 1, 1998.

27. Stickney, R. R., McGeachin, R. B., Lewis, D. H., Marks, J., Riggs, A., Sis, R. F., Robinson, E. H., and Wurts, W., Response of *Tilapia aurea* to dietary vitamin C, *J. World Maricult. Soc.*, 15, 179, 1984.

28. Shiau S. Y. and Hsu, T. S., L-ascorbyl-2-sulfate has equal antiscorbutic activity as L-ascorbyl-2-monophosphate for tilapia (*Oreochromis niloticus* × *O. aureus*), *Aquaculture*, 133, 147, 1995.

29. Shiau S. Y. and Hsu, T. S., Quantification of vitamin C requirement for juvenile hybrid tilapia, (*Oreochromis niloticus* × *O. aureus*), with L-ascorbyl-2-monophosphate-Na and L-ascorbyl-2-monophosphate-Mg, *Aquaculture*, 175, 317, 1999.

30. Shiau, S. Y. and Jan, F. L., Dietary ascorbic acid requirement of juvenile tilapia (*Oreochromis niloticus* × *O. aureus*), *Nippon Suisan Gakkaishi*, 58, 671, 1992.

31. Soliman, A. K., Jauncey, K., and Roberts, R. J., Water-soluble vitamin requirements of tilapia: ascorbic acid (vitamin C) requirement of Nile tilapia, (*Oreochromis niloticus* L.), *Aquaculture and Fish Management*, 25, 269, 1994.

32. Mahajan, C. L. and Agrawal, N. K., Nutritional requirement of ascorbic acid by indian major carp, (*Cirrhina mrigala*), during early growth, *Aquaculture*, 19, 37, 1980.

33. Sealey, W. M. and Gatlin III, D. M., Dietary vitamin C requirement of hybrid striped bass *Morone chrysops* × *M. saxatilis*, *J. World Aquacult. Soc.*, 30, 297, 1999.

34. Chavez de Martinez, M. C., Vitamin C requirement of the mexican native cichlid (*Cichlasoma urophthalmus*, Gunther), *Aquaculture*, 86, 409, 1990.

35. Martins, M. L., Effect of ascorbic acid deficiency on the growth, gill filament lesions and behavior of pacu fry (*Piaractus mesopotamicus* Holmberg, 1887), *Braz. J. Med. Biol. Res.*, 28, 563, 1995.

36. Eya, J. C., "Broken-skull disease" in african catfish *Clarias gariepinus* is related to a dietary deficiency of ascorbic acid, *J. World Aquacult. Soc.*, 27, 493, 1996.

37. Mishra, S. and Mukhopadhyay, P. K., Ascorbic acid requirement of catfish fry *Clarias batrachus* (Linn), *Indian J. Fish.*, 43, 157, 1996.

38. Gouillou-Coustans, M. F., Bergot, P., and Kaushik, S. J., Dietary ascorbic acid needs of common carp (*Cyprinus carpio*) larvae, *Aquaculture*, 161, 453, 1998.

39. Dabrowski, K., Hinterleitner, S., Sturmbauer, C., El-Fiky, N., and Wieser, W., Do carp larvae require vitamin C?, *Aquaculture*, 72, 295, 1988.

40. Reddy, H. R. V. and Ramesh, T. J., Dietary essentiality of ascorbic acid for common carp *Cyprinus carpio* L., *Ind. J. Exp. Biol.*, 34, 1144, 1996.

41. Moreau, R., Kaushik, S. J., and Dabrowski, K., Ascorbic acid status as affected by dietary treatment in the siberian sturgeon (*Acipenser baeri* Brandt): tissue concentration, mobilisation and L-gulonolactone oxidase activity, *Fish Physiol. and Biochem.*, 15, 431, 1996.

42. Ishibashi, Y., Ikeda, S., Murata, O., Nasu, T., and Harada, T., Optimal supplementary ascorbic acid level in the japanese parrot fish diet, *Nippon Suisan Gakkaishi*, 58, 267, 1992.

43. Rosenlund, G., Jorgensen, L., Waagbø, R., and Sandnes, K., Effects of different dietary levels of ascorbic acid in plaice *Pleuronectes platessa* L., *Comp. Biochem. Physiol.*, 96, 395, 1990.

44. Alexis, M. N., Karanikolas, K. K., and Richards, R. H., Pathological findings owing to the lack of ascorbic acid in cultured gilthead bream (*Sparus aurata* L.), *Aquaculture*, 151, 209, 1997.

45. Lee, K. J., Kim, K. W., and Bai, S. C., Effects of different dietary levels of L-ascorbic acid on growth and tissue vitamin C concentration in juvenile Korean rockfish, *Sebastes schlegeli* (Hilgendorf), *Aquacult. Res.*, 29, 237, 1998.

46. Kanazawa, A., Teshima, S., Koshio, S., Higashi, M., and Itoh, S., Effect of L-ascorbyl-2-phosphate-Mg on the yellowtail *Seriola quinqueradiata* as a vitamin C source, *Nippon Suisan Gakkaishi*, 58, 337, 1992.

47. Teshima, S. I., Kanazawa, A., Koshio, S., and Itoh, S., L-ascorbyl-2-phosphate-Mg as vitamin C source for the japanese flounder (*Paralichthys olivaceus*), in *Fish Nutrition in Practice*, INRA, Eds., Paris, 1993, 157.

48. Merchie, G., Lavens, P., Storch, V., Ubel, U., Nelis, H., Deleenheer, A., and Sorgeloos, P., Influence of dietary vitamin C dosage on turbot (*Scophthalmus maximus*) and european seabass *(Dicentrarchus labrax)* nursery stages, *Comp. Biochem. Physiol.*, 114, 123, 1996.

49. Aguirre P. and Gatlin III, D., M., Dietary vitamin C requirement of red drum *Sciaenops ocellatus, Aquaculture Nutr.*, 5, 247, 1999.

50. Boonyaratpalin, M., Unprasert, N., and Buranapanidgit, J., Optimal supplementary vitamin C level in seabass fingerling diet, in *The Current Status of Fish Nutrition in Aquaculture*, Takeda, M. and Watanabe, T., Eds., Tokyo University of Fisheries, Tokyo, 1989, 149.

51. Boonyaratpalin, M., Boonyaratpalin, S., and Supamataya, K., Ascorbyl-phosphate-Mg as a dietary vitamin C source for seabass (*Lates calcarifer*), presented at the third Asian Fisheries Forum, Singapore, October 26 to 30, 1992.

52. Phromkunthong, W., Boonyaratpalin, M., and Storch, V., Different concentrations of ascorbyl-2-monophosphate-magnesium as dietary sources of vitamin C for seabass, *Lates calcarifer, Aquaculture*, 151, 225, 1997.

53. Saroglia, M. and Scarano, G., Experimental induction of ascorbic acid deficiency in seabass in intensive aquaculture, *Bull. Eur. Ass. Fish Pathol.*, 12, 96, 1992.

54. Fournier, V., Gouillou-Coustans, M. F., and Kaushik, S. J., Hepatic ascorbic acid saturation is the most stringent response criterion for determining the vitamin C requirement of juvenile European sea bass (*Dicentrarchus labrax*), *J. Nutr.*, 130, 617, 2000.

55. Coustans, M. F., Guillaume, J., Métailler, R., Dugornay, O., and Messager, J. L., Effect of an ascorbic acid deficiency on tyrosinemia and renal granulomatous disease in turbot (*Scophthalmus maximus*). Interaction with a slight polyhypovitaminosis, *Comp. Biochem. Physiol.*, 97, 145, 1990.

56. Chavez de Martinez, M. C. and Richards, R. H., Histopathology of vitamin C deficiency in cichlid, *Cichlasoma urophthalmus* (Günther), *J. Fish Dis.*, 14, 507, 1991.

chapter seven

Requirements of L-ascorbic acid in a viviparous marine teleost, Korean rockfish, Sebastes schlegeli (Hilgendorf)

Sungchul Charles Bai

Contents

Abstract

This chapter reviews our three previous experiments of the L-ascorbic acid (AA) requirements in Korean rockfish, *Sebastes schlegeli* (Hilgendorf)

Figure 7.1   Korean rockfish, *Sebastes schlegeli*.

(Figure 7.1). The first experiment was conducted to develop an experimental model and a semipurified diet for AA requirement study. In the second experiment, the effects of different dietary levels of AA on growth and tissue AA concentrations were studied. In the third experiment, tissue AA concentrations of wild and cultured Rockfish averaging 400 ± 7.8g and 51 ± 6.8g were examined.

Results from the first experiment suggest that the experimental model and the semipurified diet can be useful for vitamin C requirement study, and also suggest that the dietary vitamin C requirement is greater than 39.7 mg AA, but 144.6 mg AA kg^{-1} diet is adequate for the maximum growth of juvenile Korean rockfish. In the second experiment, fish were divided into six groups and given one of six semipurified diets for 16 weeks. Fish fed the C_0 diet showed the vitamin C deficiency symptoms such as scoliosis, shortened operculae, exophthalmia, and fin hemorrhage after 12 weeks. The broken line analysis model for weight gain, specific growth rate, protein efficiency, and feed efficiency showed that optimum levels of vitamin C were 102.2, 102.1, 102.5 and 100.3 mg AA kg^{-1} diet, respectively. These findings suggest that a dietary AA level of 102.5 mg AA kg^{-1} diet is required for maximum growth of juvenile Korean rockfish. Liver and brain AA concentrations of the larger wild immature adult fish were significantly higher than those of wild and cultured young growing fish, while these AA concentrations were not significantly different from those of cultured immature adult fish. Also, the average liver and brain AA concentrations of wild young growing fish were significantly higher than those of cultured young growing fish. No significant difference existed in average muscle and gill AA concentrations of wild and cultured fish examined in this experiment.

7.1 Introduction

Many general physiological functions of L-ascorbic acid (AA) are well defined; among them, AA is known as a co-factor in the hydroxylation of proline to hydroxyproline, important for helical structure of collagen. AA is also the most powerful reducing agent available to cells, losing two hydrogen atoms to become dehydroascorbic acid, and is of general importance as an antioxidant because of its high reducing potential. All animal species require AA, but dietary essentiality is not a common characteristic of vertebrates because some can synthesize it from hexose sugars.[1] However, it has been postulated that dietary essentiality of AA was due to the absence of the enzyme L-gulonolactone oxidase which catalyzes the conversion of L-gulonolactone to AA in many species of fish, including channel catfish, *Ictalurus punctatus*, rainbow trout, *Oncorhynchus mykiss*, brook trout, *Salvelinus fontinalis*, blue tilapia, *Tilapia aurea*, Japanese eel, *Anguilla japonica*, coho salmon, *Oncorhynchus kisutch*, yellowtail, *Seriola quinqueradiata*, and shrimp, *Penaeus* sp.[2-15]

Vitamin C requirements vary among species and even within the same species maintained under different environmental conditions.[16] The recommended levels of dietary vitamin C for growing fish determined with chemically defined diets under controlled conditions include the following: 50 mg kg^{-1} diet for Salmonidae,[14, 17] 40–100 mg kg^{-1} diet for rainbow trout, *Oncorhynchus mykiss*,[14, 18] 11–60 mg kg^{-1} diet for channel catfish, *Ictalurus punctatus*,[19-21] 50 mg kg^{-1} diet for blue tilapia, *Oncorhynchus aurea*,[10] and 122 mg kg^{-1} diet for yellowtail, *Seriola quinqueradiata*.[14]

Korean rockfish, *Sebastes schlegeli* is a commercially important fish species in Korea. This species has desirable characteristics for aquaculture including high tolerance to water temperature changes, ease of seedling production due to viviparous reproductive style, and the ability to withstand high stocking density. Commercial fish culture has been rapidly developing since 1987, and its production is the second largest followed by flounder, *Paralichthys olivaceus*, in Korean mariculture. Recently in our laboratory, a series of experiments has been conducted to develop an experimental model for vitamin C requirement study, and to look into the long-term and short-term effects of different dietary levels of AA on growth and tissue AC concentrations.[22-24] Therefore, this chapter will review our previous experiments of the AA requirements in a viviparous marine teleost, Korean rockfish.

7.2   Materials and methods

7.2.1   Experimental diets

The semipurified basal diets composition used in the first and second experiments is shown in Table 7.1. Six experimental diets supplemented with 0, 25, 50, 75, 150, or 1500 mg L-ascorbic acid (AA) kg^{-1} diet (C_0, C_{25}, C_{50}, C_{75}, C_{150}, and C_{1500}, respectively) were prepared by adding appropriate amounts of vitamin C pre-mixture (10 mg AA g^{-1} cellulose) at the expense of cellulose. The experimental diets were formulated to contain 48% crude protein and 18 kJ available energy g^{-1}. The estimated available energy of the experimental diet was adjusted to have 18 kJ g^{-1} diet (16.7, 16.7, and 37.7 kJ g^{-1} for protein, carbohydrate, and lipid, respectively).[25, 26] Fish meal (10%) was added to enhance

Table 7.1   Composition of the basal diet
(dry matter basis)

Ingredient	%
Casein, vitamin free[1]	30.0
Gelatin[1]	10.0
White fish meal (defatted)[2]	10.0
Dextrin[1]	27.0
L-Arginine[3]	0.5
L-lysine·HCl[2]	0.5
DL-methionine[2]	0.5
Pollack oil	5.0
Corn oil[4]	5.0
Carboxymethylcellulose	2.0
α-Cellulose	2.5
Vitamin premix (vitamin C free)[5]	3.0
Mineral premix[6]	4.0

[1] United States Biochemical, Cleveland, OH 44122

[2] Kum Sung Feed Co., Pusan, Korea

[3] Yunsei Chemical Co., Japan

[4] Corn oil 4.9% + DHA, EPA 0.1% (25% DHA + EPA mixture)

[5] Contains (as g/100g premix) : DL-calcium pantothenate, 0.5; choline bitartrate, 10; inositol, 0.5; menadione, 0.02; niacin, 0.5; pyridoxine HCl, 0.05; riboflavin, 0.1; thiamine mononitrate, 0.05; DL-α-tocopheryl acetate, 0.2; retinyl acetate, 0.02; biotin, 0.005; folic acid, 0.018; B_{12}, 0.0002; colecalciferol, 0.008; alpha-cellulose, 87.06

[6] H-440 premix No. 5 (mineral) (NRC, 1973)

their palatability to Korean rockfish because we observed that the purified casein–gelatin based diet, low in fish meal (5%), was not readily accepted by fish in the previous feeding trial (unpublished data). Fish meal was extracted three times with a chloroform/methanol mixture (2:1, v/v) for one day and then air dried before incorporating into the experimental diets.[24] Vitamin C concentrations in the experimental diets were determined at the beginning and at the end of the feeding trial by the modified procedures of Thenen[27] as described by Bai and Gatlin.[28] Based on this analysis, the average of the vitamin C level between the beginning and the end of the first and second experiments is shown in Table 7.2. Less than 20% loss of AA activity was observed in the experimental diets during storage at −35°C for up to 4 months.

7.2.2   Experimental fish and feeding trials

Experiment 1

Prior to the start of the feeding trial, fish were fed the basal diet C_0 for 4 weeks to adjust to the semipurified diet and to deplete possible body reserves of vitamin C. The feeding trial was conducted in 60 L flow-through aquaria receiving filtered sea water at a rate of 1 L min^{-1}. Supplemental aeration also was provided to maintain dissolved oxygen near air saturation. Water temperature ranged from 15 ± 0.5°C at the beginning to 17 ± 0.5°C at the end of the experiment. Experimental fish, averaging 12.6 g, were randomly distributed in each aquarium in groups of 25 (total weight of 316.3 ± 0.58 g; mean ± S.E.M). The diets were fed twice a day at 09:00 and 16:00 for 8 weeks to triplicate groups of fish to satiation (approximately 2% of wet body weight per day).

Experiment 2

Prior to the start of the feeding trial, fish were fed the basal diet for 1 week to adjust to the semi-purified diet and to deplete possible body reserves of

Table 7.2   Proximate analysis (%) and L-ascorbic acid (AA) concentrations (µg AA g^{-1} diet) of the experimental diets (dry matter basis)[1,2]

	Diets					
	C_0	C_{25}	C_{50}	C_{75}	C_{150}	C_{1500}
Moisture (%)	30.1	29.4	30.0	32.2	32.7	30.9
Crude protein (%)	49.4	50.9	51.4	50.9	50.0	49.7
Crude lipid (%)	9.2	8.6	7.8	9.2	10.2	9.6
Ash (%)	7.2	6.3	5.7	8.2	7.6	6.9
AA (Exp. 1)[3]	39.7	64.5	88.4	98.7	144.6	1542.1
AA (Exp. 2)[3]	17.1	39.1	69.2	91.9	144.0	1395.0

[1] Values are means of duplicate samples.
[2] C_0, C_{25}, C_{50}, C_{75}, C_{150}, and C_{1500}; 0, 25, 50, 75, 150, and 1500 mg AA supplementation per kg diet.
[3] The values are the average of the AA level between the beginning and the end of the experimental period.

vitamin C. The feeding trial was conducted in 60 L flow-through aquaria receiving filtered sea water at a rate of $1\,L\,min^{-1}$. Supplemental aeration was also provided to maintain dissolved oxygen near air saturation. The temperature of water ranged from $18 \pm 0.5°C$ at the beginning to $13 \pm 0.2°C$ at the end of the experiment. Experimental fish averaging 7 g were randomly distributed (as groups of 25 fish; total weight of $178 \pm 0.41g$). The diets were fed to triplicate groups of fish to satiation (approximately 3% of wet body weight per day at the beginning and 1.5% at the end of feeding trial). Fish were fed twice a day at 09:00 and 16:00 during the first 8 weeks, and then once a day at 16:00 after 8 weeks until the end of feeding trial.

Juvenile Korean rockfish were produced at the Wando hatchery (a research station of the National Fisheries Research and Development Agency). Total fish weight in each aquarium was determined every 4 weeks following anesthesia with $100\,mg\,L^{-1}$ of MS 222 (tricaine). The amount of diet fed to fish was adjusted accordingly. The inside of the aquaria was brushed biweekly to minimize algal and fungal growth that could provide a source of vitamin C.

Experiment 3

Wild and cultured fish were supplied by Yongchang Fisheries Farm in Tong Yong, Korea. Two different size groups of fish averaging $400 \pm 7.82g$ (immature adult) and $51 \pm 6.8g$ (young growing) of live wild and cultured fish, respectively, were used.

7.2.3   Sample collection and analysis

At the end of the first feeding trial, all fish were weighed and counted to calculate weight gain (WG), feed efficiency (FE), condition factor (CF), and survival. In the second 16-week experiment (Experiment 2), WG, FE, specific growth rate (SGR), protein efficiency ratio (PER), protein productive value (PPV), hepatosomatic index (HSI), CF, and survival were measured. Blood samples were obtained from the caudal vessel. Hematocrit (PCV) was determined for three individual fish per aquarium by the microhematocrit method,[29] and hemoglobin (Hb) was also measured in the same fish by the cyanmethemoglobin procedure using Drabkin's solution. Hb standard prepared from human blood (Sigma Chemical, St. Louis, MO) was used. Tissue vitamin C concentrations of liver, muscle, gill, and brain of fish were determined in triplicates. Each sample was prepared from five fish randomly selected per aquarium. Vitamin C was analyzed by the dinitrophenylhydrazine (DNPH) spectrophotometric method[30, 31] as described by Bai and Gatlin.[28] Crude protein, moisture, and ash of whole-body were analyzed by AOAC methods.[32] Crude fat was determined by Soxtec system 1046 (Tecator AB, Sweden) after freeze drying the sample for 12 hours. All data were subjected to ANOVA using Statistix 3.1 (Analytical Software, St. Paul, MN). When a significant treatment effect was observed, a Least Significant

Difference (LSD) test was used to compare means. Treatment effects were considered with the significant level at $P < 0.05$.

7.3 Results

Experiment 1

The results are summarized in Table 7.3. Weight gain (WG) and feed efficiency (FE) of fish fed C_0 diet were significantly lower than those of fish fed diets C_{150} and C_{1500} ($P < 0.05$), while those of fish fed diets C_{25}–C_{75} were not significantly different from those of fish fed C_0, C_{150} or C_{1500} diet ($P < 0.05$). Average hematocrit (PCV) of fish fed C_0 was significantly lower than those of fish fed the other diets. No significant differences existed in survival rate (SR), hemoglobin (Hb), condition factor (CF), whole-body protein, whole-body lipid, whole-body ash, and whole-body moisture among fish fed the six experimental diets ($P > 0.05$). Muscle L-ascorbic acid (AA) concentrations (Table 7.4) of fish fed C_0 diet were lower than those of fish fed C_{150} and C_{1500} diets ($P < 0.05$), while there was no significant difference among fish fed diets C_0–C_{75} ($P > 0.05$). Liver, gill, and brain AA concentrations (Table 7.4) of fish fed C_{1500} diet were significantly higher than those of fish fed diets C_0–C_{150} ($P < 0.05$), while these values were not significantly different among fish fed diets C_0–C_{150} ($P < 0.05$).

Experiment 2

Results of growth are summarized in Table 7.5. Weight gain (WG), specific growth rate (SGR), protein efficiency ratio (PER), protein productive value (PPV), and condition factor (CF) of fish fed C_0 diet were lower than those of fish fed diets C_{25}–C_{1500} ($P < 0.05$), while these values of fish fed diets C_{25}–C_{75}

Table 7.3 Percent weight gain (WG), feed efficiency (FE), hematocrit (PCV), hemoglobin (Hb), and condition factor (CF) in Korean rockfish (*Sebastes schlegeli*) fed the experimental diets for 8 weeks (Experiment 1)[1]

	Diets						Pooled SEM
	C_0	C_{25}	C_{50}	C_{75}	C_{150}	C_{1500}	
WG (%)[2]	73.3[b]	82.6[ab]	81.8[ab]	77.5[ab]	84.4[a]	87.1[a]	1.67
FE (%)[3]	29.5[c]	55.5[b]	59.1[ab]	61.9[ab]	65.0[a]	73.8[a]	3.40
PCV (%)	34.7[b]	41.3[a]	42.0[a]	41.4[a]	41.3[a]	42.3[a]	0.35
Hb (g/dl)	8.00	8.34	8.01	8.08	8.28	8.23	0.09
CF[4]	1.79	1.84	1.80	1.80	1.81	1.81	0.01

[1] Values are means from triplicate groups of fish, and the means in each row with a different superscript are significantly different ($P < 0.05$).
[2] WG = (final wt. − initial wt.) × 100 / initial wt.
[3] FE = (wet weight gain / feed intake) × 100.
[4] CF = (wet weight / total length[3]) × 100.

Table 7.4   Four different tissue L-ascorbic acid (AA) concentrations (μg AA g^{-1} tissue) in Korean rockfish fed the experimental diets for 8 weeks[1]

	Diets						Pooled
	C_0	C_{25}	C_{50}	C_{75}	C_{150}	C_{1500}	SEM[2]
Liver	32.6^b	41.2^b	46.7^b	58.2^b	76.1^b	179.7^a	13.2
Muscle	4.84^c	10.7^{bc}	10.4^{bc}	12.4^{bc}	13.7^b	63.2^a	5.46
Gill	21.6^b	18.6^b	24.0^b	24.0^b	39.8^b	139.5^a	8.78
Brain	181.8^b	233.0^b	203.9^b	221.5^b	244.5^b	394.6^a	9.22

[1] Values are means from triplicate groups of fish, and the values in each row with a different superscript are significantly different ($P < 0.05$).
[2] Pooled standard error of mean.

Table 7.5   Percent weight gain (WG), feed efficiency (FE), specific growth rate (SGR), protein efficiency ratio (PER), protein productive value (PPV), hepatosomatic index (HSI), condition factor (CF), hematocrit (PCV), hemoglobin (Hb) and survival rate (SR) in Korean rockfish (*Sebastes schlegeli*) fed the experimental diets for 16 weeks[1]

	Diets						Pooled
	C_0	C_{25}	C_{50}	C_{75}	C_{150}	C_{1500}	SEM[2]
WG (%)[3]	61^d	144^c	151^{bc}	159^{bc}	177^{ab}	191^a	10.3
FE (%)[4]	29.5^c	55.5^b	59.1^b	61.9^b	65.0^{ab}	73.8^a	3.40
SGR (%)[5]	0.44^d	0.85^c	0.89^{bc}	0.90^{bc}	0.99^{ab}	1.08^a	0.05
PER[6]	0.50^c	1.10^b	1.18^b	1.22^b	1.31^{ab}	1.51^a	0.08
PPV (%)[7]	8.8^c	17.3^b	19.1^b	19.6^b	21.2^{ab}	24.9^a	1.23
HSI[8]	2.73^c	3.14^{bc}	3.65^{ab}	3.88^{ab}	4.04^a	4.33^a	0.15
CF[9]	1.32^c	1.43^b	1.52^b	1.47^b	1.48^b	1.61^a	0.02
PCV (%)	33.5	37.3	37.3	36.7	35.5	38.7	0.74
Hb (g/dl)	9.03	8.91	9.71	9.71	9.45	10.9	0.28
SR (%)	89^b	96^a	96^a	100^a	100^a	100^a	1.11

[1] Values are means from triplicate groups of fish where the means in each row with a different superscript are significantly different ($P < 0.05$).
[2] Pooled standard error of mean.
[3] WG = (final wt. − initial wt.) × 100/initial wt.
[4] FE = (wet weight gain/feed intake) × 100
[5] SGR = (ln final wt. − ln initial wt.)/days
[6] PER = wet weight gain/protein intake
[7] PPV = (body protein deposit/feed protein intake) × 100
[8] HSI = (liver weight/body weight) × 100
[9] CF = [fish wt. (g)/fish length (cm)^3] × 100

were lower than those of fish fed C_{1500} diet (P < 0.05). The other growth parameters also gave the similar values. Also, WG, FE, SGR, PER, PPV, and HSI of fish fed C_{150} diet were not significantly different from those of fish fed C_{1500} diet (P < 0.05). HSI of fish fed diet C_0 was lower than in fish fed diets C_{50}–C_{1500} (P < 0.05). Fish fed C_0 diet exhibited a range of external signs of vitamin C deficiency (such as scoliosis, shortened operculae, exophthalmia, and fin hemorrhage) that first appeared after 12 weeks. Survival rate (SR) of fish fed C_0 diet was lower than those of fish fed the other diets (P < 0.05). At the end of the experiment, fish fed C_0 diet exhibited spinal deformity (5.3 ± 1.08%), shortened operculae, exophthalmia, and fin hemorrhage which became apparent during the 14th week. No significant difference existed in hematocrit, hemoglobin, whole-body protein, whole-body lipid, and whole-body moisture among groups (P > 0.05). However, ash contents of fish fed C_0 and C_{25} diets were higher than those of fish fed the other diets (P < 0.05). These higher ash contents of fish fed C_0 and C_{25} diets may be related to the lower condition factor. Muscle and liver AA concentrations (Table 7.6) of fish fed C_0 diet were lower than those of fish fed C_{150} and C_{1500} diets (P < 0.05). There was no significant difference in fish fed diets C_0–C_{75} (P > 0.05). Gill AA concentrations of fish fed C_{1500} diet were higher than those of fish fed diets C_0–C_{150} (P < 0.05). AA concentrations of brain were lower in fish fed diets C_0–C_{75} than in fish fed diets C_{150} and C_{1500} (P < 0.05).

Experiment 3

Tissue AA concentrations of wild and cultured Korean rockfish are summarized in Table 7.7. Average liver and brain AA concentrations of the larger wild immature adult fish were significantly higher than those of wild and cultured young growing fish, while these AA concentrations were not significantly different from those of the larger cultured immature adult fish. Also, the average liver and brain AA concentrations of wild young growing fish were significantly higher than those of cultured young growing fish. No

Table 7.6   Four different tissue L-ascorbic acid (AA) concentrations (μg AA g^{-1} tissue) in Korean rockfish (*Sebastes schlegeli*) fed the experimental diets for 16 weeks (Example 2)[1]

	Diets						**Pooled**
	C_0	C_{25}	C_{50}	C_{75}	C_{150}	C_{1500}	**SEM**
Liver	28.3^c	33.9^{bc}	35.3^{bc}	35.7^{bc}	55.8^b	180^a	13.0
Muscle	14.9^c	18.9^{bc}	18.1^{bc}	18.8^{bc}	24.8^b	60.6^a	3.77
Gill	33.2^b	36.5^b	52.5^b	53.5^b	70.3^b	143^a	9.65
Brain	30.4^c	37.2^c	39.8^c	54.6^c	86.0^b	247^a	18.1

[1] Values are means of triplicate groups of pooled five individual fish, and where the means in each row with a different superscript are significantly different (P < 0.05).

Table 7.7 Four different tissue L-ascorbic acid (AA) concentrations (μg AA g^{-1} tissue) of wild and cultured Korean rockfish averaging 400 ± 7.8g (immature adult) and 51 ± 6.8g (young growing)[1]

	Wild		Cultured		Pooled
	400g	**51g**	**400g**	**51g**	**SEM**
Liver	79.3[a]	73.0[b]	78.5[ab]	53.3[c]	2.29
Muscle	20.8	24.6	22.0	22.8	1.15
Gill	59.3	100.7	58.5	70.3	8.59
Brain	298[a]	136[b]	287[a]	86.2[c]	19.5

[1] Values are means of three to six individual fish (three of each wild and cultured immature adult fish, five of wild, and six of cultured young growing fish), and where the values in each row with a different superscript are significantly different ($P < 0.05$).

significant difference existed in average muscle and gill AA concentrations of wild and cultured fish examined in this experiment.

7.4   Discussion

In the first experiment, the values of weight gain and growth (Table 7.3) were comparable to results obtained in the previous studies,[33] while for the second experiment the values were lower.[34] The reason for this might be attributed to the size of fish, the length of the experimental period, the water temperature and qualities, and, most importantly, the use of semipurified diets in this investigation. In the first and second experiments, the vitamin C level of 39.7 (Experiment 1) and 17.1 (Experiment 2) mg AA kg^{-1} diet from the C_0 diet might be overestimated by some interferences. Dabrowski[3] stated that the vitamin C levels determined by the dinitrophenylhydrazine (DNPH) method are likely to be overestimated. Deficiency signs such as scoliosis, shortened operculae, exophthalmia, and fin hemorrhage of the fish observed in the second experiment are known in other fish species such as channel catfish,[6, 35, 36] coho salmon,[14] rainbow trout,[18, 37, 38] carp,[39] tilapia,[10, 40, 41] or yellowtail[42] fed vitamin C-free or deficient diets. Rockfish fed C_0 diet exhibited a range of external signs of vitamin C deficiency that first appeared after 12 weeks. Lim and Lovell[35] observed AA deficiency symptoms in 2.3g channel catfish after 8–12 weeks when liver AA concentrations was < 30 μg AA g^{-1} tissue. Hilton et al.[43] found that rainbow trout given an AA-free diet suffered from deficiency symptoms such as anorexia, lethargy, and prostration when a liver concentration was < 20 μg AA g^{-1} tissue after 16–20 weeks. The AA deficiency symptoms were observed in Rockfish fed C_0 diet after 12–14 weeks when a liver concentration amounted to <30 μg AA g^{-1} (Table 7.3). There was no external sign of vitamin C deficiency in Rockfish fed the other diets. From the first experiment, it is suggested that the experimental model and the semipurified diet can be useful for the vitamin C requirement study, and the dietary AA requirement is greater than 39.7 mg AA kg^{-1} diet (C_0), whereas

144.6 mg AA kg^{-1} diet (C_{150}) is adequate for the maximum growth in Korean rockfish. From the second experiment, broken line analyses for WG, SGR, FE, and PER (Figures 7.2 and 7.3) show that the optimum levels of dietary vitamin C are 102.2, 102.1, 102.5, and 100.3 mg AA kg^{-1} diet for the maximum growth in Korean rockfish.

Liver AA concentration has been mainly used as an index of AA status of fish.[36, 43-45] Liver AA concentrations of the larger wild immature adult fish were significantly higher than those of wild and cultured young growing fish (Experiment 3; Table 7.7). Liver AA concentrations from randomly selected cultured and wild Korean rockfish showed a range averaging 53.5–79.3 (Table 7.7), whereas approximately 180 µg AA g^{-1} tissue (Tables 7.4 and 7.6)

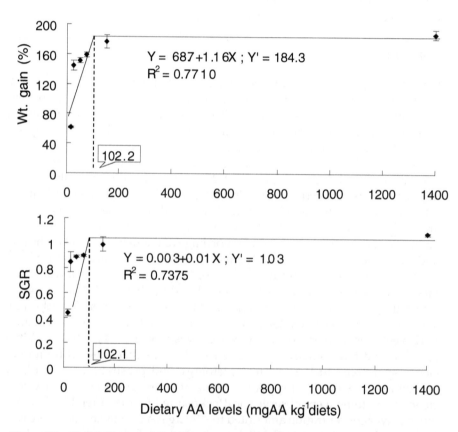

Figure 7.2 Broken line model of percent weight gain (upper) and specific growth rate (lower) in Korean rockfish fed six different levels of dietary L-ascorbic acid (AA) for 16 weeks. Values of the x-axis are the dietary AA levels in the experimental diets. Values are Means ± SEM, n = 3.

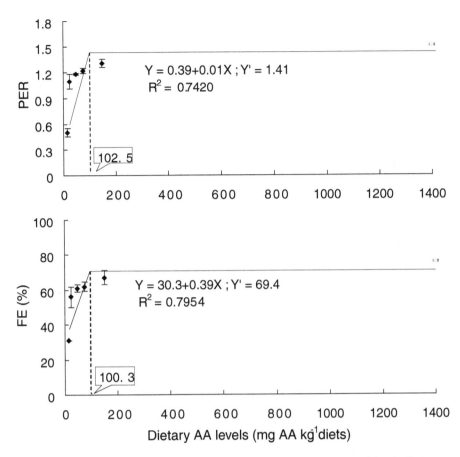

Figure 7.3 Broken line model of protein efficiency ratio (upper) and feed efficiency (lower) in Korean rockfish fed six different levels of dietary L-ascorbic acid (AA) for 16 weeks. Values of the x-axis are the dietary AA levels in the experimental diets. Values are Means ± SEM, n = 3.

can be the saturation point in this species. This assumption is also supported by our previous long-term study.[23] Juvenile Korean rockfish fed 1500 mg AA kg^{-1} diet for 28 weeks showed the liver AA concentration of 180.2 μg AA g^{-1} tissue.[23] Lim and Lovell[35] and Murai et al.[36] found a good correlation between the dietary and liver AA concentrations in channel catfish. Hilton et al.[43] and Skelbaek et al.[45] also found significant correlation in rainbow trout. In the present study, the liver AA concentrations had a positive correlation (r^2 = 0.99) with growth and with the dietary AA concentrations. However, Hilton et al.[18] found that rainbow trout maintain relatively constant liver AA levels when fed diets supplemented with 80–320 mg AA kg^{-1} diet, but when fed a diet supplemented with 1280 mg AA kg^{-1} (12 times the requirement for normal growth), liver AA increased to a level more than double of that in the fish supplemented with 320 mg AA kg^{-1} diet. Halver et al.[46] preferred anterior kidney AA concentration while Lim and Lovell[35] suggested that the AA concentrations in kidney of channel catfish did not reflect the differences among AA concentrations of the diet. The AA concentrations of four tissues in Rockfish fed the diet supplemented with 1500 mg AA kg^{-1} diet (10 times the optimum dietary level for maximum growth) were 2–3 times higher than those in fish fed the diets supplemented with 25–150 mg AA kg^{-1} diet. No prediction on the saturation point in liver AA concentration was made in the previous studies.

No significant difference existed in the average muscle AA concentrations of wild and cultured fish examined in the present study (Table 7.7). However, muscle AA concentrations had a positive correlation ($r^2 = 0.99$) with growth and with the dietary AA concentrations in the first and second experiments (Tables 7.4 and 7.6). The muscle AA concentrations could be used as an index of AA status of fish. The muscle AA concentration was lower than those in other tissues such as liver, gill, and brain in our experiments including data for wild fish. Although the unit AA concentration is the lowest in muscle, the calculated total amounts of AA in fish muscle are the highest. This was well supported by the results from Jauncey et al.,[47] Soliman et al.,[40] Al-Amoudi et al.,[48] Bai et al.,[22] and Bai and Lee.[23] Also, Al-Amoudi et al.[48] speculated that AA in muscle tissue may be in the readily available form for physiological activities.

No significant difference existed in average gill AA concentrations of wild and cultured fish examined in this experiment. Brain AA concentrations of the larger wild immature adult fish were significantly higher than those of wild and cultured young growing fish. Gill and brain AA concentrations had no positive correlation with the results of growth and with the dietary AA concentrations in the first and second experiment (Tables 7.4 and 7.6). Brain AA concentrations were not good indices of AA status in Korean rockfish, although the results of the tissue AA analyses indicated that the AA concentration was the highest in the brain.

In conclusion, based on the broken line model of growth, these findings suggest that the optimum dietary level of vitamin C is approximately 102.5 mg AA kg^{-1} diet (it can be a minimum requirement) for the maximum growth when the AA was used as the source of dietary vitamin C, but 144 mg AA kg^{-1} diet is sufficient to optimize growth and vitamin C status in juvenile Korean rockfish. The vitamin C analyses of tissues indicate that a dietary level of vitamin C greater than 1390 mg kg^{-1} diet could be necessary in order to achieve tissue vitamin C saturation. These conclusions can be well supported by the results from several studies[18, 49–52] including the third experiment which showed the increasing of the dietary requirement of AA with a decrease of fish size in the cage culture conditions. It has been well known that AA is very unstable and high losses occur during pellet manufacture and storage.[53] During feeding, further loss results when vitamin C leaches from pellets on contact with water.[44] Commercial feeds are often stored for long periods under inadequate conditions. Therefore, it is recommended that the commercial feeds for Korean rockfish should be supplemented with enough AA to maintain at least 102.5 mg kg^{-1} in the feed at the time of feeding to fish.

Acknowledgments

This research was supported in part by the funds of Woo Sung Feed Company, Ltd, Gum Sung Feed Company, the Ministry of Marine Affairs and Fisheries, the Research Center for Ocean Industrial Development (ERC designated by KOSEF), and the Seafood and Marine Bioresources Development Center at Pukyong National University.

References

1. Burns, J. J. and Conney, A. H., Metabolism of glucuronic acid and its lacton, in *Glucuronic Acid*, Dutton, G. J., Ed., Academic Press, New York, 1996, 365.
2. Wilson, R. P., Absence of ascorbic acid synthesis in channel catfish, *Ictalurus punctatus* and blue catfish, *Ictalurus frucatus*, *Comp. Biochem. Physiol.*, 46B, 636, 1973.
3. Dabrowski, K., Absorption of ascorbic acid and ascorbic sulfate and ascorbate metabolism in stomachless fish, common carp, *J. Comp. Biochem. Physiol.*, 160, 549, 1990.
4. Lovell, R. T., Essentiality of vitamin C in feeds for intensively fed caged catfish, *J. Nutr.*, 103, 134, 1973.
5. Wilson, R. P. and Poe, W. E., Impaired collagen formation in the scorbutic channel catfish, *J. Nutr.*, 103, 1359, 1973.
6. Andrews, J. W. and Murai, T., Studies on the vitamin requirements of channel catfish (*Ictalurus punctatus*), *J. Nutr.*, 105, 557, 1975.
7. Kitamura, S., Suwa, T., Ohara, S., and Nakagawa, K., Studies on vitamin requirements of rainbow trout, *Salmo gairdneri*. II. The deficiency symptoms of fourteen kinds of vitamins, *Bull. Jap. Soc. Fish.*, 33, 1120, 1965.
8. Poston, H., Effect of dietary L-ascorbic acid on immature brook trout, *N. Y. State Conserv. Dep. Fish. Res. Bull.*, 30, 46, 1967.
9. Poston, H., Effect of dietary L-ascorbic acid on immature brook trout, *N. Y. State Conserv. Dep. Fish. Res. Bull.*, 30, 46, 1967.
10. Stickney, R.R., McGeachin, R.B., Lewis, D.H., Marks, J., Riggs, R., Sis, F., Robinson, E.H., and Wurts, W., Response of Tilapia aurea to dietary vitamin C, *J. World Maricul. Soc.*, 15, 179, 1984.
11. Arai, S., Nose, T., and Hshimoto, Y., Qualitative requirements of young eels, *Anguilla japonica*, for water-soluble vitamins and their deficiency symptoms, *Bull. Freshw. Fish. Res. Lab., Tokyo*, 22, 69, 1972.
12. Guary, M., Kanazawa, A., Tanaka, N., and Ceccaldi, H. J., Nutritional requirements of prawn. III. Requirement for ascorbic acid, *Mem. Fac, Fish, Kagoshima Univ.*, 25, 53, 1976.
13. Magarelli, P. C., Jr., Hunter, B., Lightner, D. V., and Colvin, L. B., Black death and ascorbic acid deficiency disease in penaeid shrimp, *Comp. Biochem. Physiol.*, 63A, 103, 1979.

14. Halver, J. E., Ashley, L. M., and Smith, R. R., Ascorbic acid requirements of coho salmon and rainbow trout, *Trans. Am. Fish. Soc.*, 90, 762, 1969.
15. Shimeno, S., Yellowtail, *Seriola quinqueradiata*, in *Handbook of Nutrition Requirements of Finfish*, Wilson, R. P., Ed., CRC Press, Boca Raton, FL, 1991, 181.
16. Halver, J. E., Vitamin requirements of fin fish, in: *Processing from World Symposium on Fin Fish Nutrition and Fish Feed Technology*, Vol. 1, Halver, J. E. and Tiews, K., Eds., Hamburg, Heenemann, Berlin, 1979, 45.
17. Lall, S. P., Oliver, G., Weerakoon, D. E. M., and Hines, J. A., The effect of vitamin C deficiency and excess on immune response in Atlantic salmon (*Salmo salar* L.), in *Proceedings of the Fish Nutrition Meeting*, Toba, Japan, Takeda, M. and Watanabe, T., Eds., Tokyo, Japan Translation Center, 1990, 427.
18. Hilton, J. W., Cho, C. Y., and Slinger, S. J., Effect of graded level of supplemental ascorbic acid in practical diets fed to rainbow trout (*Salmo gairdneri*), *J. Fish. Res. Bd. Can.*, 35, 431, 1978.
19. Lim, C. and Lovell, R. T., Pathology of vitamin C deficiency syndrome in channel catfish *Ictalurus punctatus*, *J. Nutr.*, 108, 1137, 1978.
20. Robinson, E., Reevaluation of the ascorbic acid (vitamin C) requirements of channel catfish (*Ictalurus punctatus*), *FASEB J.*, 4, 3745, 1990.
21. El Naggar, G. O. and Lovell, R. T., L-Ascorbyl-2-monophosphate has equal antiscorbutic activity as L-ascorbic acid but L-ascorbyl-2-sulfate is inferior to L-ascorbic acid for channel catfish, *J. Nutr.*, 121, 1622, 1991.
22. Bai, S. C., Lee, K. J., and Jang, H. K., Development of an experimental model for vitamin C requirement study in Korean rockfish, *Sebastes schlegeli*, *J. Aquacul.*, 9(2), 169, 1996 (in Korean with English abstract).
23. Bai, S. C. and Lee, K. J., Long-term feeding effects of different dietary L-ascorbic acid levels on growth and tissue vitamin C concentrations in juvenile Korean rockfish, *J. Korean Fish. Soc.*, 29(5), 643, 1996 (in Korean with English abstract).
24. Lee, K. J., Kim, K. W., and Bai, S. C., Effects of dietary levels of L-ascorbic acid on growth and tissue vitamin C concentrations in juvenile Korean rockfish, *Sebastes schlegeli* (Hilgendorf), *Aquacul. Res.*, 29, 237, 1998.
25. Lee, D. J. and Putnam, G. B., The response of rainbow trout to varying protein/energy ratios in a test diet, *J. Nutr.*, 103, 916, 1973.
26. Garling, D. L., Jr. and Wilson, R. P., Effects of dietary carbohydrate-to-lipid ratios on growth and body composition of fingerling channel catfish, *Prog. Fish-Cul.*, 39, 43, 1997.
27. Thenen, S. W., Megadose effects of vitamin C on vitamin B-12 status in the rat, *J. Nutr.*, 119, 1107, 1989.
28. Bai, S. C. and Gatlin, D. M. III., Dietary rutin has limited synergistic effects on vitamin C nutrition of fingerling channel catfish (*Ictalurus punctatus*), *Fish Physiol. Biochem.*, 10(3), 183, 1992.
29. Brown, B. A., Routine hematology procedures, in *Hematology: Principles and Procedures*, Lea and Febiger, Philadelphia, 198, 71.
30. Schaffert, R. R. and Kingsley, G. R., A rapid, simple method for the determination of reduced, dehydro-, and total ascorbic acid in biological material, *J. Biol. Chem.*, 212, 59, 1955.
31. Interdepartmental Committee on Nutrition for National Defense, Serum vitamin C (ascorbic acid)-dinitrophenyl hydrazine method, in *Manual for Nutrition Surveys*, National Institutes of Health, Bethesda, MD, 1963, 117.
32. Association of Official Analytical Chemists, *Official Methods of Analysis*, 16th ed., Association of Official Analytical Chemists, Arlington, VA, 1995.

33. Lee, S. M. and Lee, J. Y., Effects of dietary α-cellulose levels on the growth, feed efficiency and body composition in Korean rockfish, *Sebastes schlegeli, J. Aquacul.*, 7(2), 97, 1994 (in Korean with English).
34. Lee, S. M., Lee, J. Y., and Hur, S. B., Eicosapentaenoic acid and docosahexaenoic acid requirement of the Korean rockfish *Sebastes schlegeli*, Book of Abstracts, *World Aquaculture '94, World Aquaculture Society Annual Meeting*, January 14–18, 1994. New Orleans, 1994, 338.
35. Lim, C. and Lovell, R. T., Pathology of the vitamin C deficiency syndrome in channel catfish (*Ictalurus punctatus*), *J. Nutr.*, 108, 1137, 1978.
36. Murai, T., Andrews, J. W., and Bauernfeind, J. C., Use of L-ascorbic acid, ethocel coated ascorbic acid and ascorbate 2-sulphate in diets for channel catfish (*Ictalurus punctatus*), *J. Nutr.*, 108, 1761, 1978.
37. Tsujimura, M., Yoshikawa, H., Hasagawa, T., Suzuki, T., Kaisai, T., Suwa, T., and Kitamura S., Studies on the vitamin C activity of ascorbic acid 2-sulfate on the feeding test of new born rainbow trout, *Vitamins* (Japan), 52, 35, 1978.
38. Sato, M., Kondo, T., Yoshinake, R., and Ikeda, S., Effect of water temperature on the skeletal deformity in ascorbic acid deficient rainbow trout, *Bull. Jap. Soc. Sci. Fish.*, 49, 443, 1983.
39. Agrawal, N. S. and Mahajan, C. L., Nutritional deficiency in an Indian major carp, *Cirrhina mrigala*, due to a vitaminosis C during early growth, *J. Fish Dis.*, 3, 231–248, 1980.
40. Soliman, A. K., Jauncey, K., and Roberts, R. H., The effect of varying forms of dietary ascorbic acid on the nutrition of juvenile tilapias (*Oreochromis niloticus*), *Aquaculture*, 52, 1, 1986.
41. Soliman, A. K., Jauncey, K., and Roberts, R. H., The effect of dietary ascorbic acid supplementation on hatchability, survival rate and fry performance in *Oreochromis mossambicus* (Peters), *Aquaculture*, 59, 197, 1986.
42. Sakaguchi, H., Takeda, F., and Tange, K., Studies on vitamin requirements by yellowtail. 1. Vitamin B6 and C deficiency symptoms, *Bull. Jap. Soc. Sci. Fish.*, 44, 1029, 1969.
43. Hilton, J. E., Cho, C. Y., and Slinger, S. J., Evaluation of ascorbic acid status of rainbow trout (*Salmo gairdneri*), *J. Fish. Res. Bd. Can.*, 34, 2207, 1977.
44. Hardie, L. J., Fletcher, T. C., and Secombes, C. J., The effect of dietary vitamin C on the immune response of the Atlantic salmon, *Aquaculture*, 95, 201, 1991.
45. Skelbaek, T., Andersen, N. G., Winning, M., and Westergaard, S., Stability in fish feed and bioavailability to rainbow trout of two ascorbic acid forms, *Aquaculture*, 84, 335, 1990.
46. Halver, J. E., Smith, R. R., Tolbert, B. M., and Baker, E. M., Utilization of ascorbic acid in fish, *Ann. N.Y. Acad. Sci.*, 258, 81, 1975.
47. Jauncey, K., Soliman, A., and Roberts, R. J., Ascorbic acid requirement in relation to wound healing in the cultured tilapia, *Oreochrmis niloticus* (Trewavas), *J. Fish. Manage.*, 16, 139, 1985.
48. Al-Amoudi, M. M., El-Nakkadi, A. M. N., and El-Nouman, B. M., Evaluation of optimum dietary requirement of vitamin C for the growth of *Oreochromis spilurus* fingerlings in water from the Red Sea, *Aquaculture*, 105, 165, 1992.
49. Ikeda, S. and Sato, M., Biochemical studies on L-ascorbic acid in carp, *Bull. Jap. Soc. Sci. Fish.*, 30, 365, 1964.

50. Sato, N., Yoshinaka, R., and Ikeda, S., Biochemical studies on L-ascorbic acid in aquatic animals. XI. Dietary ascorbic acid requirement of rainbow trout for growth and collagen formation, *Bull. Jap. Soc. Sci. Fish.*, 44, 1029, 1995.
51. Li, Y. and Lovell, R. T., Elevated levels of dietary ascorbic acid increase immune responses in channel catfish, *J. Nutr.*, 115, 123, 1985.
52. Dabrowski, K., Hinterleitner, S., Sturmbauer, C., El-Fiky, N., and Wieser, W., Do carp larvae require vitamin C?, *Aquaculture*, 72, 295, 1998.
53. Steffens, W., *Principles of Fish Nutrition*, Ellis Horwood, Chichester, 1989, 384.

chapter eight

Vitamin C requirement in crustaceans

Shi-Yen Shiau

Contents

Abstract

Vitamin C is an essential nutrient for crustaceans. Traditionally, L-ascorbic acid (AA) was the only source of vitamin C used in shrimp feed. A dietary requirement of vitamin C (in the form of AA) has been quantified for several penaeid species. However, AA is unstable, and practical diets have been shown to lose AA during processing and storage. AA derivatives with sulfate and phosphate moieties at carbon C-2 position in the lactone ring of AA have been shown to be stable and effective in satisfying the requirement of penaeid shrimp. Several derivatives, namely L-ascorbyl-2-sulfate (C2S), L-ascorbyl-2-monophosphate-Mg (C2MP-Mg), L-ascorbyl-2-monophosphate-Na (C2MP-Na) and L-ascrobyl-2-polyphosphate (C2PP), have been used to quantify the vitamin C requirements for penaeid shrimps. The requirement of these derivatives as vitamin C sources for penaeid shrimp was more markedly reduced than the value estimated earlier with AA. Caution needs to be taken, however, when comparing the requirements of each of these ascorbate derivatives from the published data.

8.1   Introduction

Although most land animals do not require a dietary source of vitamin C, most aquatic animals, including crustaceans, are extremely sensitive to vitamin C deficiency. Although Lightner et al.[1] suggested the possibility of limited synthesis of vitamin C in some species of penaeid shrimp, all species of shrimp and prawns (freshwater) tested to date require a dietary source of vitamin C. Thus, even if some limited ability to synthesize vitamin C exists in some shrimp species, it is insufficient to meet metabolic requirements.

8.2   Vitamin C deficiency symptom

In shrimp, a dietary lack of vitamin C leads to impaired collagen synthesis.[2] The nutritional disease referred to as "black death" is an external manifestation of the reduction in collagen synthesis. This deficiency syndrome, characterized by melanized lesions distributed throughout the collagenous tissue underlying the exoskeleton, was originally described in *Farfantepenaeus californiensis*, *Farfantepenaeus aztecus*, and *Litopenaeus stylirostris*.[1] Similar lesions have been described in other penaeid shrimp, *Penaeus monodon*[3] and *Litopenaeus vannamei*,[4] as well as the freshwater prawn, *Macrobrachium rosenbergii*.[5] Symptoms of vitamin C deficiency are somewhat different in *Marsupenaeus japonicus*, being a decolorization and development of abnormal grayish-white color on the carapace margins, lower abdomen, and tips of walking legs.[6]

8.3   L-ascorbic acid requirement

A quantitative requirement of AA for maximum growth has been determined for several penaeid species including *M. japonicus*,[6, 7] *F. californiensis*,[8] and *P. monodon*.[9] Kuruma shrimp *M. japonicus* was shown to require 3000 mg/kg diet[6] or 10,000–20,000 mg/kg diet[7] of AA for optimal growth. The AA requirement for maximum growth of *F. californiensis* was 2000 mg/kg diet.[8] For *P. monodon*, a requirement of 2000 mg/kg diet was suggested.[9] In all these studies, AA was used as the source of vitamin C. Because of the instability of AA in nature, consequently these estimates were too high.

8.4   Instability of L-ascorbic acid

AA is the source of vitamin C traditionally used to feed fish and shrimp, but it is a water-soluble, thermolabile vitamin. It is also easily oxidized to an inactive form, diketogulonic acid, during processing and storage due to exposure to high temperature, oxygen, and light.[10–12] In our previous study, even the process used to make experimental diets in the laboratory exposes the feed ingredients to relatively mild conditions compared to those used in feed manufacturing; nevertheless, the process destroyed approximately 75% of the initial amount of supplemental AA in shrimp feeds.[13] Also, a steady decline in AA activity down to 85.7% was found when extracted by water and allowed to stand for 60 min in the laboratory before high performance liquid chromatography (HPLC) analysis.

Slow feeding behavior of shrimp further reduces availability due to leaching of the hydrophilic acid. Attempts have been made to increase retention of vitamin C activity in shrimp feeds by using alternative forms of AA, such as L-ascorbyl-2-monophosphate-Mg.

8.5   Ascorbic acid derivatives

AA derivatives with sulfate and phosphate moities at the unstable carbon C-2 position in the lactone ring are highly resistant to oxidation.[14] Up to the present time, a number of AA derivatives have been developed (Table 8.1) and used to quantify the vitamin C requirement for penaeid shrimp. They have been shown to be effective in satisfying the requirement of penaeid shrimp. For example, the estimated requirement for AA estimated earlier as 2000 mg/kg diet for AA[9] was reduced to 210 mg/kg for L-ascorbyl-2-polyphosphate (C2PP),[15] to 40 mg/kg for L-ascorbyl-2-monophosphate-Mg (C2MP-Mg).[16] The requirements of *M. japonicus* for C2MP-Mg was 215–430 mg/kg[17] and 500 mg/kg[18] and of *L. vannamei* for C2PP was 90–120 mg/kg.[19] Requirements of penaeid shrimp for AA and its derivatives in meeting vitamin C are summarized in Table 8.2.

Strict comparison of the potency of each source of AA from these studies is perhaps misleading because the duration of the study, the initial body weight of the shrimp, the water temperature, and other experimental

Table 8.1   Commonly used ascorbic acid derivatives in shrimp feed

Ascorbic acid	Abbreviation
L-ascorbic acid	AA
L-ascorbyl-2-sulfate	C2S
L-ascorbyl-2-monophosphate-Mg	C2MP-Mg
L-ascorbyl-2-monophosphate-Na	C2MP-Na
L-ascorbyl-2-polyphosphate	C2PP

Table 8.2   Ascorbic acid and its derivatives in meeting vitamin C requirement of penaeid shrimp

Vitaminer	Species	Ascorbic acid activity (%)	Requirement (mg/kg diet)			Reference
			Supplementation	Analyzed	Equivalent	
AA	P. monodon	100	2000–2500	–	–	9
C2MP-Mg	M. japonicus	?	215–430	?	?	17
C2MP-Mg	M. japonicus	?	500	?	?	18
C2PP	L. vannamei	11.61	(775–1034)*	?	90–120	19
C2S	P. monodon	48	–	156.97	75.35	16
C2MP-Mg	P. monodon	46.46	–	40.25	18.70	16
C2MP-Mg	P. monodon	(50)*	100–200	?	50–100	3
C2PP	P. monodon	15	210	?	?	15
C2PP	P. monodon	25	–	(117.08)*	29.27	20
C2S	P. monodon	48	–	(153.81)*	73.83	20
C2MP-Mg	P. monodon	46.46	–	48.40	22.47	21
C2MP-Na	P. monodon	25.20	–	106.07	26.73	21

* Figures in parentheses were calculated based on information of this paper.

conditions were not the same. Furthermore, the data used to quantify the vitamin C requirements vary because some results are based on the supplementation level, while others are based on dietary levels that are actually determined. For example, it is very difficult to obtain a comparison between the requirement of C2PP in Chen and Chang's study[15] and the requirement of C2MP-Mg in Catacutan and Lavilla-Pitogo's study[3] in meeting the vitamin C requirement of *P. monodon*. Other than the different experimental conditions of the two studies, the actual analyzed dietary ascorbate concentrations of the two studies were not given. Thus, a comparison between different AA derivatives within a study for an individual aquatic species is preferred to a cross-comparison of the potency of each AA derivative from different studies. Recently, a series of studies has been conducted in our laboratory comparing the biopotency of each ascorbate derivative in meeting the vitamin C requirements for juvenile *P. monodon*.

8.5.1   C2MP-Mg vs. C2S

A growth experiment was conducted to compare C2MP-Mg and C2S with AA for providing the dietary source of vitamin C activity for juvenile *P. monodon* (mean weight 1.06 ± 0.05 g).[16] C2MP-Mg (46.46% AA activity, Showa Denko K. K., Tokyo, Japan), C2S (52.85% AA activity, Pfizer Inc., New York), and AA (100% activity, Merck Co., Germany) were each added to the basal diet formula to obtain concentrations of 0, 30, 50, 200, 500, 1000, and 2000 mg/kg diet. However, the actual concentrations as determined by analysis differed from the above and are presented in Table 8.3. Each diet was fed to

Table 8.3   Ascorbic acid concentrations (mg/kg diet) in diets containing L-ascorbyl-2-monophosphate-Mg (C2MP-Mg), L-ascorbyl-2-sulfate (C2S) and L-ascorbic acid (AA)

	Added C2MP-Mg, C2S, and C1 in basal diet (mg/kg diet)						
	0	30	50	200	500	1000	2000
C2MP-Mg group							
Analyzed C2MP-Mg	–	22.5	37.6	150.4	377.8	756.5	1549.7
Ascorbic acid equivalent[a]	–	10.5	17.5	69.9	175.5	351.5	720.0
C2S group							
Analyzed C2S	–	24.3	39.5	156.0	386.3	778.6	1566.5
Ascorbic acid equivalent[a]	–	11.7	18.9	74.9	185.4	373.7	751.9
AA group							
Analyzed AA	–	7.5	12.6	53.5	128.7	261.5	523.7
Ascorbic acid equivalent[a]	–	7.5	12.6	53.5	128.7	261.5	523.7

[a] Calculated ascorbic acid equivalency of ascorbyl ester based on manufacturer's declarations of the source materials. (From Shiau, S. Y. and Hsu, T. S., Vitamin C requirement of grass shrimp, *Penaeus monodon*, as determined with L-ascorbyl-2-monophosphate. *Aquaculture*, 122, 347-357, 1994. Reproduced with permission.)

three replicate groups of shrimp, and the study was carried out for 8 weeks. Results indicated that shrimp fed diets supplemented with C2MP-Mg , C2S, or AA had significantly ($P < 0.05$) higher weight gains and better feed conversion ratio (FCR) than those fed the unsupplemented control diet. Survival rates were significantly higher ($P < 0.05$) in shrimp fed diets supplemented with >37.6, >386.3 or >261.5 mg/kg of C2MP-Mg, C2S, or AA, respectively, than those fed the unsupplemented control diet. Broken-line regression analysis indicated that the requirement was 40.25 mg C2MP-Mg/kg diet (equivalent of 18.7 mg AA/kg diet) or 156.97 mg C2S/kg diet (equivalent of 82.95 mg AA/kg diet), indicating that C2S was only about 25% as effective as C2MP-Mg in meeting the vitamin C requirement.

8.5.2   C2PP vs. C2S

Comparison of C2PP with C2S in meeting vitamin C requirements of juvenile *P. monodon* was conducted[20] by feeding the shrimp (mean weight 0.79 ± 0.08 g) purified diets with 6 levels of either supplemented C2PP (0, 120, 200, 800, 2000, and 4000 mg/kg diet, 25% vitamin C activity, Roche Rovimix Stay-C 25, Switzerland) or C2S (0, 30, 50, 200, 500, and 1000 mg/kg diet, 48% vitamin C activity, Pfizer Inc., New York) for 8 weeks. The actual AA concentrations in experimental diets are shown in Table 8.4. Results indicated that shrimp fed diets containing ≥ 22.81 mg of C2PP/kg or ≥ 72.41 mg of C2S/kg diet had significantly ($P < 0.05$) higher weight gain than shrimp fed the unsupplemented control diet. Feed conversion ratio (FCR) was poor in shrimp fed the diet lacking supplementary vitamin C. Shrimp fed diets supplemented with 126.9 and 292.8 mg of C2PP/kg or ≥ 72.4 mg of C2S/kg diet had a higher survival than shrimp fed the unsupplemented control diet. Weight gain analyzed by broken-line regression indicated that the dietary level of AA from

Table 8.4   Ascorbic acid concentrations (mg/kg diet) in diets containing L-ascorbyl-2-polyphosphate (C2PP) and L-ascorbyl-2-sulfate (C2S)

	Added C2PP in basal diet (mg/kg diet)					
	0	120	200	800	2000	4000
Analyzed C2PP	–	66.36	91.24	50.76	1171.36	2382.24
Ascorbic acid equivalent[a]	–	16.59	22.81	126.90	292.84	595.56
	Added C2S in basal diet (mg/kg diet)					
	0	30	50	200	500	1000
Analyzed C2S	–	22.54	36.83	150.85	398.17	756.27
Ascorbic acid equivalent[a]	–	10.82	17.68	72.41	191.12	363.01

[a] Calculated ascorbic acid equivalency of ascorbyl ester based on manufacturer's declarations of the source materials. (From Hsu, T. S. and Shiau, S. Y., Comparison of L-ascorbyl-2-polyphosphate with L-ascorbyl-2-sulfate in meeting vitamin C requirements of juvenile grass shrimp *Penaeus monodon. Fisheries Sci.*, 63, 958–962, 1997. Reproduced with permission.)

each source for juvenile *P. monodon* was 29.27 mg of AA/kg diet for C2PP and 73.83 mg of AA/kg diet for C2S, suggesting that C2S is about 40% as effective as C2PP in meeting the vitamin C requirements.

This comparison of the growth response does not reflect bioavailability of the vitamin directly, as the secondary factor may impact the growth (see Section 8.5.3).

8.5.3   C2MP-Na vs. C2MP-Mg

Hsu and Shiau[21] conducted another experiment to quantify the level of C2MP-Na (25.20% vitamin C activity, Showa Denko K. K., Tokyo, Japan) needed to satisfy the dietary vitamin C requirement for juvenile *P. monodon*. C2MP-Mg (46.46% AA content, Showa Denko K. K.) was also included in this study for comparison. Purified diets with 7 levels of AA equivalents (0, 30, 70, 150, 300, 600, and 1200 mg/kg diet) from either supplemental C2MP-Na or C2MP-Mg were each fed to *P. monodon* (mean weight 0.55 ± 0.04 g) for 8 weeks. The actual AA concentrations in diets are shown in Table 8.5. Results of the broken-line regression analysis indicated that the adequate dietary AA from each source for shrimp was 106.1 mg of C2MP-Na/kg (equivalent to 26.7 mg of AA/kg) diet and 48.4 mg of C2MP-Mg (equivalent to 22.5 mg AA/kg) diet, and it also indicated that C2MP-Na was about 84% as effective as C2MP-Mg in meeting the vitamin C requirement for *P. monodon*.

The dietary C2MP-Na and C2MP-Mg levels required for *P. monodon* to maximize body AA concentrations are likely to be higher than the values obtained from the growth data. When the broken-line model analysis of hepatopancreatic AA concentrations was employed, values of 162.2 mg/kg diet of C2MP-Na and 144.2 mg/kg diet of C2MP-Mg were estimated to be

Table 8.5   Ascorbic acid concentrations (mg/kg diet) in diets containing
L-ascorbyl-2-monosphosphate-Na (C2MP-Na) and
L-ascorbyl-2-monophosphate-Mg (C2MP-Mg)

	Added C2MP-Na and C2MP-Mg in basal diet (mg/kg diet)						
	0	30	70	150	300	600	1200
C2MP-Na group							
Analyzed C2MP-Na	–	22.95	53.76	118.65	232.50	486.67	946.83
Ascorbic acid equivalent[a]	–	5.78	13.55	29.89	58.59	122.64	238.60
C2MP-Mg group							
Analyzed C2MP-Mg	–	22.41	52.56	112.94	227.40	457.21	923.72
Ascorbic acid equivalent[a]	–	10.39	24.36	52.36	105.42	211.96	428.24

[a] Calculated ascorbic acid equivalency of ascorbyl ester based on manufacturer's declarations of the source materials. (From Hsu, T. S. and Shiau, S. Y., Comparison of vitamin C requirement for maximum growth of grass shrimp, *Penaeus monodon*, with L-ascorbyl-2-monophosphate-Na and L-ascorbyl-2-monophosphate-Mg. *Aquaculture*, 163, 203–213, 1998. Reproduced with permission.)

adequate for *P. monodon*.[21] Ideally, the estimates derived from weight gain and hepatopancreatic tissue content analysis should support one another. Animals sometimes manifest a nutrient requirement to maximize body concentration that is higher than that to maximize growth.

8.6   Conclusion

The unstable nature of L-ascorbic acid (AA) requires the use of more stable forms of AA derivatives in shrimp feed. The relative ability to utilize the various protected forms of vitamin C is likely the result of differences in absorption. Accordingly, the biopotency of each ascorbate source is critical in determining the supplemental dietary level for crustaceans. Caution should be taken when comparing the requirements of each of the ascorbate derivatives from the published data. Cross-comparison of C2S, C2PP, C2MP-Mg, and C2MP-Na in meeting the vitamin C requirement of *P. monodon* indicates that the biopotencies of these compounds are C2MP-Mg (1) > C2MP-Na (84%) > C2PP (64%) > C2S (25%). More information is needed on the biopotency of these derivatives in various species of penaeid shrimp.

References

1. Lightner, D. V., Colvin, L. B., Brand, C., and Donald, D. A., Black death, a disease syndrome of penaeid shrimp related to a dietary deficiency of ascorbic acid. *Proc. World Maricult. Soc.*, 8, 611–623, 1977.
2. Hunter, B., Magarelli, P. C., Jr., Lightner, D. V., and Colvin, L. B., Ascorbic acid-dependent collagen formation in penaeid shrimp. *Comp. Biochem. Physiol.*, 64B, 381–385, 1979.
3. Catacutan, M. R. and Lavilla-Pitogo, C. R., L-ascorbyl-2-phosphate Mg as a source of vitamin C for juvenile *Penaeus monodon*. *Israeli J. Aquacult., Bamidgeh*, 46, 40–47, 1994.
4. Montoya, N. and Molina, C., Optimum supplemental level of L-ascorbyl-2-phosphate-Mg to diet for white shrimp *Penaeus vannamei*. *Fisheries Sci.*, 61, 1045–1046, 1995.
5. Heinen, J. M., Nutritional studies on the giant Asian prawn, *Macrobrachium rosenbergii*. Ph.D. dissertation, Boston University, Boston, 1984.
6. Deshimaru, O. and Kuroki, K., Studies on a purified diet for prawn. VII: Adequate dietary levels of ascorbic acid and inositol. *Bull. Jpn. Soc. Sci. Fish.*, 42, 571–576, 1976.
7. Guary, M., Kanazawa, A., Tanaka, N., and Ceccaldi, H. J., Nutritional requirements of prawn. VI. Requirement for ascorbic acid. *Mem. Fac. Fish.* Kagoshima University, 25, 53–57, 1976.
8. Lightner, D. V., Hunter, B., Magarelli, P. C., Jr., and Colvin, L. B., Ascorbic acid: nutritional requirement and role in wound repair in penaeid shrimp. *Proc. World Maricult. Soc.*, 10, 513–519, 1979.
9. Shiau, S. Y. and Jan, F. L., Ascorbic acid requirement of grass shrimp, *Penaeus monodon. Nippon Suisan Gakkaishi*, 58, 363, 1992.

10. Hilton, J. W., Cho, C. Y., and Slinger, S. T., Factors affecting the stability of supplemental ascorbic acid in practical trout diets. *J. Fish. Res. Board Can.*, 34, 683–687, 1977.

11. Lovell, R. T. and Lim, C., Vitamin C in pond diets for channel catfish. *Trans. Am. Fish. Soc.*, 107, 321–325, 1978.

12. Soliman, A. K., Jauncey, K., and Roberts, R. T., Stability of ascorbic acid (vitamin C) and its forms in fish feeds during processing, storage and leaching. *Aquaculture*, 60, 73–83, 1987.

13. Shiau, S. Y. and Hsu, T. S., Stability of ascorbic acid in shrimp feed during analysis. *Nippon Suisan Gakkaishi*, 59, 1535–1537, 1993.

14. Tolbert, B. M., Downing, M., Carlson, R. W., Knight, M. K., and Bakre, E. M., Chemistry and metabolism of ascorbic acid and ascorbate sulfate. *Ann. N.Y. Acad. Sci.*, 258, 48–69, 1975.

15. Chen, H. Y. and Chang, C. F., Quantification of vitamin C requirements for juvenile shrimp (*Penaeus monodon*) using polyphosphorylate L-ascorbic acid. *J. Nutr.*, 124, 2033–2038, 1994.

16. Shiau, S. Y. and Hsu, T. S., Vitamin C requirement of grass shrimp, *Penaeus monodon*, as determined with L-ascorbyl-2-monophosphate. *Aquaculture*, 122, 347–357, 1994.

17. Shigueno, K. and Itoh, S., Use of Mg-L-ascorbyl-phosphate as a vitamin C source in shrimp diets. *J. World Aquacult. Soc.*, 19, 168–174, 1988.

18. Alava, V. R., Kanazawa, A., Teshima, S., and Koshio, S., Effect of dietary L-ascorbyl-2-phosphate magnesium on gonadal maturation of *Penaeus japonicus*. *Nippon Suisan Gakkaishi*, 59, 691–696, 1993.

19. He, H. and Lawrence, A. L., Vitamin C requirements of the shrimp *Penaeus vannamei*. *Aquaculture*, 114, 305–316, 1993.

20. Hsu, T. S. and Shiau, S. Y., Comparison of L-ascorbyl-2-polyphosphate with L-ascorbyl-2-sulfate in meeting vitamin C requirements of juvenile grass shrimp *Penaeus monodon*. *Fisheries Sci.*, 63, 958–962, 1997.

21. Hsu, T. S. and Shiau, S. Y., Comparison of vitamin C requirement for maximum growth of grass shrimp, *Penaeus monodon*, with L-ascorbyl-2-monophosphate-Na and L-ascorbyl-2-monophosphate-Mg. *Aquaculture*, 163, 203-213, 1998.

chapter nine

Dietary requirements for ascorbic acid by warmwater fish

R.T. Lovell

Contents

9.1 Early research

The first report that warmwater fish required a dietary source of vitamin C (L-ascorbic acid) was made by Ikeda and Sato,[1] who reported that common carp (*Cyprinus carpio*) could synthesize the vitamin, but not in sufficient quantity for optimum growth. Subsequently, Kitamura et al.[2] demonstrated spinal deformities in common carp fed diets devoid of vitamin C, much like those in rainbow trout (*Oncorhynchus mykiss*) fed similar diets, which confirmed that young carp required a dietary source of the vitamin. Dupree[3] was unable to demonstrate a need for dietary vitamin C by channel catfish (*Ictalurus punctatus*) fed purified diets in aquaria for 36 weeks. His report indicated a slow rate of growth, possibly due to low water temperature of 23°C, which may explain the lack of response. However, later Lovell[4] and Wilson and Poe[5] almost simultaneously demonstrated that a dietary deficiency of vitamin C would cause the economically important "broken back" syndrome in channel catfish, and associated this with malsynthesis of collagen in the matrix of the vertebrae. Wilson and Poe[5] showed that young channel catfish did not possess the enzyme L-gluonolactone oxidase activity required to synthesize ascorbic acid from glucose. Subsequently, deficiency signs such as structural deformities in vertebrae, fins, gills, and eye cartilage, lethargy, and poor growth were reported in Indian carp[6] and blue tilapia (*Oreochronis aureus*).[7] Lim and Lovell[8] reported that 30–50 mg kg^{-1} of dietary vitamin C was sufficient for normal growth and to prevent spinal deformities in channel catfish with an initial weight of 2.3 grams. This was similar to the requirement reported for this species by Andrews and Murai,[9] although they did not observe spinal damage in vitamin C deprived fish. Stickney et al.[7] also reported that 25 to 50 mg kg^{-1} was the vitamin C requirement for small blue tilapia. Martins[10] found that tropical characinoid fish, pacu (*Piaractus mesopotamicus*) juveniles required 50 mg kg^{-1} in the dry diet. This range of dietary concentration has generally been the recommended range for warmwater fishes as documented in the NRC[11] publication *Nutrient Requirements of Fish*.

9.2   Variation in requirements

The dietary requirement for vitamin C for a fish species varies with the size of the fish and metabolic function. Li and Lovell[12] reported that the requirement for maximum weight gain for channel catfish weighing less than 10 grams was 50 mg kg^{-1}, but 30 mg kg^{-1} was sufficient for fish weighing 50 grams and above. Dabrowski et al.[13] showed that ascorbic acid concentration in tissues of channel catfish declined with ontogeny, indicating that the dietary requirement changes with fish size. The study of Li and Lovell[12] also indicated that 30 mg kg^{-1} was sufficient for prevention of scoliosis and lordosis, although 50 mg kg^{-1} was required for maximum weight gain in the small fish. Lim and Lovell[8] found that 30 mg kg^{-1} of dietary vitamin C was satisfactory for optimum growth of channel catfish, but 90 mg kg^{-1} provided for optimum rate of wound healing. Higher than normal requirements have been reported for maximum resistance against bacterial infection in warm and cold water fishes. Li and Lovell[12] showed that increasing the dietary allowance of vitamin C to 10 times the requirement for growth reduced mortality and increased specific and nonspecific immune responses in channel catfish challenged with *Edwardsiella ictaluri*.

Blom and Dabrowski[14] provided evidence that the highest quality of eggs and spermatozoa were produced in rainbow trout when a dietary level of vitamin C eight- to ten-fold higher than the growth requirement was fed. Japanese parrot fish *(Oplegnathus fasciatus)* showed greater tolerance to intermittent hypoxic stress when fed a diet with 300 mg than 75 mg of vitamin C per kg which was sufficient for preventing deficiency. Nitrite-induced methemoglobin concentrations were lower in channel catfish fed a diet containing 8000 mg ascorbic kg^{-1} than in fish fed a diet containing 63 mg ascorbic acid kg^{-1}.

9.3   Deficiency signs and metabolic roles

A number of vitamin C deficiency signs and metabolic functions have been described in warmwater fish. These include structural deformities in fins, vertebrae, gill operculum and support cartilage, and blood vessels. Conversion of iron and folic acid to reduced forms for metabolism, and detoxification of various xenobiotics have been associated with vitamin C. Most of these reactions can be associated with the role of vitamin C as a reducing compound. For example, structural deformities can be associated with collagen metabolism where vitamin C is involved in mixed function oxidase activity affecting the hydroxylation of lysine and proline when procollagen is converted to collagen. Vitamin C is necessary for maximum immunity against bacterial infections. Vitamin C is assumed to be necessary in the synthesis of collagenous components of the complement system[12] and for protecting oxidation-sensitive membrane phospholipids, such as eicosanoids, which are precursors of immunostimulators.[15] Vitamin C was necessary for successful reproduction in rainbow trout, apparently in protecting oxidation sensitive genetic material in gametes,[14] and probably has the same function in warmwater fish.[16]

9.4   Indicators of vitamin C status in fish

Subclinical indicators used to evaluate vitamin C status in warmwater fish include tissue levels of vitamin C and collagen content of bones. Lim and Lovell[8] reported that tissue (liver, kidney, serum) levels of vitamin C increased as dietary intake increased to 10 to 20 times the dietary requirement for optimum growth and, thus, were not an accurate measure of the optimum dietary requirement. They found, however, that liver concentration of vitamin C below 30 mg kg^{-1} of wet tissue was indicative of a scorbutic condition in channel catfish. Hilton et al.[17] found that a similar concentration in coldwater rainbow trout indicated dietary deficiency of vitamin C.

Lim and Lovell[8] and Wilson and Poe[5] found that collagen content of vertebrae is a useful index of dietary requirement and vitamin C status of channel catfish. Both research teams reported that collagen content of vertebrae did not increase appreciably when the dietary level of vitamin C increased above that for optimum growth. Lim and Lovell[8] showed that vertebral collagen content of 25% of dry bone indicated a dietary deficiency and 30% indicated a dietary adequacy of vitamin C in channel catfish.

9.5   Reevaluation of vitamin C requirements

The listed NRC dietary requirement for vitamin C for young channel catfish and carp, and also for two salmonid species, is 50 mg kg^{-1}. These values are probably overestimations; for channel catfish there is good evidence that this diet concentration is too high. The early studies from which these values were derived used nonstabilized sources of ascorbic acid and the concentrations in the diets when fed were 20–50% lower than the designed levels. El Naggar and Lovell[18] showed that up to 50% of the supplemented ascorbic acid in supplemental diets was lost, even when the diets were stored at temperature below freezing. When they used a stabilized source of vitamin C, L-ascorbyl-2-monophosphate, in the experimental diets, the vitamin C requirement for maximum growth and prevention of scurvy signs in small (less than 13 g) channel catfish was 11 mg kg^{-1}. Robinson[19] used L-ascorbyl-2-polyphosphate as the vitamin C source and found that approximately 15 mg kg^{-1} of vitamin C was sufficient for diets of young channel catfish. Both of these values determined with the stabilized vitamin C source, 11 and 15 mg kg^{-1}, are well below the NRC listed requirement and suggested that the early requirements for catfish, and possibly other species where nonstabilized ascorbic acid was the dietary source, may be overestimates.

9.6   Vitamin C and immunity

A number of studies have shown that teleost fishes deprived of dietary vitamin C are more sensitive to bacterial infections. Li and Lovell[12] found reductions in antibody production, phagocytic activity of macrophages and complement activity in channel catfish fed diets low or devoid in vitamin C, indicating that the vitamin affected both cellular and humoral immune responses. This was different from responses of rainbow trout where only phogacytosis and complement activity were suppressed in vitamin C deprived fish while antibody production was reduced in vitamin E deficient fish.[20]

Several studies with channel catfish have shown higher mortalities from *Edwardsiella ictaluri* infection in fish deprived of vitamin C.[12, 21, 22] Li and Lovell[12] and Duncan and Lovell[21] found that higher than normal dietary levels enhanced resistance against this pathogen in small catfish, although Li and Robinson[22] failed to repeat these findings. Studies with rainbow trout and Atlantic salmon *(Salmo salar)* have also shown that feeding higher levels of vitamin C than is required for normal growth increases several nonspecific immune responses.[23, 20]

9.7   Bioavailability of sources of vitamin C

L-ascorbic acid is relatively sensitive to oxidation at carbon 2 in the lactone ring in its molecular structure. Thus, moisture, heat, and exposure to prooxidants reduces its vitamin C activity. Lovell and Lim[24] and El Naggar and Lovell[18] showed that approximately 50% of the vitamin potency of ascorbic acid was lost during extrusion processing of fish feeds, and that its half-life in the finished feed was less than 90 days.

Because of the large losses of ascorbic acid during processing and storage, new compounds with vitamin C activity have been investigated for use in aquatic feeds. For a compound to replace ascorbic acid, information must be obtained on its efficacy (does it have the same potency as L-ascorbic acid) and stability during processing and storage. Compounds that have been presented for use in fish feeds are phosphate, sulfate, and glucose derivatives of ascorbic acid. These compounds are relatively stable during processing and storage of the feeds and have been reported to have vitamin C activity for salmonids,[25] channel catfish,[9] tilapia,[26] and other species. However, the vitamin C activity of these derivatives apparently varies among fish species. Therefore, to be used effectively in fish feeds the molar activity, or vitamin potency, of the phosphate, sulfate, or glucose derivatives relative to L-ascorbic acid must be known.

El Naggar and Lovell[18] fed channel catfish ascorbic acid, ascorbic acid phosphate, or ascorbic acid sulfate at equimolar ascorbic acid levels ranging from below the dietary requirement to an excess of the requirement. They regressed response data (weight gain, bone collagen content, and overt signs of scurvy) against dietary level of ascorbic acid, and used slope ratio analysis to determine the vitamin C potency of ester derivatives relative to L-ascorbic acid. Their data showed that generally L-ascorbyl-2-phosphate and L-ascorbic acid had equal vitamin C potency, but that L-ascorbyl-2-sulfate had less than 10% of the vitamin activity of L-ascorbic acid. Previous reports had indicated that catfish could use the sulfate derivative as a source of vitamin C; however, an excess of vitamin C at only one dietary level was used in these studies providing only a qualitative and not quantitative comparison. In order to prescribe dietary allowances of vitamin C in warmwater fish feeds, the vitamin activity of the vitamin C source must be known.

Shiau and Hsu,[26] using broken-line regression analysis, showed that hybrid tilapia (*Oreochromis niloticus* X *O. aureus*) could use ascorbic acid sulfate equally as well as ascorbic acid phosphate for growth and collagen synthesis, but both were less efficacious than ascorbic acid. Abdelghany[27] also found with Nile tilapia (*O. niloticus*) that ascorbic acid sulfate and ascorbic acid phosphate had equimolar vitamin activity. The glucose derivative, ascorbate-2-glucose, was found to have equal potency as ascorbic acid for growth and prevention of scurvy in hybrid *Clarias* catfish by Khajaren and Khajaren.[28]

Buddington et al.[29] demonstrated that in channel catfish the phosphate was released from ascorbyl-2-mono (or poly) phosphate by hydrolysis prior to absorption of ascorbic acid from the intestine, and concluded that production of hydrolases by brush-border membranes in the intestine is essential to utilization of ascorbate esters. This suggested that activity of sulfatase enzyme in the intestine may influence the ability of various fish to use the sulfate derivative as a source of vitamin C.[30]

9.8   Interaction of vitamin C with other nutrients

Because of its reducing properties, vitamin C may influence the bioavailability and, thus, the dietary requirement of other nutrients. For example, iron and other elements that are usually ingested in oxidized forms must be reduced to more soluble forms for absorption and metabolism; vitamin C has been demonstrated to be functional in this process. Another example is the effect of dietary vitamin C on folate requirement. Duncan and Lovell[21] found that the dietary folate requirement of channel catfish fed a marginal level of vitamin C, mg kg^{-1} (or the minimum requirement for normal growth) required approximately 4 mg kg^{-1} of folate, but when fed an excess of vitamin C, 200 mg kg^{-1}, folate requirement was reduced to 0.4 mg kg^{-1}. Vitamin C is functioning in reduction of folate to the active coenzyme, 5-methyl tetrahydrofolate, and apparently a higher requirement of vitamin C was necessary for optimal activation of folate than is required for growth. This revelation indicated that the dietary concentration of vitamin C and other micronutrients can influence the dietary requirement of another nutrient.

9.9   Summary

The several species of cichlids, catfishes, and cyprinids investigated require a dietary source of vitamin C for normal function. Quantitative requirements of dietary vitamin C have been reported for several species and appear to be relatively similar. Early reports indicted that approximately 50 mg kg^{-1} was the recommended dietary allowance for young fish of warmwater species; however, recent studies using vitamin C sources stabilized against oxidation loss, suggested that the requirement may be reduced by 50% or more. Several overt signs and metabolic anomalies have been well identified in vitamin C deficient warmwater fishes, generally associated with its role as a reducing compound. Vitamin C is necessary for maximum resistance against bacterial infections, and its role has been identified in specific and nonspecific immune responses. L-ascorbyl-2-phosphate is stable against oxidation loss and has equimolar potency to conventional L-ascorbic acid for most fish; however, L-ascorbyl-2-sulfate is used poorly by channel catfish but relatively well by some tilapias.

References

1. Ikeda, S. and Sato M., Biochemical studies of L-ascorbic acid by carp, *Bull. Jpn. Soc. Sci. Fish.*, 30, 365, 1964.
2. Kitamura, S., Ohara, S., Suwa, T., and Nakagawa, K., Studies on vitamin requirements of rainbow trout, *Salmo gairdneri.* 1. On the ascorbic acid, *Bull. Jpn. Soc. Sci. Fish.*, 31, 818, 1965.
3. Dupree, H. K., Vitamins essential for growth of channel catfish, *Ictalurus punctatus*, Technical Paper No. 7. Washington, D.C., Bureau of Sport Fisheries and Wildlife, 1966.
4. Lovell, R. T., Essentiality of vitamin C in feeds for intensively fed caged channel catfish, *J. Nutr.*, 103, 134, 1973.
5. Wilson, R. P. and Poe, W. E., Impaired collagen formation in the scorbutic channel catfish, *J. Nutr.*, 103, 1359, 1973.
6. Agrawal, N. K. and Mahajan, C. L., Nutritional deficiency disease in an Indian major carp, *Cirrhina mrigala* Hamilton, due to avitaminosis C during early growth, *J. Fish. Dis.*, 3, 231, 1980.
7. Stickney, R. R., McGeachin, R. B., Lewis, D. H., Marks, J. , Riggs, A., Sis, R. F., Robinson, E. H., and Wurts, W. , Response of *Tilapia aurea* to dietary vitamin C, *J. World Maricult. Soc.*, 15, 179, 1984.
8. Lim, C. and Lovell, R. T. Pathology of the vitamin C deficiency syndrome in channel catfish (*Ictalurus punctatus*), *J. Nutr.*, 108, 1137, 1978.
9. Andrews, J. W. and Murai, T., Studies on the vitamin C requirements of channel catfish (*Ictalurus punctatus*), *J. Nutr.*, 105, 557, 1974.
10. Martins, M. L., Effect of ascorbic acid deficiency on pocu fry (*Piaractus mesoptomicus*), *Braz. J. Med. Biol. Res.*, 28, 563, 1995.
11. National Research Council, *Nutrient Requirements of Fish*, National Academy Press, Washington, D.C., 1993.
12. Li, Y. and Lovell, R. T., Elevated levels of dietary ascorbic acid increase immune responses in channel catfish, *J. Nutr.*, 115, 123, 1985.
13. Dabrowski, K., Moreau, R., El-Saidy, D., and Ebeling, J., Ontogenetic sensitivity of channel catfish to ascorbic acid deficiency, *J. Aquatic Animal Health.*, Vol. 8. No. 1. 22, 1996.
14. Blom, J. and Dabrowski, K., Reproductive success of female rainbow trout in response to graded dietary ascorbyl monophosphate levels, *Biol. Reprod.*, 52, 1073, 1995.
15. Fracalossi, D. M., Craig-Schmidt, M., and Lovell, R. T., Effect of dietary lipid sources on production of leukotriene B by head kidney of channel catfish held at different water temperatures, *J. Aquatic Animal Health.*, 6, 242, 1994.
16. Santiago, C. B. and Gonzel, A. C., Effect of prepared diet and vitamins A, E and C supplementation on the reproductive performance of cage-reared bighead carp *Aristichthys nobilis* (Richardson), *J. App. Ichth.*, 16, 8, 2000.
17. Hilton, J. W., Cho, C. Y., and Slinger, S. J., Effect of graded levels of supplemental ascorbic acid in practical diets fed to rainbow trout (*Salmo gairdneri*), *J. Fish. Res. Board Can.*, 35, 431, 1978.
18. El Naggar, G. O. and Lovell, R. T., L-Asorbyl-2-monophosphate has equal antiscorbutic activity as L-ascorbic acid but L-ascorbyl-2-sulfate is inferior to L-ascorbic acid for channel catfish, *J. Nutr.*, 121, 1622, 1991.

19. Robinson, E. H., Vitamin C studies with channel catfish, *Tech. Bull., Miss. Agri. Exp. Sta.*, No. 182, 8pp., 1992.
20. Verlhac, V., Doye, A., Gabaudan, J., Troutland, D., Deschaux, P., Kaushik, S. J., and (ed) Luquet, P., Vitamin nutrition and fish immunity influence of antioxidant vitamins (C and E) on immune response of rainbow trout. Fish nutrition in practice: 4th international symposium on fish nutrition and feeding. Biarritz, France, *Les Colloques*, No. 61, 167, 1991.
21. Duncan, P. L. and Lovell, R. T., Influence of vitamin C on the folate requirement of channel catfish, *Ictalurus punctatus*, for growth, hematopoiesis, and resistance to *Edwardsiella ictaluri* infection, *Aquaculture*, 127, 233, 1994.
22. Li, M. H. and Robinson, E. H., Effect of dietary vitamin C on tissue vitamin C concentration in channel catfish, *Ictalurus punctatus*, and clearance rate at two temperatures—a preliminary investigation, *J. Applied Aquacult.*, 4, 59, 1994.
23. Waagob, R., Glette, J., Raa-Nilsen, E., and Sandnes, K., Dietary vitamin C, immunity and disease resistance in Atlantic salmon (*Salmo salar*), *Fish Physio. Biochem.*, 12, 61, 1993.
24. Lovell, R. T., and Lim, C., Vitamin C in pond diets for channel catfish, *Trans. Am. Fish. Soc.*, 107, 321, 1978.
25. Tucker, B. W. and Halver, J. E., Ascorbate-2-sulfate metabolism in fish, *Nutr. Rev.*, 42(5), 173, 1984.
26. Shiau, S. Y. and Hsu, T. S., L-ascorbyl-2-sulfate has equal antiscorbutic activity as L-ascorbyl-2-phosphate for tilapia, *Aquaculture*, 133, 147, 1995.
27. Abdelghany, A. E., Growth response of Nile tilapia to dietary L-ascorbic acid, L-ascorbyl sulfate and L-ascorbyl-2-polyphosphate, *Aquaculture*, 150, 449, 1996.
28. Khajaren, J. and Khajaren, S., Stability and bioavailability of vitamin C-glucose in hybrid *Clarius* catfish, *Aquaculture*, 151, 219, 1997.
29. Buddington, R. K., Puchal, A. A. , Houpe, K. L., and Diehl W. J., III, Hydrolysis and absorption of two monophosphate derivatives of ascorbic acid by channel catfish, *Aquaculture*, 114, 317, 1993.
30. Matusiewicz, M. and Dabrowski, K., Characterization of ascorbyl esters hydrolysis in fish, *Comp. Bioch. Physiol.*, 110B, 739, 1995.

chapter ten

The impact of micronutrients on the requirement of ascorbic acid in crustaceans and fish

Rune Waagbø, Kristin Hamre, and Amund Maage

Contents

Abstract

Teleost fish lack the enzymes for endogenous ascorbic acid (AA) synthesis and therefore have a qualitative requirement for AA. The minimum dietary requirement for AA in most fish species has been estimated in the range of 10 to 60 mg kg^{-1} dry diet when free AA was used. Below this level of supplementation the fish show dramatic deficiency symptoms such as reduced growth, vertebrae deformity, anemia, increased mortality, and reduced resistance to infections. Most ingredients used in feed for fish and crustaceans do not contain AA, due to type and pretreatment (heat) of the raw materials. Consequently, AA must be supplemented in the feeds to cover requirements for growth, health, and reproduction. By using mono- and polyphosphate derivatives of AA with high stability during processing and storage of feeds and high bioavailability, the requirement for AA has been determined to be around 20 mg kg^{-1} in several fish, and between 40 and 210 mg kg^{-1} in shrimp species.

Interactions between AA and several micronutrients in the diet and in the fish tissues affect the requirement for AA supplementation and the AA status. The strongest interactions have been found for antioxidant vitamins and astaxanthin, as well as iron and copper. High dietary levels of AA may improve fish immunity and resistance to stress and infectious diseases. Contradictory scientific results on this area may partly be explained by nutrient and mineral interactions.

This chapter covers micronutrient interactions that may affect the requirement of AA in fish and crustaceans to support optimal growth and health.

10.1   Introduction

Vitamin C, or ascorbic acid (AA) is an essential nutrient for teleost fish,[1-5] due to the missing enzymatic activity of L-gulonolactone oxidase (E.C. 1.1.3.8) needed for AA biosynthesis. It has been suggested that shrimps have a limited ability to synthesize vitamin C,[6] although endogenous AA synthesis does not seem to meet the requirement in young crustaceans.[7] The vitamin C nutrition of aquacultural species has been a focus of many research works due the low stability of AA during fish feed production and storage, the essential and multifunctional biochemical roles of AA in fish and crustaceans, as well as its potential stress-ameliorating and immunostimulating effects. These aspects have been discussed in detail in several recent reviews.[8-15]

The exact magnitude of vitamin C-related disorders in aquaculture is not known, but vitamin C deficiency has caused significant losses in practical fish farming, especially during the sensitive start feeding period.[9] The exact requirement of AA depends on interactions between AA and other nutrients and feed ingredients, as well as other factors such as age, health status, and exposure to unfavorable environmental conditions. High dietary levels of AA may improve fish immunity and resistance to infectious diseases.[12, 16] However, several studies do not support immunostimulating or stress modulating effects of high dietary levels of AA, and contradictory scientific results in this area may partly be explained by nutrient and environmental interactions. Even though interactions between AA and other nutrients have been observed in many studies on fish, few studies have been specifically designed to study effects of other nutrients and environmental factors on AA metabolism, and few reviews have focused on such interactions in fish.[9, 17, 18] The purpose of this review is to present the current knowledge of micronutrients that modify the amount of AA needed to support optimal growth and health in fish and crustaceans, including interactions at the dietary, intestinal, and metabolic levels.

10.2   Vitamin C deficiency symptoms and established minimum requirements

Under experimental conditions, fish and crustaceans fed vitamin C levels below the requirement show dramatic deficiency symptoms. Reduced growth, vertebrae deformity, histopathological changes, anemia, increased mortality, and reduced resistance to infections are commonly observed in fish.[19–25] The symptoms reflect the many essential roles of AA in the body, but their sensitivity with respect to detection of suboptimal AA nutrition varies.

   AA is an essential cofactor for the proline and lysine hydroxylases, which catalyse hydroxylation of the respective protein-bound amino acids in procollagen. Lower proportions of hydroxy (OH)-proline and OH-lysine (making cross-links between collagen subunits) result in impaired connective tissue development and function, giving symptoms such as spinal deformities (lordosis and scoliosis), other bone deformities (jaws, fins), and hemorrhages. Although some studies failed to demonstrate the essentiality of AA in crustacean species,[26] others have shown that diets devoid of AA affect growth and collagen formation. Further, suboptimal AA status has been related to the "black death" syndrome in *Penaeus stylirostris* and *P. californiencis*.[27, 28] In a study on *P. japonicus*, Kanazawa[29] found a clear dose-related preventive effect of AA on mortality.

   Besides the role as cofactor in enzymatic reactions, AA serves as a water-soluble reducing agent, or antioxidant in the tissues, and participates in a series of less specified biochemical reactions.[9, 30] Thus, the symptoms at AA deficiency and suboptimal intakes vary between species.[24, 25] Despite large differences in analytical methods, critical tissue AA levels have been used as a clinical sign of deficiency. Liver AA concentrations below 20 and 30 μg g^{-1} were suggested to indicate AA deficiency in rainbow trout[21] and channel catfish juveniles,[31] while deficiency appears to occur below 10 μg g^{-1} in juvenile

Atlantic salmon.[23] Establishment of reliable tissue AA threshold values for AA-related biochemical processes, defined for fish and shrimp species according to their age, would be valuable parameters in future AA research.

The indicators used in AA requirement studies differ in their sensitivity and can be divided into three categories according to their usefulness in AA nutrition: 1. absolute AA deficiency, indicated by deficiency symptoms anorexia, lethargy, reduced growth, mortality, impaired connective tissue manifested as bone/skin hydroxyproline, deformation and hyperplasia of gill epithelial cell layer, fin erosion, hemorrhage exophthalmia, granulomatous hypertyrosinemia, ascites, and muscular hemorrhage; 2. suboptimal AA status indicated by anemia, suboptimal tissue AA concentration, reduced immunity and secondary infections; and 3. high or pharmacological doses of dietary AA indicated by tissue AA saturation and manifested with positive effects on detoxification, resistance to infections and stress, specific immunity, and wound healing.

Minimum requirement levels for AA in different fish and crustacean species have been determined under variable experimental rearing conditions and using different diet compositions. The status of AA is, however, a result of dietary AA level and chemical form, as well as the feed composition and intake and site-dependent environmental conditions. Most of the dietary AA in formulated diets originates from the supplemented vitamin mixtures, since most of the AA from the feed raw materials are lost during their processing to meals or during feed production. Until recently, stable and bioavailable AA derivatives were not available, and requirement studies were performed using crystalline-free AA. The low stability of crystalline AA in practical fish feeds[6, 27, 32–34] has complicated accurate requirement studies. Oxidation of AA in the experimental diets during production and storage, and leaching into the water when feeding, has made it difficult to know the exact amount ingested by the fish or crustacean. Thus, many of the interactions observed between AA and feed nutrients have been related to the instability of AA in the feed, and this has probably led to too high estimates of the minimum requirements of AA in diets for fish (60 mg kg^{-1};[31] 100 mg kg^{-1};[35] approx. 50 mg kg^{-1})[25] and shrimp species (up to 10,000 mg kg^{-1}).[36]

10.2.1 Vitamin C sources

The vitamin C requirement clearly depends on the stability and availability of the vitamin C source used, and several AA compounds (coated products, derivatives) have been tested with respect to feed stability during processing and storage, as well as to vitamin C bioactivity.[9, 25, 37–39] So far, the phosphate derivatives of AA (ascorbic acid mono phosphate, AAmP and ascorbic acid polyphosphate, AApP) seem to be the most effectively utilized vitamin C sources for fish and crustaceans, due to both superior feed stability and bioactivity. AA-2-sulfate (AAS) exhibits far lower bioactivity both in fish (approximately 15%)[37, 40, 42] and crustaceans (approximately 25%).[39, 43] The AA derivatives have to be hydrolyzed to AA in the intestine before absorption through a Na-mediated process.[44, 45] Different rates of hydrolysis by intestinal phosphatases and sulphatases may explain the difference in bioactivity between AA-phosphate and -sulphate derivatives.[41, 46–48] Of the AA phosphate derivatives, *in vitro* intestinal hydrolysis of AApP to AA and phosphate seemed less efficient than that of AAmP.[48] A similar difference was observed in rapid-growing *P. monodon*, where AApP showed 64% of the bioactivity of AAmP.[39] Despite similar trends, liver AA levels measured between 12 and 24 weeks after start feeding were similar in Atlantic salmon juveniles fed equimolar amounts (10 and 100 mg kg^{-1}) of either crystalline AA, AAmP, or AApP (Figure 10.1).

Between 10 and 20 mg AA equivalents kg^{-1} feed approximated the minimum requirement in several salmonid species, including Atlantic salmon, *Salmo salar*[23, 49] and rainbow trout, *Oncorhynchus mykiss* juveniles,[50] using fishmeal-based diets and AA phosphate derivatives. Diets supplemented with 10 mg AA equiv. kg^{-1}, using crystalline AA, AAmP, or AApP, supported growth and collagen formation, but blood hemoglobin concentrations were suppressed compared to fish fed 100 mg AA equiv. kg^{-1}. Levels of 10 mg kg^{-1} are therefore considered as a suboptimal AA supplementation (Table 10.1). Another species, sea bass (*Lates calcarifer*) seemed to cover its requirement for vitamin C by 30 mg AAmP kg^{-1}.[51]

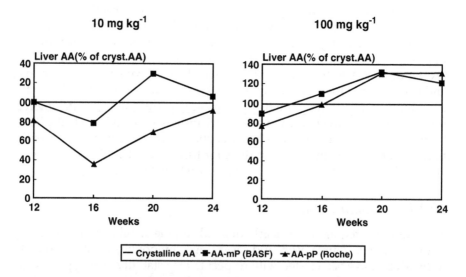

Figure 10.1 The phosphate derivatives AAmP (BASF, Germany) and AApP (Roche, Switzerland) showed similar bioactivity at suboptimal (10 mg AA equiv. kg^{-1}) and optimal (100 mg AA equiv. kg^{-1}) inclusion levels in Atlantic salmon fed from first feeding, as compared to crystalline AA. After 20 weeks the phosphate derivatives at 100 mg AA equiv. kg^{-1} significantly improved liver AA levels compared to crystalline AA. The suboptimal level caused anemia (Waagbø and Maage, unpublished data).

Table 10.1 Liver total ascorbic acid (AA), liver iron, and blood hemoglobin
concentrations in Atlantic salmon fed diets with 0, 10, and 100 mg AA,
AA-monophosphate (AamP) or AA-polyphosphate (AApP) kg^{-1} for 24 weeks
(Waagbø and Maage, unpublished results)

Dietary AA form and level	Liver AA (μg g^{-1})	Liver iron (μg g^{-1})	Blood Hb (g 100 ml^{-1})
AA 0 (n=2)	1 a	104	3,4 a
AA 10	4,0 (0,6) a	82 (1)	5,6 (0,3) b
AA 100	48 (1) b	80 (5)	9,2 (0,2) c
AA-mP 10	4,3 (0,3) a	89 (14)	4,8 (0,3) ab
AA-mP 100	59 (2) bc	87 (3)	8,9 (0,3) c
AA-pP 10	3,7 (0,3) a	75 (2)	4,4 (0,3) ab
AA-pP 100	64 (4) c	75 (1)	8,5 (0,1) c

Different letters indicate significant differences within a column (n=3, p < 0.05).

Very high AA recommendations have been suggested for several shrimp species (ranging from 1000 to 20,000 mg AA kg^{-1}) to correct for both vitamin instability as well as leaching losses from the feed during long exposure in water.[6, 8, 27, 29] However, by using AApP and AAS as dietary AA sources, Hsu and Shiau[39] recently found 30 and 75 mg AA equiv. kg^{-1}, respectively, to meet the vitamin C requirement in juvenile grass shrimp (*Penaeus monodon*). Other authors have determined the AA requirements in shrimps to be between 40 and 210 mg kg^{-1}, depending on species (*P. vannamei, P. monodon*) and AA phosphate source.[52, 7, 43, 53, 54] No quantitative data on the AA requirement in freshwater prawn species have been gathered.

The discrepancy between minimum AA requirements and the amount needed to maintain optimal health seemed to be wide. Practical requirements are based upon the minimum requirements, with a broad safety margin, taking into consideration such factors as the availability and stability of the AA source and variation in farming conditions and environment. AA supplementations of diets for aquaculture should cover requirements for growth, health, and reproduction. The target levels of AA in commercial fish feeds are most often above 100 mg kg^{-1}, even when using stable phosphate derivatives. Dabrowski[55] suggested that the optimal dietary AA concentration should maintain steady state tissue concentrations of AA in fish larvae. In Atlantic salmon juveniles fed practical diets from start feeding, 10 to 20 mg kg^{-1} supported a stable liver AA concentration (from weeks 14 to 23) of approximately 10 μg g^{-1}.[23] However, 10 μg g^{-1} of liver is probably below desired level. Dietary levels of AA affect the liver AA concentration, however, with different kinetics depending on the dietary level[56] and feeding regime.[45] The metabolism of AA probably includes regulation of intestinal uptake and excretion of surplus AA. Fluctuations in AA status related to AA metabolism, feed intake, and interactions with dietary, physical, and environmental fac-

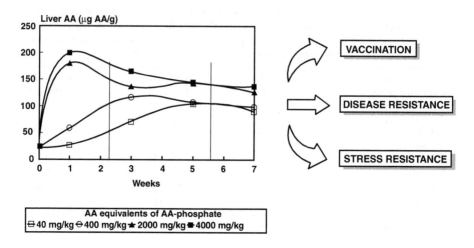

Figure 10.2 Fluctuations in ascorbic acid (AA) status (dietary level, feed intake, metabolism, and excretion rate) may impact physiological and immunological systems temporarily, depending on time of feeding (indicated by the vertical bars, where groups of fish show different AA status).

tors may result in time-dependent and temporary beneficial effects on immunity, resistance to infections and stress response (Figure 10.2), and may partly explain contradictory scientific results in this area.

10.2.2 Changes in AA requirement in life cycle

Studies have indicated that the AA requirement decreases with age,[17, 21, 57–59] probably related to a decreasing metabolic rate with increasing size[60] and increased AA storage capacity. This seemed to be true also for shrimps.[7]

Feeding fish and shrimp larvae with live feed (natural zooplankton, i.e., *Temora longicornis, Artemia* spp., etc.) represent a link between crustacean and fish nutrition with respect to AA.[61] Natural zooplankton fed to halibut larvae contains considerable amounts (600 to 1000 μg AA g^{-1} DW; equivalent of 60 to 100 μg AA g^{-1} wet weight) of AA.[62] AA-enriched *Artemia* spp. *nauplii* are commonly used as first feed in larviculture,[60, 61] and depending on the regime of enrichment they may contain between 500 and 2300 μg AA g^{-1} DW.[61] Accordingly, the dietary AA concentrations of fish and crustacean larvae seem to exceed the estimated AA requirements observed in fish fed formulated feeds from the onset start feeding (salmonids). From the high AA concentrations found in fish eggs,[9, 11, 62-64] natural zooplankton and non-enriched *Artemia*, it is speculated that the fish larvae need elevated dietary AA to pass the critical weaning period.[61, 63] Studies indicate superior availability of AA from *Artemia* compared to dry feeds.[65] However, halibut larvae fed on grown *Artemia* (770 μg AA g^{-1} DW) or dry feed containing

AApP (277 μg AA equiv. g^{-1} DW) during 20 days, grew equally well and retained similar amounts of AA in the whole body.[63] Interactions between AA and other micronutrients in the live food or larval body may justify elevated requirements of AA in the fish and crustacean larvae. A constant weight-specific AA content in halibut larvae weighing above 10 mg dry weight indicated a saturation of whole-body AA by using live food with 600 to 1000 μg AA g^{-1} DW.[62] The status of AA is probably determined through regulation of intestinal AA uptake[45] or excretion of excess AA, as was suggested for Atlantic salmon fed elevated levels of AAmP[56] (Figure 10.2).

Despite the possible important roles of vitamin C in smoltification processes in salmonids (synthesis of collagen, hormones, and neurotransmittors, as well as change in enzyme activities and general metabolism), a study on Atlantic salmon did not support the anticipated effect of an increased requirement of dietary vitamin C during this phase.[66]

Vitamin C plays important roles in fish reproduction, affecting hormone synthesis,[58] vitellogenesis,[58] and results in variable oocyte AA retention and gamete quality.[9, 64, 67, 68] Peroxidative damage to rainbow trout spermatozoa during the reproduction phase may be prevented by adequate broodstock AA nutrition, thereby preserving sperm quality.[69]

Several studies have shown that a higher dietary AA input is required in fish exposed to unfavourable conditions such as infections[70, 71] and environmental pollutants.[9, 72–75, 146] Similarly, higher dietary AA has been shown to have positive effects on immune functions and resistance to diseases and stress.[12, 16, 76]

10.3 Micronutrient interactions

The interaction between AA and other micronutrients has been shown to affect the AA status and requirement for AA supplementation.[9, 18] The strongest interactions have been found for antioxidant vitamins and astaxanthin, as well as minerals such as iron and copper.

10.3.1 Interactions with vitamin E

While AA acts as a radical scavenger in the aqueous phase, vitamin E is regarded as the most important lipid soluble antioxidant in animal tissue.[77] Interactions between the two may be coupled to the simultaneous protection of both water- and lipid soluble molecules against free radicals attack. Further, a hypothesis first proposed by Tappel[78] states that AA may reduce the vitamin E radicals generated in the process where vitamin E scavenges lipid peroxyl radicals. Oxidized AA is reduced by glutathione *in vivo*, and oxidized glutathione is in turn reduced by NADPH.[79, 80] Thus, it appears that the red-ox defense is coupled with the generation of reducing equivalents through energy metabolism (Figure 10.3).

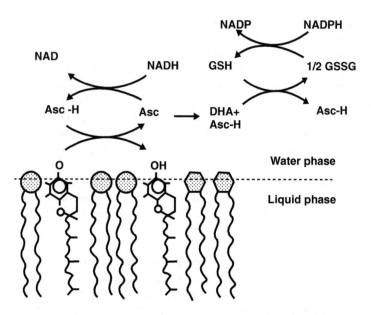

Figure 10.3 Proposed mechanism for the regeneration of α-tocopherol from the tocopheroxyl radical (Tappel, 1962). Ascorbic acid (Asc) is oxidized in the process, but can be regenerated by glutathione either chemically or enzymatically (Winkler et al., 1994; Meister, 1994) or enzymatically by NADH (Meister, 1994). Oxidized glutathione (GSSG) is reduced by glutathione reductase at the expense of NADPH generated in the pentose phosphate shunt (Meister, 1994).

The reduction of the vitamin E radical by AA has been shown in several *in vitro* systems, i.e., in homogenous solution,[81] liposomes,[82] organelle preparations,[83] and low density lipoprotein, LDL.[84] Although experimental results are conflicting,[85, 86] it is now commonly accepted that the process also proceeds *in vivo*. The standard potential difference of the reaction makes it thermodynamically feasible at physiological pH,[87] but for the reaction to proceed there must be a possibility for the components to interact. It is assumed that the active and polar phenol group of α-tocopherol resides at the interface between lipid and aqueous compartments, for example, at the membrane surface.[88] In this position it can react with water-soluble molecules such as AA. Further, the concentrations of reactants and products affect the rate and direction of red-ox reactions, according to the Nernst equation.

Poston and Combs[89] reported on interactions of AA and vitamin E with selenium in Atlantic salmon. They found that both vitamins prevented symptoms of selenium deficiency and concluded that knowledge of the interdependence of these nutrients may help explain some of the discrepancies in their reported quantitative requirements. In spite of this, very few studies focused on the interactions of AA with vitamin E (and selenium) in fish.

One way in which vitamin E might affect the requirement of AA is through modulation of the AA retention. In Atlantic salmon, juveniles supplemented with 0 to 60 mg kg^{-1} AA and 0 to 220 mg kg^{-1} α-tocopherol,[49] we found that high dietary α-tocopherol led to an increase in AA uptake in the liver at dietary AA concentrations above 15 mg kg^{-1} (Figure 10.4). Similar results were reported by White et al.,[90] also in Atlantic salmon. However, in postsmolt Atlantic salmon, an increase in dietary α-tocopherol from 80 to 400 mg kg^{-1} had no or a slightly negative effect on liver AA levels in fish supplemented with 60 mg AA kg^{-1}, using AApP (Figure 10.5). This discrepancy may have to do with the tendency for vitamin E to act as a pro-oxidant when present in high concentrations, especially when the AA concentration is low.[91, 92] Alternatively, this may reflect differences in developmental stage of the fish, since the efficiency of both α-tocopherol and AA uptake in salmon changes during development[49] (see previous discussion).

Rønnestad et al.[62] studied the mass balance of AA and α-tocopherol in Atlantic halibut (*Hippoglossus hippoglossus*, L.) larvae and found that 95% of AA but only 30% of the α-tocopherol in the yolk, had been transferred to the body at the onset of the first exogenous feeding. This resulted in a low and stable body level of α-tocopherol (≈25μg g^{-1} DW) during the yolk sac stage and increasing concentration of α-tocopherol in the remaining yolk as other nutrients were utilized. Concomitant with final yolk absorption, α-tocopherol concentrations in the larval body increased five-fold, but the excess of vitamin E was rapidly excreted. The larval α-tocopherol then stabilized between 50 and 100 μg g^{-1} DW. Slightly delayed compared to the peak for α-tocopherol, AA concentrations increased two-fold. This increase could

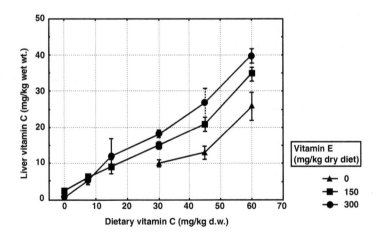

Figure 10.4   Liver ascorbic acid concentration in Atlantic salmon juveniles fed graded levels of AA (as AAmP) and α-tocopherol (as acetate ester) for 22 weeks. Mean ± SEM, n = 5 (adapted from Hamre et al., 1997 with permission from *Free Radical Biology and Medicine*, Elsevier Science Inc.)

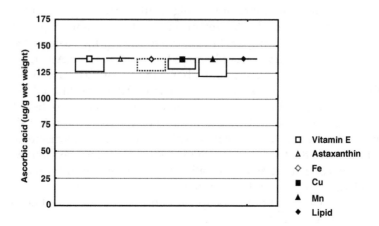

Figure 10.5 Concentration of ascorbic acid (μg g^{-1}) in the liver of Atlantic salmon postsmolts fed diets supplemented with 60 mg kg^{-1} ascorbic acid and two levels of vitamin E (80 or 400 mg kg^{-1}), astaxanthin (10 or 50 mg kg^{-1}), lipid (170 or 330 mg kg^{-1}), iron (70 or 1200 mg kg^{-1}), copper (8 or 110 mg kg^{-1}), and manganese (10 or 200 mg kg^{-1}) in a 2^{7-3} reduced factorial design. The diets were fed for 22 weeks. Points represent the average liver AA concentration and bars indicate the magnitude and direction of the significant effects of feeding high levels of the nutrient variables. Regression coefficients ($p < 0.05$) were calculated using multiple linear regression (extracted from Hamre et al., unpublished data).

not be explained by AA transfer in connection with final yolk absorption or by AA levels in the live feed. It was therefore suggested that the high vitamin E levels of the larvae at final yolk absorption stimulated the uptake of AA in a similar manner to Atlantic salmon.[49]

High vitamin E did not enhance liver uptake of AA at low dietary AA concentrations (Figure 10.4) and vitamin C deficiency developed slightly earlier in fish fed the highest vitamin E levels than in those fed intermediate levels.[49] The difference was too small to have any major influence on the AA requirement, but it supports the hypothesis that vitamin E radical is reduced by AA in the fish tissues. If recovery is to proceed, there should be enough AA present to prevent accumulation of vitamin E radicals, as these radicals may act as pro-oxidants and promote lipid oxidation.[92]

Vitamin C-deficient Atlantic salmon juveniles showed a large drop in liver vitamin E concentration in parallel with lowered vertebrae hydroxyproline concentration (Figure 10.6), indicating a vitamin E "sparing" by feeding an AA-adequate diet as opposed to an AA-deficient diet. Raising dietary AA to a higher level than the minimum requirement did not have any further effect on the uptake of vitamin E in the fish body. Several studies have indicated that AA "spares" vitamin E in terrestrial animals,[85] whereas other authors could not detect this effect.[86] These discrepancies may be partly explained by satisfactory AA levels, whereas AA has to be deficient in order to have a negative impact on vitamin E status.

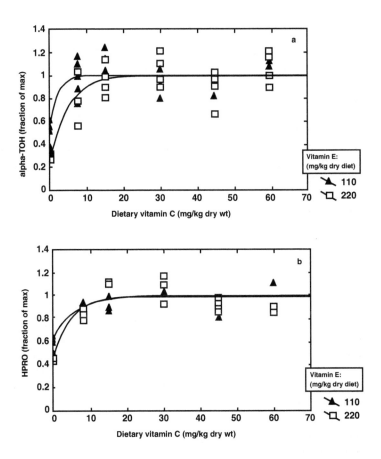

Figure 10.6   Liver α-tocopherol (A) and vertebrae hydroxyproline (B) concentrations in response to dietary AA in Atlantic salmon juveniles fed graded levels of AAmP and α-tocopherol-acetate for 22 weeks. Results are given as the fraction of average concentrations of α-tocopherol and hydroxyproline when AA was adequate. Each point represents one pooled sample from 10 fish, and there are 5 samples per treatment. Vertebrae and livers were dissected from the same fish (adapted from Hamre et al., 1997; with permission from *Free Radical Biology and Medicine,* Elsevier Science, Inc.).

Vitamin C supplementation above the minimum requirement has a marked effect on the vitamin E requirement in salmonids, as shown by Frischknecht et al.[93] and Hamre et al.[49] In both studies, fish fed diets without vitamin E supplementation were protected by AA against mortality, growth depression, anemia, and erytrocyte fragility (spontanous hemolysis) in a dose-dependent manner. Hamre et al.[49] also found that AA inhibited lipid oxidation in the liver and accumulation of autofluorescent granules in erythrocytes. Since AA did not affect the vitamin E status of these fish, it was suggested that the protective role of AA was to lower the oxidative challenge on the organism by scavenging water-soluble radicals. Based on these studies, it can be concluded that AA supplemented at levels above the minimum requirement may prevent vitamin E deficiency in salmonids.

In contrast to the experiments with salmonids, Gatlin et al.[94] did not induce vitamin E deficiency in channel catfish (*Ictalurus punctatus*) fed a diet deficient in both vitamins C and E. Results from several studies suggested that fish fed low levels of predominantly saturated lipids, are remarkably resistant to vitamin E deficiency.[95–97] The oxidative challenge of an organism, and thereby the vitamin E requirement, will increases in response to higher levels of polyunsaturated fatty acids (PUFA) in the diet.[98–100] In the experiment of Gatlin et al.[94] the dietary lipid source was 4% lard, and a possible explanation for the lack of vitamin E deficiency symptoms may therefore be that exposure of the fish to oxidative stress was quite low in this experiment.

10.3.2   Retinoids and carotenoids

Hypervitaminosis A did not affect the AA requirement in rainbow trout, as evaluated by the occurrence of AA deficiency symptoms with dietary vitamin A up to 124000 IU or 37.2 mg kg^{-1}.[101] Several studies have shown that excess dietary vitamin A may cause backbone deformities in fish, but this pathology seems not to be related to vitamin C deficiency.[102] Carotenoids and retinoids may act synergistically against lipid peroxidation, as components in the integrated cellular antioxidant protection system.[103] In a feeding experiment with purified diets, Christiansen et al.[104] showed that the dietary supplementation and status of astaxanthin interacted with the status of several antioxidant vitamins, including AA in Atlantic salmon. This effect seems to be related to a general antioxidant-sparing action at low dietary levels of astaxanthin, resulting in deficiency. Our results indicated that dietary astaxanthin levels (10 and 50 mg kg^{-1}) in practical diets do not significantly affect AA status in Atlantic salmon reared in sea water (Figure 10.5).

10.3.3   Other vitamins

Studies on interactions between AA and other water-soluble vitamins are sparse, and most studies are concerned with the impact of high dietary AA on vitamin stability and metabolism.[105]

Graded dietary AA (0, 20, 200 mg AA kg^{-1}, as AAmP) modulated the response of channel catfish to dietary folate, and vice versa, as illustrated by red blood cell pathologies in a study by Duncan and Lovell.[106] Vitamin C deficiency signs recorded as survival rates, and changes in hematological parameters were reversed by increasing folic acid supplementation (0, 0.4, 4.0 mg kg^{-1}). However, the deficiency-related growth retardation and bone deformities were not affected. By adding 20 and 200 mg AA kg^{-1} to the feed, the authors showed that dietary AA reduced the requirement for folate in channel catfish.[106] Increased dietary AA alleviated the symptoms of folate deficiency, and thus one may conclude that the requirement for AA to sustain growth, survival, and fish health increases when folate supplementation is low. The interaction between AA and folate may be related to the role of AA as a reducing agent and antioxidant, protecting reduced forms of folates in metabolism.

Flavonoids provide antioxidant protection, and AA may affect these properties. Dabrowski et al.[38] suggested that the better retention observed for AA derived from *Artemia* than AA from dry feeds may be related to interactions of flavonoids in *Artemia* with AA utilization in fish larvae. Inclusion of the flavonoid rutin in diets for channel catfish did, however, not improve AA status.[107] The potential for using flavonoids as natural antioxidants in fish diets should be further evaluated.

Antagonisms between high dietary levels of AA and other water-soluble vitamins have been suggested in man, due to elevated excretion or destruction, but newer studies do not support these interactions.[105]

Kitamura et al.[108] referred to a study on guinea pigs where the requirement of AA was increased during tryptophan deficiency, which could explain the development of scoliosis in tryptophan-deficient rainbow trout.[109]

10.3.4   *AA interactions with minerals*

AA is a strong two-electron reducing agent, and is easily oxidized in one-electron steps by metal ions and metal complexes. As a reducing and chelating agent, AA interacts with minerals in the diet as well as in the tissues. In general, AA enhances the intestinal absorption of iron and selenium, reduces the absorption of copper, nickel, and manganese, has minor effects on calcium, zinc, and cobalt absorption, and has no effect on the heavy metals such as cadmium and mercury.[30] Further, mineral intake from water represents an important source of essential elements for aquatic organisms, but may also represent toxic challenges from environmental pollutants. In the following section this will be discussed for fish. Little information is available for crustaceans.

10.3.5   *Iron*

The interaction of AA and iron is a classic research area in human and animal nutrition. Due to the general high incidences of iron malnutrition among humans, this interaction is most often studied and discussed with respect to how AA positively affects iron nutrition and metabolism, and it is suggested that iron absorption should be considered in setting AA requirement in humans.[110, 111] In fish nutrition, however, the dietary content and elemental form of iron may impact the requirement of AA through the pro-oxidant properties, and elevated dietary iron has been claimed to compromise fish health (see Blazer,[16] Waagbø,[12] and Andersen[112]).

Suboptimal AA supplementation (1–10 mg kg^{-1}) or body reserve AA status lead to symptoms of an iron deficiency-like anemia in young and broodstock salmonids, despite adequate dietary iron supplementation[58, 40, 113] (Table 10.1). Broodstock rainbow trout fed a diet devoid of AA turned anemic during vitellogenesis, as compared to fish fed adequate AA.[58] The blood hemoglobin concentration in the vitamin C deficient fish was strongly negatively correlated to the gonadosomatic index, indicating that maternal haem-Fe was used to support the ovary with iron. The lack of dietary AA or low AA status, therefore, seemed to affect iron absorption and metabolism in fish, probably related to the lack of suitable reducing power needed in iron mobilization from storage.[114]

Dietary AA facilitates the intestinal absorption of nonheme iron in humans by reducing ferric iron to the ferrous form as well as increasing the solubility of iron in alkaline pH by chelating iron.[110, 115] The iron absorption in rainbow trout was positively affected by a diet containing crystalline AA as opposed to an AA-free diet, but not in diets with AS;[47] however, several other studies did not show convincing data on this effect.[41, 113] By using liver iron concentrations, minor effects on absorption may be masked by time and general iron metabolism. Serum iron concentrations after three weeks increased significantly (from 0.7 to 1.15 µg ml^{-1}) with increasing dietary AApP (20 to 4000 mg AA equiv. kg^{-1}) indicating a positive, but time dependent effect on absorption, since the corresponding liver AA concentration did not reflect this interaction (Waagbø and Verlhac, unpublished data). Increasing levels of AA phosphate derivatives (AAmP or AApP) in the diet did not seem to affect liver iron status (reflecting both absorption and/or metabolism) in Atlantic salmon juveniles or smolts[116] (Table 10.1). Elevated dietary iron may increase the need for AA supplementation, when using unstable AA forms or highly oxidized lipid (rancid) diets.[116, 117] Increasing dietary iron concentrations fed to Atlantic salmon reduced the dietary AA utilization levels and liver AA status in a dose-dependent manner.[118] Different chemical forms of iron also have varied effects on the liver AA status. In fish meal diets supplemented above 100 mg heme bound iron kg^{-1}, the supplemented AA (150 mg ethyl cellulose-coated AA) was nearly completely lost compared to acceptable AA levels analyzed in diets and fish fed equimolar amounts of iron sulfate and elementary iron.[119] The interaction of iron with the AA requirement seems to occur mainly at the dietary and/or intestinal levels. Despite some controversies on iron regulation in fish, the absorption of non-heme iron seems to be well regulated at the intestinal level in Atlantic salmon.[112] Thus, AA may facilitate iron absorption,[47] however, this may only be of biological significance at low dietary iron concentrations and body iron status. Heme-iron is absorbed through a different route which may be less dependent on iron status, probably similar to the mechanisms of heme absorption in humans.[111, 119–120] Thus, both dietary iron concentrations and chemical form affect AA requirement, mainly by oxidizing AA in the diet. The AA phosphate derivatives seem to escape oxidation in the gastointestinal tract,[44] thus using

AA-phosphate derivatives and adequate feed antioxidants, dietary iron up to 600 mg kg^{-1} did not seem to increase the need for AA supplementation.[116] In an experiment with adult Atlantic salmon specifically designed to study interactions between dietary pro- and antioxidants (multivariate approach), the high dietary iron concentration (1200 mg ferrous sulfate kg^{-1}) caused a minor reduction in AA status (approx. 7.5 %, $p < 0.05$) in fish fed diets with 30 to 60 mg AA kg^{-1} by use of AApP (see Figure 10.5). Thus, high inclusion levels of dietary iron sulfate seem to impact the metabolic requirement of AA to a minor extent.

Increased liver iron status was observed in salmon fed increasing dietary heme iron, and this may elevate the need for AA to protect against *in vivo* oxidation.[119] High iron status has been considered to compromise fish health (see Blazer[16] and Waagbø[12]) due to both increased risk of oxidation and bacterial and parasitic invasions and infections. In view of the hypothesis that iron regulation in fish is inferior to that of terrestrial vertebrates, elevated levels of dietary iron have been claimed to challenge fish health by overriding the capacity of transferrin binding. Unspecifically bound iron may cause hazardous oxidations, favoring rapid growth of pathogenic bacteria, causing increased incidence of infections, microthrombosis, and secondary skin ulcers.[121–122] It has been suggested that the iron acquisition ability of several bacteria is closely related to their virulence.[123] Again, when using adequate levels of dietary AA-phosphate derivatives and feed antioxidants, high dietary iron did not have detectable effects on disease resistance and immunity.[116–117, 124–125]

The dietary requirement for iron of *P. vannamei* seemed to be covered by unsupplemented diets which contained approximately 12 mg iron kg^{-1} and 3000 mg AA equiv. kg^{-1} as AApP.[126] The low iron requirement probably reflects the use of copper-containing hemocyanin as respiratory pigment in shrimp instead of hemoglobin as in vertebrates. To our knowledge, no information on interactions between the two nutrients in crustacean nutrition is available.

10.3.6   Copper

From the literature on copper nutrition in fish, tissue Cu retention does not always reflect dietary copper levels up to 100 mg kg^{-1},[127–131] and this differs from data on the retention of waterborne copper.[132–135] AA affected the retention and toxicity of waterborne copper in carp and rainbow trout.[132, 133] In a 9-week study on carp, the toxicity of 0.05 ppm waterborne copper (measured as tissue Cu accumulation) was reduced by supplementing the diet with 2000 mg AA kg^{-1}, with a concomitant reduction in liver AA compared to fish fed no AA. Liver L-gulonolactone oxidase activity was reduced with increasing Cu accumulation.[132] In a similar 18-week study with rainbow trout, nearly identical results were obtained.[133] This shows clearly that the requirement of AA increases with high Cu load and that excess AA is needed to depress waterborne Cu accumulation in the fish liver. The uptake of dietary Cu appears to be much more restricted than the uptake over the gills.[130, 135] Dietary Cu seems to be strictly regulated at the intestinal level, and based on traditional biological responses, dietary levels up to 500 mg Cu kg^{-1} may be permitted in Atlantic salmon.[131,136] AA levels as high as 10,000 mg kg^{-1} did not prevent dietary Cu (800 mg kg^{-1})-induced lesions.[127] While AA enhances the availability of iron in the diet, intestinal Cu absorption is decreased by AA in some animals,[137] including fish.[41] These interactions may be related to the reducing capacity of AA under the physiological conditions found in the intestine, keeping iron in a favorable state, while copper is kept in an unfavorable state for absorption. The AA-induced reduction in body Cu may also be related to increased excretion of Cu through the liver and bile. In a multivariate interaction study with adult Atlantic salmon, high dietary Cu concentrations (110 mg kg^{-1}) as compared to dietary Cu near the requirement (8 mg kg^{-1}), reduced liver AA status by 6% in groups fed moderate dietary AA levels (between 30 and 60 mg AA kg^{-1}, as AApP). The liver Cu concentration was not affected by increasing dietary Cu (Figure 10.5). These data indicated that elevated dietary Cu concentrations have minor effects on the dietary requirement of AA, and probably mediated through both intestinal and metabolic interactions. Extremely high levels of Cu lead to peroxidation in liver and intestine of Atlantic salmon, probably arising from oxidation in the diet or target tissues.[131] It is not known whether supplementation of antioxidant vitamins could alleviate this pro-oxidant effect of Cu.

10.3.7   Other minerals

With the exception of iron, interactions between AA and minerals that may cause vitamin C deficiency have not been observed in our studies of Atlantic salmon [113] (Waagbø, and Maage, unpublished data). In an indirect absorption study, Dabrowski and Köck[47] suggested that both iron and zinc absorption were facilitated by AA, as compared to AAS. Decreased body Cu and Zn concentrations were found in rainbow trout fed 500 mg AA kg^{-1} for 25 weeks, as compared to fish fed equimolar amounts of AAS or no AA.[41] In the study by Yamamoto et al.,[133] where groups of rainbow trout were fed diets with or without 2000 mg AA kg^{-1} and exposed to waterborne Cu for 18 weeks, elevated Zn levels were found in the gills but not in the livers, kidneys, or intestines of fish fed no AA. Maage et al.[113] and Lanno et al.[127] did not find reduction in hepatic Zn and Cu concentrations in Atlantic salmon or rainbow trout fed graded AA levels up to 5000 and 10000 mg AA kg^{-1}, respectively. In the roach, *Rutilus rutilus* L. larvae fed zinc levels between 142 and 483 mg kg^{-1} dry matter, whole body Zn between 334–1604 µg g^{-1} was positively correlated to the body levels of AA ranging between 11–82 µg g^{-1}. These results indicated a metabolic interaction between AA and Zn.[129]

In fish fed an AA-deficient diet, calcium absorption both from freshwater and diet seemed to be disturbed. This was shown as reduced Ca in gill, skin, bone, and muscle tissues in snake heads, *Channa punctatus*,[138] and whole body Ca in rainbow trout[41] fed AA-deficient compared to AA-supplemented diets. In the first experiment there seemed to be a compensatory intestinal Ca uptake, while in the latter experiment the Ca reduction was time-dependent and disappeared the second half of the experiment.[41] The high mortality in the latter experiment may, however, have selected for survivors with higher Ca. Ishibashi et al.[139] found somewhat lower blood Ca in AA deficient Japan parrot fish (*Oplegnathus fasciatus*) juveniles. A reduction in several other clinical blood parameters, including protein, indicated that this may be related to an osmotic disturbance. In an unpublished study on Atlantic salmon juveniles we found minor effects of dietary AA (0, 10, and 100 mg AA, AApP, and AAmP kg^{-1}) and the AA status on bone Ca and P concentrations (Waagbø and Maage, unpublished data). To our knowledge, effects of dietary Ca on AA requirement have not been observed, and interactions between AA and Ca therefore seem to arise as a secondary to AA deficiency lesions of epithelial cells. Indirect effects of dietary AA on Ca and bone metabolism in farmed animals, by modulating the conversion of vitamin D to its active 1,25 dihydroxy vitamin D_3,[140] cannot be excluded in intensive aquaculture.

Dietary megadoses of crystalline AA (5000 mg kg^{-1}), but not equivalent doses of AAS, lowered the hepatic selenium concentrations in Atlantic salmon, probably by reducing Se and making it less available for intestinal absorption.[113] Sufficient Se status is necessary for maximal glutathion peroxidase activity, and low dietary Se levels may be compensated by increased supplementation of other components in the integrated antioxidant system, such as vitamins E and C.[103, 141]

In Atlantic salmon fed diets with low AA diets (30–60 mg AA kg^{-1} as AApP), among the dietary variables included (vitamin E, astaxanthin, Fe, Cu Mn and lipid level), manganese (10 and 200 mg kg^{-1}) had the largest effect on liver AA concentration (−12%, see Figure 10.5). In an *in vitro* study with Atlantic salmon head kidney phagocytes suspended in plasma, an increase of plasma Mn from 0.035 to 0.194 µg ml^{-1} was shown to inhibit the production of reactive oxygen species from stimulated phagocytes.[142] The inhibitory effects were only seen in plasma with high AA concentrations (approx. 37 µg ml^{-1}). A possible interaction between dietary Zn, Mn, and AA (as AAS) in the development of bacterial kidney disease in sockeye salmon (*Oncorhynchus nerka*) was not clearly demonstrated by Bell et al.[143] Our recent studies indicate both dietary and metabolic interactions between AA and Mn which warrant further research.

A study on the effects of gradually increasing dietary magnesium (Mg) at two different protein levels on the vitamin C status in common carp (*Cyprinus carpio* L.) showed that high protein level (44%) reduced the AA status compared to low protein levels (25%), while no correlation between dietary Mg inclusion levels and AA concentrations in hepatopancreas, kidney, and brain was observed.[144] However, significant losses (95% of 1680 mg kg^{-1}) of the supplemented AA during diet processing may have masked potential Mg-AA interactions.

Several authors have been concerned about interactions between AA and unfavorable environmental factors (reviewed by Sandnes[9]), and although Cd is not a micronutrient, some comments on Cd toxicity are included for comparison with other minerals. High dietary levels of AA (6000 mg kg^{-1}) reduced the toxicity of waterborne Cd (0.14 ppm for 24 days) in rainbow trout,[75] while mullet (*Mugil cephalus* L.) exposed to 10 mg Cd l^{-1} for six weeks showed a 60% reduction of liver AA accompanied by accumulation of Cd in liver and gill tissue.[73] Dietary AA has minor effects on Cd accumulation at low (0.11 mg kg^{-1}) dietary Cd levels[113] but may explain relatively low additional Cd accumulation in fish fed five-fold this level.[145] Recent data on Atlantic salmon parr suggested that very high dietary Cd levels (up to 250 mg kg^{-1}) may be tolerated, and do not have an impact on liver AA status (A.K. Lundebye, personal communication).

To summarize, the information available on AA-mineral interactions seems to concentrate on: 1) disturbed mineral metabolism in a state of AA deficiency, which may arise as secondary to the lack of AA and be related to its essential functions; 2) differences in AA interactions beween water- and feed-borne minerals, showing most pronounced effects of water-borne minerals on dietary AA requirement; 3) intestinal interactions where AA facilitate (possibly Fe, Zn) or prevent uptake (Se, Cu); and 4) metabolic interactions (Fe, Cu, Zn, Mn, Cd). Some of the weak interactions between AA and minerals may be secondary to other micronutrients strongly influenced by AA. The cited and several older experiments dealing with AA and minerals in fish have led to a set of indications that should be verified in specially well-designed interaction experiments under both suboptimal and favorable environmental conditions.

10.4 Perspectives

Recommendations for supplementation of AA in diets to fish and crustaceans are based upon minimum requirements obtained in carefully performed feeding experiments. By using stable phosphate derivatives of AA, most of the micronutrient interactions with AA in the diet have been eliminated. The present knowledge on factors that may alter the AA requirement by interactions occuring during diet preparation, absorption, and *in vivo* utilization are scarce, and still it is difficult from a scientific point of view to recommend AA supplementations for practical farming situations, as well as during stressful conditions, outbreak of infectious diseases, and during vaccination. Increased knowledge on the role of AA in biochemical and cellular mechanisms and its interactions with other essential micronutrients is clearly required.

Acknowledgments

Some of the referred unpublished studies using phosphate derivatives of AA have been kindly supported and performed in collaboration with BASF, Germany (AA-Ca-mono-phosphate), F. Hoffmann–La Roche, Switzerland (AA-mono-and polyphosphates), and Bioakva Innovation AS, Norway.

References

1. Yamamoto, Y., Sato, M., and Ikeda, S., Existence of L-gulonolactone oxidase in some teleosts, *Bull. Jap. Soc. Sci. Fish.*, 44, 775, 1978.

2. Soliman, A. K., Jauncey, K., and Roberts, R. J. Qualitative and quantitative identification of L-gulonolactone oxidase activity in some teleosts, *Aquacul. Fish. Manage.*, 1, 249, 1985.

3. Dabrowski, K., Primitive actinopterigian fishes can synthesize ascorbic acid, *Experientia*, 50, 745, 1994.

4. Moreau, R. and Dabrowski, K., The primary localization of ascorbate and its synthesis in the kidneys of acipenserid (*Chondrostei*) and teleost (*Teleostei*) fishes, *J. Comp. Physiol. B.*, 166, 178, 1996.

5. Maeland, A. and Waagbø, R., Examination of the qualitative ability of some cold water marine teleosts to synthesise ascorbic acid, *Comp. Biochem. Physiol.*, 121A, 249, 1998.

6. Lightner, D. V., Hunter, B., Magarelli, P. C., Jr., and Colvin, L. B., Ascorbic acid: nutritional requirement and role in wound repair in penaeid shrimp, *Proc. World Maricult. Soc.*, 10, 513, 1979.

7. He, H. and Lawrence A. L., Vitamin C requirements of the shrimp *Penaeus vannamei*, *Aquaculture*, 114, 305, 1993.

8. Chuang, J. L., Nutrient requirements, feeding and culturing practices of *Penaeus monodon*: a review. *F. Hoffmann–La Roche Ltd., Animal Nutrition and Health*, Switzerland, 1990.

9. Sandnes, K., Vitamin C in fish nutrition—a review, *Fisk. Dir. Skr. Ser. Ern.*, 4, 3, 1991.

10. Dabrowski, K., Ascorbate concentration in fish ontogeny, *J. Fish Biol.*, 40, 273, 1992.

11. Dabrowski, K. and Blom, J., The effect of ascorbic acid in rainbow trout broodstock diets on deposition of vitamin C in eggs and and survival of embryos, *Comp. Biochem. Physiol.*, 108A, 129, 1994.

12. Waagbø, R., The impact of nutritional factors on the immune system in Atlantic salmon, *Salmo salar* L.: a review, *Aquacult. Fish. Man.*, 25, 175, 1994.

13. Woodward, B., Dietary vitamin requirements of cultured young fish, with emphasis on quantitative estimates for salmonids, *Aquaculture*, 124, 133, 1994.

14. Boonyaratpalin, M., Nutrient requirements of marine food fish cultured in Southeast Asia, *Aquaculture*, 151, 283, 1997.

15. Merchie, G., Lavens, P., and Sorgeloos, P., Optimization of dietary vitamin C in fish and crustacean larvae: a review, *Aquaculture*, 155, 165, 1997.

16. Blazer, V. S., Nutrition and disease resistance in fish, *Annu. Rev. Fish Dis.*, 2, 309, 1992.

17. Hilton, J. W., Ascorbic acid-mineral interactions in fish, in *Ascorbic Acid in Domestic Animals*, Wegger, I., Tagewerker, F. J., and Moustgaard, J., Eds., The Royal Danish Agriculturist Society, Copenhagen, 1984, 218.

18. Hilton, J. W., The interaction of vitamins, minerals and diet composition in the diet of fish, *Aquaculture*, 79, 223, 1989.

19. Poston, H. A., Effect of dietary L-ascorbic acid on immature brook trout, in *Fisheries Research Bulletin No. 31*, State of New York Conservation Department, Albany, 45, 1967.

20. Halver, J. E., Smith, R. R., Tolbert, B. M., and Baker, E. M., Utilization of ascorbic acid in fish, *Ann. N.Y. Acad. Sci.*, 258, 81, 1975.

21. Hilton, J. W., Cho, C. Y. , and Slinger, S. J., Effect of graded levels of supplemental ascorbic acid in practical diets fed to rainbow trout (*Salmo gairdneri*), *J. Fish Res. Board Can.*, 35, 431, 1978.
22. Meier, W. and Wahli, T., *Vitamin C deficiency in the rainbow trout,* Animal Nutrition Events # 2242, Animal Nutrition and Health, F. Hoffmann–La Roche Ltd., Switzerland, 1990.
23. Sandnes, K., Torrissen, O. J., and Waagbø, R., The minimum dietary requirement of vitamin C in Atlantic salmon (*Salmo salar*) fry using Ca ascorbate-2-monophosphate as dietary source, *Fish Physiol. Biochem.*, 10, 315, 1992.
24. Tacon, A. G. J., *Nutritional Fish Pathology,* United Nations Development Programme, Food and Agriculture Organization of the United Nations, Rome, Italy, 1992.
25. NRC, *Nutrient requirements of fish,* National Research Council, Academy Press, Washington, D.C., 1993.
26. Desjardins, L. M., Castell, J. D., and Kean, J. C., *Can. J. Fish Aquat. Sci.*, 42, 370, 1995.
27. Lightner, D. L., Colvin, L. B., Brand, C., and Danald, D. A., Black death, a disease syndrome of penaeid shrimp related to dietary deficienct of ascorbic acid, *Proc. World Maricult. Soc.*, 8, 611, 1977.
28. Magarelli, P. C., Jr., Hunter, B., Lightner, D. V., and Colvin, L. B., Black death: an ascorbic acid deficiency disease in penaeid shrimp, *Comp. Biochem. Physiol.*, 63A, 103, 1979.
29. Kanazawa, A., *Prawn Feeds,* American Soybean Association, Taipei, Taiwan, 1985.
30. Tsao, C. S., An overview of ascorbic acid chemistry and biochemistry, in *Vitamin C in Health and Disease,* Packer, L. and Fuchs, J., Eds., Marcel Dekker, New York, 1997, 25.
31. Lim, C. and Lovell, R. T., Pathology of the vitamin C deficiency syndrome in channel catfish (*Ictalurus punctatus*), *J. Nutr.*, 108, 1137, 1978.
32. Hilton, J. W., Cho, C. Y., and Slinger, S. J., Evaluation of the ascorbic acid status of rainbow trout (*Salmo gairdneri*), *J. Fish. Res. Board Can.*, 34, 2207, 1977.
33. Sandnes, K. and Utne, F., Processing loss and storage stability of ascorbic acid in dry fish feed, *Fisk. Dir. Skr. Ser. Ern.*, 11, 39, 1982.
34. Soliman, A. K., Jauncey, K., and Roberts, R. J., Stability of L-ascorbic acid (vitamin C) and its forms in fish feeds during processing, storage and leaching, *Aquaculture*, 60, 73, 1987.
35. NRC, *Nutrient requirements of coldwater fishes,* 16, National Academy Press, Washington, D.C., 1981.
36. NRC, *Nutrient requirements of warmwater fishes,* National Academy Press, Washington, D.C., 1983.
37. Waagbø, R., Øines, S., and Sandnes, K., The stability and biological availability of different forms of vitamin C in feed for Atlantic salmon (*Salmo salar*), *Fisk. Dir. Skr. Ser. Ern.*, 4, 95, 1991.
38. Dabrowski, K., Matusiewicz, M., and Blom, J. H., Hydrolysis, absorption and bioavailability of ascorbic acid esters in fish, *Aquaculture*, 124, 169, 1994.
39. Hsu, T. -S. and Shiau, S. -Y., Comparison of L-ascorbyl-2-polyphosphate with ascorbyl-2-sulfate in meeting vitamin C requirements of juvenile grass shrimp *Penaeus monodon, Fisheries Science*, 63, 958, 1997.
40. Sandnes, K., Hansen, T., Killie, J.-E. A., and Waagbø, R., Ascorbate-2-sulfate as a dietary vitamin C source for Atlantic salmon (*Salmo salar*): 1. Growth,

bioactivity, haematology and humoral immune response, *Fish Physiol. Biochem.,* 8, 419, 1990.

41. Dabrowski, K., El-Fiky, N., Frigg, M., and Wieser, W., Requirement and utilization of ascorbic acid and ascorbic sulfate in juvenile rainbow trout, *Aquaculture,* 91, 317, 1990.

42. Tsujimura, M., Thirty-year history of research about vitamin C activity in a naturally occuring ascorbic acid derivative; L-ascorbic acid 2-sulfate, *J. Kagawa Nutr. Univ.,* 28, 31, 1997.

43. Shiau, S. Y., and Hsu, T.-S., Vitamin C requirements of grass shrimp *Penaeus monodon,* as determined with L-ascorbyl-2-monophosphate, *Aquaculture,* 122, 347, 1994.

44. Buddington, R. K., Puchal, A. A., Houpe, K. L., and Diehl,W. J., Hydrolysis and absorption of two monophosphate derivatives of ascorbic acid by channel catfish *Ictalurus punctatus* intestine, *Aquaculture,* 114, 317, 1993.

45. Blom, J. H. and Dabrowski, K., Continuous or "pulse-and-withdraw" supply of ascorbic acid in the diet: a new approach to altering the bioavailability of ascorbic acid, using teleost fish as a scurvy-prone model, *Internat. J. Vit. Nutr. Res.,* 68, 88, 1998.

46. Sandnes, K. and Waagbø, R., Enzymatic hydrolysis of ascorbate-2-monophosphate and ascorbate-2-sulphate *in vitro,* and bioactivity of ascorbate-2- monophosphate in Atlantic salmon (*Salmo salar*), *Fisk. Dir. Skr. Ser. Ern.,* 4, 33, 1991.

47. Dabrowski, K. and Köck, G., Absorption and interaction with minerals of ascorbic acid and ascorbic sulfate in digestive tract of rainbow trout, *Can. J. Fish. Aquat. Sci.,* 46, 1952, 1987.

48. Matusiewicz, M. and Dabrowski, K., Characterization of ascorbyl esters hydrolysis in fish, *Comp. Biochem. Physiol.,* 110B, 739, 1995.

49. Hamre, K., Berge, R. K., Waagbø, R., and Lie, Ø., Vitamins C and E interact in juvenile Atlantic salmon (*Salmo salar* L.), *Free Rad. Biol. Med.,* 22, 137, 1997.

50. Cho, C.Y. and Cowey, C. B. Utilization of monophosphate esters of ascorbic acid by rainbow trout (*Oncorhynchus mykiss*), in *Proceedings from Fish Nutrition in Practice,* Biarritz, France, 1993, 149.

51. Phromkunthong, W., Boonyaratpalin, M., and Storch, V., Different concentrations of ascorbyl-2-monophosphate-magnesium as dietary sources of vitamin C for seabass, *Lates calcarifer, Aquaculture,* 151, 225, 1997.

52. Shigueno, K. and Itoh, S., Use of Mg-L-ascorbyl-2-phosphate as a vitamin C source in shrimp diets, *J. World Aquacult. Soc.,* 19, 168, 1988.

53. Chen, H. Y., and Chang, C. F., Quantification of vitamin C requirements for juvenile shrimps (*Penaeus monodon*) using polyphosphorylated L-ascorbic acid, *J. Nutr.,* 124, 2033, 1994.

54. Catacutan, M. R. and Lavilla-Pitogo, C. R., L-Ascorbyl-2-phosphate-Mg as a source of vitamin C for juvenile *Penaeus monodon, Isr. J. Aquacult.-Bamidgeh,* 43, 35, 1994.

55. Dabrowski, K., Gulonolactone oxidase is missing in teleost fish. The direct spectrophotometric assay, *Biol. Chem. Hoppe-Seyler,* 371, 207, 1990.

56. Waagbø, R., Glette, J., Nilsen, E. R., and Sandnes, K., Dietary vitamin C, immunity and disease resistance in Atlantic salmon (*Salmo salar*), *Fish Physiol. Biochem.,* 12, 61, 1993.

57. Sato, M., Yoshinaka, R., and Ikeda, S., Dietary ascorbic acid requirement of rainbow trout for growth and collagen formation, *Bull. Jap. Soc. Scient. Fish.,* 44, 1029, 1978.

58. Waagbø, R., Thorsen, T., and Sandnes, K., Role of dietary ascorbic acid in vitellogenesis in rainbow trout (*Salmo gairdneri*), *Aquaculture*, 80, 301 1989.
59. Boonyaratpalin, M., Unprasert, N., and Buranapanidgit, J., Optimal supplementary vitamin C level in seabass fingerling diet, in *The Current Status of Fish Nutrition in Aquaculture*, Takeda, M. and Watanabe, T., Eds., Tokyo University of Fisheries, Tokyo, 1989, 149.
60. Dabrowski, K., Administration of gulonolactone does not evoke ascorbic acid synthesis in teleost fish, *Fish Physiol. Biochem.*, 9, 215, 1991.
61. Merchie, G., Lavens, P., Radull, J., Nelis, H., De Leenheer, A., and Sorgeloos, P., Evaluation of vitamin C-enriched *Artemia* nauplii for larvae of the giant freshwater prawn, *Aquacult. Int.*, 3, 355, 1995.
62. Rønnestad, I., Lie, Ø., Hamre, K., and Waagbø, R., L-Ascorbic acid and α-tocopherol in larvae of Atlantic halibut before and after exogenous feeding, *J. Fish Biol.*, 55, 720, 1999.
63. Maeland, A., Rosenlund, G., Stoss, J., and Waagbø, R., Weaning of Atlantic halibut, *Hippoglossus hippoglossus* L., using formulated diets with various levels of ascorbic acid, *Aquacult. Nutr.*, 5, 211, 1999.
64. Mangor-Jensen, A., Holm, J. C., Rosenlund, G., Lie, Ø., and Sandnes, K., Effect of dietary vitamin C on maturation and egg quality of cod (*Gadus morhua* L.), *J. World Aquacult. Soc.*, 25, 30, 1994.
65. Dabrowski, K., Assay of ascorbic phosphates and *in vitro* availability assay of ascorbic mono- and polyphosphates, *J. Sci. Food and Agric.*, 52, 409, 1990.
66. Waagbø, R. and Sandnes, K., Effects of dietary vitamin C on growth and parr-smolt transformation in Atlantic salmon, *Salmo salar* L., *Aquacult. Nutr.*, 2, 65, 1996.
67. Soliman, A. K., Jauncey, K., and Roberts, R. J., The effects of dietary ascorbic acid supplementation on hatchability, survival rate and fry performance in *Oreochromis mossambicus* (Peters), *Aquaculture*, 59, 197, 1986.
68. Blom, J. H. and Dabrowski, K. Reproductive success of female rainbow trout (*Oncorhynchus mykiss*) in response to graded dietary ascorby mono-phosphate levels, *Biol. Reprod.*, 52, 1073, 1995.
69. Liu, L., Ciereszko, A., Czesny, S., and Dabrowski, K., Dietary ascorbyl monophosphate depresses lipid peroxidation in rainbow trout sperm, *J. Aquatic Animal Health*, 9, 249, 1997.
70. Li, Y. and Lovell, R. T., Elevated levels of dietary ascorbic acid increase immune responses in channel catfish, *J. Nutr.*, 115, 123, 1985.
71. Wahli, T., Meier, W., and Pfister, K., Ascorbic acid induced immune-mediated decrease in mortality in *Ichtyopthirius multifiliis* infected rainbow trout (*Salmo gairdneri*), *Acta Tropica*, 43, 287, 1986.
72. Agrawal, N. K., Juneja, C. J., and Mahajan, C. L., Protective role of ascorbic acid in fishes exposed to organochlorine pollution, *Toxicology*, 11, 369, 1978.
73. Thomas, P., Bally, M., and Neff, J. M., Ascorbic acid status of mullet, *Mugil cephalus* Linn., exposed to cadmium, *J. Fish Biol.*, 20, 183, 1982.
74. Thomas, P. and Neff, J. M., Effects of a pollutant and other environmental variables on the ascorbic acid content of fish tissues, *Marine Environ. Res.*, 14, 489, 1984.
75. Yamamoto, Y. and Inoue, M., Effects of dietary ascorbic acid and dehydro-ascorbic acid on the acute cadmium toxicity in rainbow trout, *Bull. Jap. Soc. Sci. Fish.*, 51, 1299, 1985.

76. Fletcher, T. C., Dietary effects on stress and health, in *Fish Stress and Health in Aquaculture*, Iwama, G. K., Pickering, A. D., Sumpter, J. P., and Schreck, C. B., Eds., Cambridge University Press, Cambridge, 223, 1997.

77. Packer, L. and Kagan, V. E., Vitamin E: The antioxidant harvesting center of membranes and lipoproteins, in *Vitamin E in Health and Disease*, Packer, L. and Fuchs, J., Eds., Marcel Dekker, New York, 1993, 179.

78. Tappel, A. L., Vitamin E as the biological lipid antioxidant, *Vit. Horm.*, 20, 493, 1962.

79. Meister, A., Glutathione-ascorbic acid antioxidant system in animals, *J. Biol. Chem.*, 269, 9397, 1994.

80. Winkler, B. S., Orselli, S. M., and Rex, T. S., The redox couple between glutathione and ascorbic acid: a chemical and physiological perspective, *Free Rad. Biol. Med.*, 17, 333, 1994.

81. Packer, J. E., Slater, T. F., and Willson, R. L., Direct observation of a free radical interaction between vitamin E and vitamin C, *Nature*, 278, 737, 1979.

82. Niki, E., Antioxidants in relation to lipid peroxidation, *Chem. Phys. Lipids*, 44, 227, 1987.

83. Wefers, H. and Sies, H., The protection by ascorbate and glutathione against microsomal lipid peroxidation is dependent on vitamin E, *Eur. J. Biochem.*, 174, 353, 1988.

84. Kagan, V. E., Serbinova, E. A., Forte, T., Scita, G., and Packer, L., Recycling of vitamin E in human low density lipoproteins, *J. Lipid Res.*, 33, 385, 1992.

85. Chen, L. H., Interaction of vitamin E and ascorbic acid, *In vivo*, 3, 199, 1989.

86. Burton, G. W., Wronska, U., Stone, L., Foster, D. O., and Ingold, K. U., Biokinetics of dietary RRR-α-tocopherol in the male guinea pig at three dietary levels of vitamin C and two levels of vitamin E. Evidence that vitamin C does not "spare" vitamin E *in vivo*, *Lipids*, 25, 199, 1990.

87. Buettner, G. R., The pecking order of free radicals and antioxidants: lipid peroxidation, alpha-tocopherol, and ascorbate, *Arch. Biochem. Biophys.*, 300, 535, 1993.

88. Kagan, V. E., Zhelev, Z. Z., Bakalova, R. A., Serbinova, E. A., Ribarov, S. R., and Packer, L., Intermembrane transfer of alpha-tocopherol and its homologs, in *Vitamin E in Health and Disease*, Packer, L. and Fuchs, J., Eds., Marcel Dekker, New York, 1993, 171.

89. Poston, H. A. and Combs, G. F., Jr., Interrelationships between requirements for dietary selenium, vitamin E and L-ascorbic acid by Atlantic salmon (*Salmo salar*) fed a semipurified diet, *Fish-Health-News*, 8, 6, 1979.

90. White, A., Fletcher, T. C., Secombes, C. J., and Houlihan, D. F., The effect of different dietary levels of vitamins C and E and their tissue levels in Atlantic salmon, *Salmo salar* L., in *Fish Nutrition in Practice. Proc. IV International Symposium on Fish Nutrition and Feeding*, Kaushik, S. J. and Luquet, P., Eds., Les Colloques 61, INRA Editions, Biarritz, France, 1993, 203.

91. Frankel, E. N., Antioxidants in foods and their impact on food quality, *Food Chemistry*, 57, 51, 1996.

92. Bowry, V. W., Ingold, K. U. and Stocker, R., Vitamin E in human low-density lipoprotein. When and how this antioxidant becomes a pro-oxidant, *Biochem. J.*, 288, 341, 1992.

93. Frischknecht, R., Wahli, T., and Meier, W., Comparison of pathological changes due to deficiency of vitamin C, vitamin E and combinations of vitamins C and E in rainbow trout, *Oncorhynchus mykiss* (Walbaum), *J. Fish Dis.*, 17, 31, 1994.

94. Gatlin, D. M., Poe, W. E., Wilson, R. P., Ainsworth, A. J., and Bowser, P. R., Effects of stocking density and vitamin C status on vitamin E-adequate and vitamin E-deficient fingerling channel catfish, *Aquaculture,* 56, 187, 1986.
95. Cowey, C. B., Adron, J. W., Walton, M. J., Murray, J., Youngson, A., and Knox, D., Tissue distribution, uptake and requirement for alpha-tocopherol of rainbow trout (*Salmo gairdneri*) fed diets with a minimal content of unsaturated fatty acids, *J. Nutr.,* 111, 1556, 1981.
96. Cowey, C. B., Adron, J. W., and Youngson, A., The vitamin E requirement of rainbow trout (*Salmo gairdneri*), given diets containing polyunsaturated fatty acids derived from fish oil, *Aquaculture,* 30, 85, 1983.
97. Gatlin, D. M., Poe, W. E., and Wilson, R. P., Effect of singular and combined dietary deficiencies of selenium and vitamin E on fingerling channel catfish (*Ictalurus punctatus*), *J. Nutr.,* 116, 1061, 1986.
98. Watanabe, T., Takeuchi, T., Wada, M., and Uehara, R., The relationship between dietary lipid levels and alpha-tocopherol requirement of rainbow trout, *Bull. Jap. Soc. Sci. Fish.,* 47, 1463, 1981.
99. Watanabe, T., Takeuchi, T., and Wada, M., Dietary lipid levels and the alpha-tocopherol requirement of carp, *Bull. Jap. Soc. Sci. Fish.,* 47, 1585, 1981.
100. Horwitt, H. K., Data supporting supplementation of humans with vitamin E, *J. Nutr.,* 82, 424, 1991.
101. Hilton, J. W., Cho, C. Y., and Slinger, S. J., Effect of hypervitaminosis A on the development of ascorbic acid deficiency in underyearling rainbow trout (*Salmo gairdneri* R.), *Aquaculture,* 13, 325, 1978.
102. Hilton, J. W., Hypervitaminosis A in rainbow trout (*Salmo gairdneri*): Toxicity signs and maximum tolerable level, *J. Nutr.,* 113, 1737, 1983.
103. Cowey, C.B., The role of nutritional factors in the prevention of peroxidative damage to tissues, *Fish Physiol. Biochem.,* 2, 171, 1986.
104. Christiansen, R., Glette, J., Lie, Ø., Torrissen, O. J., and Waagbø, R., Antioxidant status and immunity in Atlantic salmon (*Salmo salar* L.) fed semi-purified diets with or without astaxanthin supplementation, *J. Fish Dis.,* 18, 317, 1995.
105. Rivers, J. M., Safety of high-level vitamin C ingestion, *Ann. New York Acad. Sci.,* 498, 445, 1987.
106. Duncan, P. L. and Lovell, R. T., Influence of vitamin C on the folate requirement of channel catfish, *Ictalurus punctatus,* for growth, hematopoiesis, and resistance to *Edwardsiella ictaluri* infection, *Aquaculture,* 127, 233, 1994.
107. Bai, S. C. and Gatlin, D. M., III, Dietary rutin has limited synergistic effects on vitamin C nutrition in channel catfish (*Ictalurus punktatus*), *Fish Physiol. Biochem.,* 10, 183, 1992.
108. Kitamura, S., Ohara, S., Suwa, T., and Nakagawa, K., Studies on the vitamin requirements of rainbow trout, *Salmo gairdneri* L., *Bull. Jap. Soc. Sci. Fish.,* 31, 818, 1965.
109. Halver, J. E. and Shanks, W. E., Nutrition of salmonoid fishes. VIII. Indispensable amino acids for sockeye salmon. *J. Nutr.,* 72, 340, 1962.
110. Lynch, S. R. and Cook, J. D., Interaction of vitamin C and iron, in *Micronutrient interactions: Vitamins, minerals and hazardous elements,* Levander, O. A. and Cheng, L., Eds., *Ann. New York Acad. Sci.* 355, 1980, 32.
111. Hallberg L., Brune, M., and Rossander-Hulthen, L., Is there a physiological role of vitamin C in iron absorption?, *Ann. N.Y. Acad. Sci.,* 498, 324, 1987.

112. Andersen, F., Studies on iron nutrition in Atlantic salmon (*Salmo salar*) with respect to requirement, bioavailability, interactions and immunity. Ph.D. thesis, University of Bergen, Norway, 126 pp., 1997.

113. Maage, A., Waagbø, R., Olsson, P. E., Julshamn, K., and Sandnes, K., Ascorbate - 2-sulfate as a dietary vitamin C source for Atlantic salmon (*Salmo salar*): 2. Effects of dietary levels and immunization on the metabolism of trace elements, *Fish Physiol. Biochem.*, 8, 429, 1990.

114. Mazur, A., Mechanism of plasma iron incorporation into hepatic ferritin, *J. Biol. Chem.*, 135, 595, 1960.

115. Han, O., Failla, M. L., Hill, A. D., Morris, E. R., and Smith J. C., Jr., Reduction of Fe(III) is required for uptake of non-heme iron by Caco-2 cells, *J. Nutr.*, 125, 1291, 1995.

116. Andersen, F., Lygren, B., Maage, A., and Waagbø, R., Interaction between two dietary levels of iron and two forms of ascorbic acid and the effect growth, antioxidant status and some non-specific immune parameters in Atlantic salmon (*Salmo salar*) smolts, *Aquaculture*, 161, 437, 1998.

117. Desjardins, L. M., Hicks, B. D., and Hilton, J. W., Iron atalyzed oxidation of trout diets and its effect on the growth and physiological response of rainbow trout, *Fish Physiol. Biochem.*, 3, 173, 1987.

118. Andersen, F., Maage, A., and Julshamn, K., An estimation of dietary iron requirement of Atlantic salmon, *Salmo salar* L. parr, *Aquacult. Nutr.*, 2, 41, 1996.

119. Andersen, F., Lorentzen, M., Waagbø, R., and Maage, A., Bioavailability and interactions with other micronutrients of three dietary iron sources in Atlantic salmon, *Salmo salar*, smolts, *Aquacult. Nutr.*, 3, 239, 1997.

120. Hultèn, L., Gramatkovski, R., Gleerup, A., and Hallberg, L., Iron absorbtion from the whole diet. Relation to meal composition, iron requirements and iron stores, *Eur. J. Clin. Nutr.*, 49, 794, 1995.

121. Rørvik, K. A., Salte, R., and Thomassen, M., Effects of dietary iron and n-3 unsaturated fatty acids (omega-3) on health and immunological parameters in farmed salmon. Abstract no. 71, *Proc. Eur. Assoc. Fish Pathol.*, Budapest, Hungary, 1991.

122. Salte, R., Rørvik, K. A., Røed, E., and Nordberg, K., Winter ulcers of the skin in Atlantic salmon, *Salmo salar* L.: pathogenesis and possible aetiology, *J. Fish Dis.*, 17, 661, 1994.

123. Ida, T. and Wakabayashi, H., Relationship between iron acquisition ability and virulence of *Edwardsiella tarda*, the etiological agent of paracolo disease in Japanese eel *Anguilla japonica*, in *Proceedings of the Second Asian Fisheries Forum*, Hirano, R. and Hanyu, I., Eds., Tokyo, 1990, 667.

124. Rasmussen, K.J., Spray-dried blood in diets to Atlantic salmon (*Salmo salar*), *Fisk. Dir. Skr. Ser. Ern.*, 6, 151, 1994.

125. Lall, S. P., Naser, N., Olivier, G., and Keith, R., Influence of dietary iron on immunity and disease resistance in Atlantic salmon, *Salmo salar*, *VII International Symposium on Nutrition and Feeding of Fish*, abstract, 11–15. Aug. 1996.

126. Davis, D. A., Lawrence, A. L., and Gatlin, D. M., III, Evaluation of the dietary iron requirement of *Penaeus vannamei*, *J. World Aq. Soc.*, 23, 15, 1992.

127. Lanno, R. P., Slinger, S. J., and Hilton, J. W., Effects of ascorbic acid on dietary copper in rainbow trout (*Salmo gairdneri* Richardson), *Aquaculture*, 49, 269, 1985.

128. Julshamn, K., Andersen, K. J., Ringdal, O., and Brenna, J., Effects of dietary copper on hepatic concentration and subcellular distribution of copper and zinc in the rainbow trout (*Salmo gairdneri*), *Aquaculture*, 73, 143, 1988.

129. Dabrowski, K., Segener, H., Dallinger, R., Hinterleintner, S., Sturmbauer, C., and Wieser, W., Rearing of roach larvae: the vitamin C, minerals interrelationship and nutritional-related histology of the liver and intestine, *J. Anim. Physiol. Anim. Nutr.*, 62, 188, 1989.

130. Lorentzen, M., Maage, A., and Julshamn, K., Supplementing copper to a fish meal based diet fed to Atlantic salmon parr affects liver and selenium concentrations, *Aquacult. Nutr.*, 4, 67, 1998.

131. Berntssen, M.H.G., Lundebye, A.-K., and Hamre, K., Tissue lipid peroxidative responses in Atlantic salmon (*Salmo salar* L.) parr fed high levels of dietary copper and cadmium, *Fish Physiol. Biochem.*, (2000, in press).

132. Yamamoto, Y., Ishii, T., Sato, M., and Ikeda, S., Effect of dietary ascorbic acid on the accumulation of copper in carp, *Bull. Jap. Soc. Sci. Fish.*, 43, 989, 1977.

133. Yamamoto, Y., Hayama, K., and Ikeda, S., Effect of dietary ascorbic acid on the copper poisoning in rainbow trout, *Bull. Jap. Soc. Sci. Fish.*, 47, 1085, 1981.

134. Ikeda, S., Importance of ascorbic acid to fish farming, in *Ascorbic Acid in Domestic Animals, Proc. of the 2nd Symposium*, Kartause, Ittingen, Switzerland 1992, 378.

135. Miller, P. A., Lanno, R. P., McMaster, M. E., and Dixon, D. G., Relative contributions of dietary and waterborne copper to tissue copper burdens and water-copper tolerance in rainbow trout (*Oncorhynchus mykiss*), *Can. J. Fish. Aquat. Sci.*, 50, 1683, 1993.

136. Berntssen, M. H. G., Lundebye, A. K., and Maage, A., Effects of elevated dietary copper concentrations on growth, feed utilisation and nutritional status of Atlantic salmon (*Salmo salar* L.) fry, *Aquaculture*, 174, 167, 1999.

137. Solomons N. W. and Viteri, F. E., Biological interactions of ascorbic acid and mineral nutrients, *Adv. Chem. Series*, 200, 551, 1982.

138. Mahajan, C. L. and Agrawal, N. K., The role of vitamin C in calcium uptake by fish, *Aquaculture*, 19, 287, 1980.

139. Ishibashi, Y., Ikeda, S., Murata, O., Nasu, T., and Harada, T., Optimal supplementary ascorbic acid level in the Japanese parrot fish diet, *Nippon Suisan Gakkaishi*, 58, 267, 1992.

140. Weiser, H., Schlachter, M., Probst, H. P., and Kormann, A. W., The relevance of ascorbic acid for bone metabolism, in *Ascorbic Acid in Domestic Animals. Proceedings of the 2nd Symposium (9th–12th October 1990)*, Wenk, C., Fenster, R., and Völker, L., Eds., Kartause, Ittingen, Switzerland, 1992, 73.

141. Winston, G. W. and Di Giulio, R. T., Pro-oxidant and antioxidant mechanisms in aquatic organisms, *Aquatic Toxicol.*, 19, 137, 1991.

142. Lygren, B. and Waagbø, R., Nutritional impacts on the chemiluminescent response of Atlantic salmon (*Salmo salar* L.) head kidney phagocytes, *in vitro*, *Fish Shellfish Immunol.*, 9, 445, 1999.

143. Bell, G. R., Higgs, D. A., and Traxler G. S., The effect of dietary ascorbate, zinc, and manganese on the development of experimentally induced bacterial kidney disease in sockeye salmon (*Oncorhynchus nerka*), *Aquaculture*, 36, 293, 1984.

144. Dabrowska, H. and Dabrowski, K., Influence of dietary magnesium on mineral, ascorbic acid status and glutathione concentrations in tissues of a freshwater fish, the common carp, *Magnesium and Trace Elements*, 9, 101, 1990.

145. Maage, A., Comparison of cadmium concentrations in Atlantic salmon (*Salmo salar*) fry fed different commercial feeds, *Bull. Environ. Contam. Toxicol.*, 44, 770, 1990.

146. Fox, M. R. S., Protective effects of ascorbic acid against toxicity of heavy metals, *Ann. N. Y. Acad. Sci.*, 335, 144, 1975.

chapter eleven

The role of ascorbic acid and its derivates in resistance to environmental and dietary toxicity of aquatic organisms

Leif Norrgren, Hans Börjeson, Lars Förlin, and Nina Åkerblom

Contents

11.1   Background

The aquatic environment often constitutes a terminal sink for many pollutants used or produced in various industrial processes. The relationship
between environmental factors, i.e., temperature, salinity, oxygen, metals,
pesticides, and aquatic animal health, has been the subject of experimental
studies and environmental monitoring programs over recent decades.
Aquatic organisms include a large variety of species representing different
trophic levels that respond differently and thereby reflect numerous aspects
of environmental pollution. Different species of invertebrates frequently represent the first consumer trophic level and are active filter feeders or actively
feed on organic matter in sediments, which means that they are exposed to
and absorb non-desirable toxic compounds. Invertebrates, i.e., mussels and
crustaceans, are important food items for many fish species and thereby act
as transmitters of both essential nutrients and nonessential compounds that
can be toxic and/or biomagnified in aquatic food webs.

The involvement of ascorbic acid during toxic stress on aquatic organisms
is poorly investigated, and the purpose of this chapter is to summarize our present knowledge of ascorbic acid and its derivates, with special emphasis on the
relation between ascorbic acid and toxic stress in aquatic animals.

11.2   Ascorbic acid in aquatic invertebrates

Feeding experiments conducted on the freshwater prawn *Macrobrachium
rosenbergii,* showed increased survival of juvenile individuals fed diets containing ascorbic acid at concentrations up to 200 mg/kg food compared with
groups fed lower concentrations.[1] Very high mortality was recorded in
groups fed less than 50 mg/kg, which is associated with the prawns' inability to extricate itself successfully from the old exoskeleton during ecdysis. In
another study, grass shrimp (*Penaeus monodon*) were fed diets supplemented
with ascorbic acid in different forms, e.g., L-ascorbyl-2-sulphate, L-ascorbyl-
2-monophosphate, and L-ascorbic acid, in concentrations ranging between 0
and 2000 mg/kg food for eight weeks. Shrimps fed diets supplemented with
ascorbic acid had significantly higher weight gain and better food conversion
ratio than those fed an ascorbic acid-deficient diet. It was concluded that
L-ascorbyl-2-sulphate was only about 25% as effective as L-ascorbyl-2-
monophosphate in meeting the ascorbic acid requirement.[2] The role of ascor-

bic acid for aquatic invertebrate health in relation to oxidative stress was poorly investigated. However, ascorbic acid seems to have essential functions for normal growth and ecdysis.[1]

11.3 Parameters affecting the ascorbic acid concentrations in fish

The impact of hyposalinity, shallow water, handling, temperature, and nitrite on ascorbic acid concentrations in various fish organs has been studied by a number of authors.[3,4,5,6,7]

11.3.1 Temperature

The concentration of ascorbic acid in different organs in fish fluctuates over the year. Thomas et al.[3] showed that the ascorbic acid concentration is maximal in liver during the summer when the temperature is high, and minimal during the winter. This may be due to a decreased dietary intake of the vitamin at low water temperature. The highest ascorbic acid concentration was detected in the brain. In contrast to the liver, the concentration in the brain declined during the summer. Thomas[5] showed that increasing water temperature caused a decline in brain ascorbic acid content in juvenile mullet (*Mugil cephalus* L). A temperature-dependent generation of reactive O^{2-} due to lipid peroxidation has been shown in freshwater catfish (*Heteropneustes fossilis*).[6] Furthermore, hepatic superoxide dismutase (SOD) activity increased, and the ascorbic acid concentration in the liver was significantly reduced at increased water temperatures. There were also declines in major phospholipids, which might be due to reactive metabolites formed as a result of lipid peroxidation. Elevated temperatures also caused decreased concentrations of ascorbic acid in both gill and air sac tissue of freshwater catfish.[7] The increased SOD activity and depletion of ascorbic acid in different tissues suggest that SOD and ascorbic acid act as effective oxyradical scavengers as a result of temperature stress.

11.3.2 Salinity

Gilthead seabream (*Sparus aurata*) kept under hyposaline conditions showed no tissue alteration in either ascorbate or ascorbate-2-sulfate concentrations.[4] However, the ascorbic acid concentration in gills of mullet increased rapidly under hyposaline conditions, indicating an inverse relationship between gill ascorbic acid concentration and salinity.[5] The inhibiting effect of ascorbic acid on Na+, K+-ATPase in gill was examined *in vitro*. The results suggested that ascorbic acid participates in regulating of oubain-sensitive Na+, K+-ATPase activity in the gills of euryhaline teleosts. Hyposalinity caused a rapid and profound depletion of ascorbic acid concentrations in the kidney, while a marked increase was initially obtained in the brain.[5]

11.3.3   Confinement stress

Shallow water stress led to significant decrease in liver, kidney, and spleen ascorbic acid concentrations and a decrease in the ascorbate-2-sulfate concentration in the kidney of gilthead seabream.[4] Gilthead seabream subjected to handling responded with a marked increase in splenic ascorbic acid and ascorbate-2-sulfate concentrations, but no effects on concentrations of these two substances in other tissues were found.[4] Thompson et al.[8] investigated whether high levels of ascorbic acid can ameliorate the down-regulation of the immune system in connection with stress. Juvenile Atlantic salmon (*Salmo salar*) were maintained on low, normal, and high ascorbic acid diets (0.082, 0.44, and 3.17 g/kg food) and subjected to a 2 h confinement stress. Different concentrations of ascorbic acid in the diet did not, however, affect hyperglycemic response, leukocyte respiratory burst activity, bactericidal activity, leukocyte migration, or plasma bactericidal activity. Production of specific antibodies following immunization with *Aeromonas salmonicida* was higher in fish fed low ascorbic acid than in fish fed high levels. The brain ascorbic acid content was unaffected by capture stress. On the other hand, capture stress caused a rapid and profound depletion of ascorbic acid concentrations in the kidney.[5]

11.3.4   Ammonia

Nitrite is an intermediate product of nitrification and may reach toxic concentrations during certain conditions, i.e., anoxia. The toxicity of nitrite is related to its ability to oxidize hemoglobin to methemoglobin, a form not capable of carrying oxygen. To determine the ability of ascorbic acid to alleviate nitrite-induced methemoglobinemia, Wise et al.[9] fed channel catfish (*Ictalurus punctatus*) with different concentrations of ascorbic acid in connection with exposure to nitrite. A significant reduction in the rate of methemoglobin formation was recorded in fish fed 805 and 7720 mg ascorbic acid/kg food. In addition, a positive correlation between fish size, low temperature, and reduced nitrite toxicity has been shown.[10]

11.4   Interactions between organic toxicants and ascorbic acid in fish

Studies of feral fish species are often considered an important tool in ecotoxicological studies. One frequently used parameter to estimate physiological alterations in fish living in industrial recipient waters is analysis of cytochrome P450-dependent enzymes, which are responsible for biotransformation of compounds such as many organochlorines.[11] Furthermore, various types of pathological lesions such as cranial and skeletal deformities have been described in species living in the vicinity of industrial point sources, i.e., pulp mills.[12] Skin lesions and fin erosion are also commonly reported effects

as a result of exposure for toxic components.[13] Ascorbic acid has a fundamental role in the metabolism of certain xenobiotics, i.e., chloroorganics
which are biotransformed through the cytochrome P450 enzyme system.
Ascorbic acid also has essential functions during synthesis of collagen, which
is the major protein component in cartilage and bone tissues.[14] It has been
proposed that two hydroxylative processes in collagen formation and detoxification might compete for available ascorbic acid and that overutilization in
one metabolic process may have negative impact on others.[14]

11.4.1   Pesticides

Channel catfish fed different concentrations of ascorbic acid during
toxaphene exposure got reduced ascorbic acid concentration in vertebrae.[14]
In addition, a high frequency of vertebrae lesions was recorded in groups fed
diets containing 63 and 670 mg/kg of ascorbic acid, but no such lesions were
observed in the group fed 5000 mg/kg. This indicates that toxaphene may
cause ascorbic acid deficiency in bone tissue. However, the finding that the
hepatic ascorbic acid concentration was not affected by toxaphene is contradictory to our knowledge on a relocalization of ascorbic acid from liver to
other tissues. Furthermore, the hepatic AHH activity was higher in fish
exposed to toxaphene, indicating detoxification and, consequently, utilization of ascorbic acid resources in the hepatocytes. The role of ascorbic acid in
connection with pesticide-induced hematotoxicity in *Clarias batrachus*
showed that exposure to the pesticide DDT (0.05 ppm) results in reduced
numbers of red and white blood cells, depressed hemoglobin concentration,
and a decline in oxygen-carrying capacity of the erythrocytes. Fish fed a diet
supplemented with ascorbic acid (5 g/100 g body weight) showed less effects
of DDT, which suggests the role of ascorbic acid as a protective factor during
DDT-induced haematotoxicity in fish.[15] In a similar study, Agrawal et al.[16] fed
the air-breathing fish (*Channa punctatus*) a diet containing aldrin or a diet containing both aldrin and ascorbic acid (5000 mg/kg). Exposure to aldrin alone
caused high mortality and significantly increased the number of erythrocytes, hemoglobin, and hematocrit concentrations. The protective role of
ascorbic acid was evident and supplementation to the diet resulted in a ten-
fold reduction in the death rate and a significantly lowered hematological
response due to the aldrin exposure. In another study, rainbow trout
(*Oncorhynchus mykiss*) were injected with the organochlorine insecticide
lindane (10 or 50 mg/kg body weight).[17] Fish were fed ascorbic acid as
ascorbate-2-polyphosphate at concentrations of 60 and 2000 mg/kg of food 1
month before lindane exposure and throughout the experiment. Effects of
lindane appeared approximately one month after injection and influenced
lysozyme level and ceruloplasmin activity. In addition, a reduction in the
proportion of B-lymphocytes in head kidney was found. Dietary ascorbic
acid at a high dose caused enhancement of phagocytosis, mitogen-induced
proliferation, and antibody response. The proportion of B-lymphocytes was

also modified by ascorbic acid, whereas the level of lysozyme was lower in fish fed the high dose. These results confirmed the immunostimulating effects of ascorbic acid.

11.4.2   Polychlorinated biphenyls

Exposure of lake trout (*Salvelinus namaycush*) for a pentachlorobiphenyl (PCB #126) caused a dose-dependent activity increase of the hepatic cytochrome P450 dependent enzyme etoxyresorufin-O-deethylase (EROD).[18] In addition, significantly elevated liver membrane breakdown products (TBARS) were recorded, indicating oxidative stress. The liver concentration of ascorbic acid was not different between the groups. Synergistic effects between metals and organic pollutants have been studied in Atlantic croaker (*Micropogonias undulatus*).[19] Fish were exposed to 5 mg Cd/l water for 6 days followed by 1 mg Cd/l for 33 days or fed a diet contaminated with the technical PCB blend Arochlor 1254 at a concentration of 0.42 mg/100 g body weight. Both Cd and PCB caused lipid peroxidation. Exposure to Cd resulted in decreased levels of ascorbic acid in liver and ovary. No differences in ascorbic acid concentration of these two tissues were recorded after PCB exposure. The decrease in ascorbic acid concentration caused by Cd might stimulate lipid peroxidation. In contrast, exposure to PCB, which caused no difference in ascorbic acid concentration, suggests that PCB stimulated lipid peroxidation in these tissues by a different mechanism.

11.4.3   Oil

Tissue concentrations of ascorbic acid in mullet after exposure to stress factors including salinity, capture, and temperature were compared with tissue concentrations after 1 week's exposure to a 20% water-soluble fraction of a fuel oil.[20] The ascorbic acid concentration in the brain was not affected by oil exposure. The concentration in gill decreased as a consequence of oil exposure, and there was also a marked depletion in the posterior kidney. Also, liver ascorbic acid reserves declined by 45% due to the oil exposure. Depletion of liver ascorbic acid concentration may be a characteristic response of fish to certain types of pollutants. Data, collected over several seasons, suggested that juvenile mullet may be most susceptible to the depleting effects of pollutants on ascorbic acid at the end of the winter, when hepatic stores reach a minimum level.[20] Striped mullet (*Mugil cephalus*) were exposed to water-soluble fractions of two fuel oils and a crude oil.[21] Depleted ascorbic acid levels were primarily recorded in gills and kidney. It is possible that the ion- or osmo-regulatory dysfunction that are observed in fish exposed to oil and other adverse environmental stimuli are associated with a loss of ascorbic acid reserves in the gills. After 1 week of exposure, the liver ascorbic acid concentration decreased, which may reflect an increased ascorbic acid requirement for detoxification and protective functions. A significant decline in brain ascorbic acid concentration was only observed after chronic exposure to high concentrations of oil. A chronic depletion may cause impaired neural function. The only tissue not responding with depletion of ascorbic acid was muscle, illustrating that tissue ascorbic acid reserves are severely depleted in fish exposed to oil.

11.5   Interactions between metals and ascorbic acid in fish

11.5.1   Lead

Newly hatched rainbow trout fry exposed to different concentrations of lead in the water and fed diets deficient in ascorbic acid are affected by similar symptoms described in association with ascorbic acid deficiency, including spinal curvatures.[22] Lead concentrations in fish blood vary proportionally with the concentrations in surrounding water. No correlation between lead in tissues and the concentration of ascorbic acid in the diet has been recorded. Furthermore, hematocrit, blood iron concentrations, red blood cell counts, and red blood cell volumes were unaffected by dietary ascorbic acid concentrations, waterborne lead or the interaction between these treatments. Consequently, there are no indications that the etiology of lead toxicity in trout is directly related to the metabolism or function of ascorbic acid.[22]

11.5.2   Copper

In a study addressing copper toxicity in fish at various ascorbic acid status, juvenile rainbow trout were fed diets unsupplemented or supplemented with ascorbic acid (0, 100, 1000, or 10000 mg/kg) and copper (0 or 800 mg/kg) for 16 weeks.[23] Supplementation with ascorbic acid did not appear to markedly affect either the absorption or the metabolism of copper. Fish fed diet with copper had significantly lower body weight compared with fish fed a diet without copper. A supplementation of ascorbic acid at a level of 10,000 mg/kg caused a small, but significant increase in body weight compared with the groups fed a diet containing copper. Dietary copper did not affect the ascorbic acid concentration in head kidney or liver. Similarly, dietary ascorbic acid did not affect the copper levels in kidney or liver. Plasma copper levels were not affected by dietary ascorbic acid supplementation, indicating that ascorbic acid was not inhibiting the uptake of dietary copper into the blood or affecting the plasma clearance rate of copper. This is in direct contrast to the effect of dietary ascorbic acid on waterborne copper toxicity in rainbow trout and carp and indicates that dietary copper is metabolized in a manner different to that of waterborne copper.[24, 25] Ascorbic acid has been shown to decrease the uptake of waterborne copper in carp and rainbow trout.[24, 25] Prevention of copper accumulation by ascorbic acid intake was observed in hepatopancreas, gills, kidney, intestine, liver, and vertebrae. Dietary ascorbic acid supplementation also prevented the anemia caused by copper accumulation.

11.5.3 Cadmium

Respiratory parameters of an animal are important to assess toxic stress, as they are valuable indicators of the functions of all vital life-sustaining processes of the body. Several authors have studied the interrelation between cadmium and ascorbic acid. In a study by Sastry and Shukla[26] the freshwater fish *Channa punctatus* was exposed to Cd (11.2 mg/l) for 96 h and effects of Cd on the rate of uptake of oxygen were investigated. Acute exposure to Cd caused a decrease in oxygen uptake. Ascorbic acid failed to reduce the toxicity, measured as whole body oxygen uptake. There was, however, an increase of oxygen uptake in gill, muscle, and liver in fish exposed to Cd in combination with ascorbic acid compared with fish exposed to Cd alone. It has been shown that Cd increases the production of reactive oxygen species in the organism and inhibits the activity of some enzymes of the antioxidative system.[27, 28] Goldfish (*Carassius auratus gibelio*, Bloch.) were exposed to 20 mg Cd/l water for 1, 4, 7, and 15 days.[29] After exposure, the activities of superoxide dismutase (SOD) and catalase (CAT) were significantly decreased. At the same time, liver ascorbic acid was increased, indicating that, although ascorbic acid plays a role in scavenging reactive oxygen species in the liver of goldfish, by diminishing the effects of lipid peroxidation caused by Cd, the liver is supplied with ascorbic acid from other tissues. Thomas et al.[30] investigated ascorbic acid status in mullet (*Mugil cephalus*) exposed to Cd (1 or 10 mg Cd/l). Hepatic ascorbic acid decreased by 60% after 6 weeks of exposure, which might be due to increased utilization or decreased transport of the vitamin to the liver during Cd exposure. The accumulation of Cd in the liver was extremely high (1262 µg Cd/g dry weight). Accumulation in the gills was also high (approximately 350 µg Cd/g d.w.) and the ascorbic acid concentration in the gills was decreased to 50% of the control after 42 days of exposure. This suggested that there was a great requirement for ascorbic acid in tissues which accumulated large amounts of Cd. Very low levels of Cd accumulated in the brain. Cd caused, however, marked fluctuations in brain ascorbic acid levels, which may lead to neurochemical imbalances. During the last two weeks of Cd exposure, also the kidney ascorbic acid concentrations were reduced to 38% of the control group. Thus, it may be concluded that the overall resistance of fish and the ability to maintain further stresses may be severely decreased during marginal ascorbic acid deficiency due to Cd exposure. Rainbow trout fed diets deficient in ascorbic acid, tocopherol, or both or neither, were exposed to one of three Cd concentrations (0, 2, 4 µg Cd/l).[31] Fish fed a diet deficient in ascorbic acid accumulated more hepatic Cd than fish fed the reference diet. Rambeck et al.[32] suggested that ascorbic acid inhibits Cd accumulation by increasing absorption of iron, which competes with Cd for cellular binding and transport sites. In conclusion, fish exposed to Cd had significantly depleted ascorbic acid concentrations compared with fish fed the reference diet. Chronic exposure to Cd-induced hyperglycemia in striped mullet. Because high glucose concentrations can reduce uptake of ascorbic acid, tissue stores of ascorbic acid may be further depleted in fish exposed to Cd.[30] Metallothionein concentrations were lower in fish fed the diet deficient in ascorbic acid and exposed to 4 μg Cd/l and fish fed diet deficient in both vitamins. This may indicate a role of ascorbic acid in metallothionein synthesis.[31] Glutathione peroxidase (GPx) activity was lower in all fish fed a diet deficient in both vitamins. More research is required to determine if ascorbic acid and tocopherol are directly involved in GPx synthesis or activation, or whether these compounds affect the activity of the enzyme indirectly through cellular control of antioxidant equivalents. Zikic et al.[29] showed that exposure to Cd in goldfish caused a increase in liver ascorbic acid content.

11.5.4   Zinc

Toxic effects of zinc in relation to ascorbic acid in *Channa punctatus* were evaluated by Sen et al.[33] It was reported that zinc in the concentrations of 13.18, 17.5, 21.9, and 23.07 mg/l caused a decline in concentrations of ascorbic acid in both brain and liver tissue. The decreased levels indicated alteration of NADP, either due to stress of hypoxia, acidosis, or zinc toxicity. The lowered ascorbic acid concentrations may result in failure to tolerate chronic stress.

11.6   Interactions between ascorbic acid and other vitamins

To determine the possible synergistic effects between the bioflavonoid rutin and ascorbic acid, channel catfish (*Ictalurus punctatus*) were fed diets with different concentrations of the two compounds for 16 weeks. Already after 10–12 weeks of feeding diet deficient in ascorbic acid, fish with deformed spinal columns, external hemorrhages, and fin erosion were described. The fish also had diminished body weight gain, poor food conversion, low hematocrit, decreased hepatosomatic index, as well as reduced liver, muscle, and plasma ascorbic acid concentrations. In conclusion, the study indicated only limited synergistic effects between dietary rutine, isoflavonoid with antioxidant capacity, and ascorbic acid in channel catfish.[34] In another study, rainbow trout were fed diets containing different combinations of ascorbyl-monophosphate and all-rac-alpha-tocopherol acetate (vitamin E).[35] Diet deficient in ascorbic acid caused reduced growth rate, hemorrhages, gill lesions, severe deformations, and fractures of the vertebral column. Diet deficient in vitamin E caused splenic hemosiderosis, decreased hematocrit and hemoglobin content, reduced red blood cell numbers and increased spontaneous erythrocyte hemolysis. Fish fed diet deficient in both vitamins (C and E) exhibited high mortality and were anemic after 8–12 weeks of feeding. Examination of fish deficient in vitamins C and E revealed a severe muscular dystrophy and splenic hemosiderosis. The results show clearly that there is a strong interaction between ascorbic acid and tocopherol in rainbow trout and that ascorbic acid can prevent myodegeneration occurring as a consequence of tocopherol deficiency. Several studies have shown the beneficial effect of dietary ascorbic acid on resistance of fish to bacterial, viral, and parasitic diseases. The influence of combined vitamins C and E on nonspecific immunity and disease resistance has been studied in rainbow trout subjected to different infectious agents and fed different combinations of vitamins C and E.[36] They also found the highest macrophage oxidative burst activity, measured by chemiluminescence, in fish receiving both vitamins at high levels and observed the best survival rates in trout infected with VHS virus, among those given both vitamins at high levels.

11.7   Ascorbic acid and reproductive failures in fish

The impact of ascorbic acid on male gametogenesis and fertility was investigated in rainbow trout fed diets containing different concentrations of ascorbyl monophosphate. No dose-dependent effects on fish growth or gonadosomatic index were recorded. However, the ascorbic acid concentration in seminal fluid was affected directly by the dietary ascorbic acid concentration. Ascorbic acid deficiency reduced both sperm concentration and motility, and consequently also the fertility, which indicated importance of ascorbic acid for gamete quality in male fish.[37] The role of dietary ascorbic acid in females has been studied in tilapia (*Oreochromis mossambicus*) fed diets deficient or supplemented with ascorbic acid (1250 mg/kg diet).[38] The results showed that the hatchability of eggs originating from individuals fed unsupplemented food was significantly lower and the percentage of deformed fry was significantly higher. Furthermore, fry in these groups performed poorly in respect of growth, food utilization, and survival rate. The importance of ascorbic acid in the diet during early development was confirmed by supplementation of ascorbic acid, which improved performance, growth, food utilization, and survival rate compared with fish suffering from ascorbic acid deficiency. Similar results have been obtained in experiments with rainbow trout given different diets containing 0 or 1000 mg ascorbic acid/kg.[39] Eggs from fish fed a diet without added ascorbic acid had a decreased hatchability compared with the control group. In addition, egg mortality before the eyed stage was significantly higher in the ascorbic acid deficient group. Fecundity was not, however, statistically different between the groups, nor were significant differences observed between the egg diameters among the groups. In another experiment, rainbow trout were fed diets devoid of ascorbic acid or supplemented with ascorbate-monophosphate at a concentration of 300 mg/kg during a period of 8 months.[40] Maturing females in the group fed diets devoid of ascorbic acid had a significantly lower increase in body weight than those in the group supplemented with ascorbate-monophosphate. Eggs were obtained from all females in the ascorbate-monophosphate group, whereas only half of the females matured in the group fed ascorbic acid-free diet. The percentage of hatched embryos and the ascorbic acid concentration in eggs and newly hatched fry were lowest in the group fed the ascorbic-free diet. Low hatchability in ascorbic acid deficient groups has been described in earlier studies.[38, 39] Blom and Dabrowski[41] examined the relative importance of both maternal and offspring ascorbic acid intake on offspring performance in rainbow trout. Fry from females fed diet containing ascorbic acid and kept on a diet low in ascorbic acid content had reduced growth rate, but the mortality was not increased until after week 15. Feeding a high ascorbic acid diet to ascorbic acid deficient offspring showed that a severe deficiency in ascorbic acid during embryonic development and yolk-sac absorption did not result in irreversible damage to most of the fry, and those that survived the initial recovery phase were capable of normal growth. The dietary ascorbic acid intake in juvenile fish is therefore more important to the post-hatching performance than the initial amount of ascorbic acid supplied in the egg. In summary, this suggested that supplementation of ascorbic acid results in maternal transfer of the vitamin to the eggs, and consequently in amelioration of ascorbic acid deficiency during early stages of life.

During the 1990s, high mortalities in early-life stages have been recorded in different fish species in the Great Lakes and Baltic Sea. These types of disorders, affecting several species of salmonids, have been designated Early Mortality Syndrome (EMS) and the M74 syndrome, respectively, in these two locations. Both EMS and M74 are associated with the progeny of certain females.[42] Low thiamin concentration in roe and yolk-sac fry is the most significant feature of both EMS and M74. This thiamin deficiency is supported by the fact that both EMS and M74 are treatable if the offspring are immersed in thiamin-enriched water.[43, 44] However, the primary cause of EMS and M74 is still not known. Other factors considered to be involved in the etiology of M74 are oxidative stress due to organic pollutants, i.e., PCBs.[45, 46] Also other vitamins, i.e., astaxanthin, tocopherol (vitamin E) and ubiquinon (vitamin Q), have been considered as involved in the etiology.[47] The involvement of ascorbic acid in the etiology of M74 has been studied in Baltic salmon (*Salmo salar*) alveins by comparing healthy family groups with progeny developing M74. Since ascending spawners sometimes are injected with thiamin one month before ovulation to prevent development of M74, a third group was included.[48] Ascorbic acid status in Baltic salmon yolk-sac fry was shown to be significantly reduced in sac-fry affected by the M74 syndrome (A) compared to healthy sac-fry (B).[48] Progeny of females treated with intra peritoneal injection of thiamin before ovulation showed a low to intermittent ascorbic acid status without suffering from M74 (C) (Table 11.1).

As shown by Börjeson et al.[49] thiamin treatment of females gives high thiamin status of eggs and yolk-sac fry and the result may indicate that a deficiency in ascorbic acid to a certain extent may be inflicted by thiamin. Perhaps the mechanism involved is related to the protection against oxidative stress.[48] The results indicated that ascorbic acid may, in the same way as several other vitamins, play a role in the etiology of M74.

Table 11.1   Mean ascorbic acid concentration and standard deviation
(μmol/g wet weight) in Baltic salmon yolk-sac fry with or without M74.
P-values calculated against the M74 groups

	M74	Healthy	Thiamin injected
Mean ± SD	2.5 ± 0.5	3.4 ± 0.9	3.2 ± 0.9
Range	1.8–3.4	2.3–4.9	1.9–4.3
P-value		0.004	0.016

References

1. D'Abramo, L.R.D., Moncreiff, C.A., Holcomb, F.P., Montanez, J.L., and Buddington, R.K., Vitamin C requirement of the juvenile freshwater prawn, *Macrobrachium rosenbergii. Aquaculture*, 128, 269–275, 1994.
2. Shiau, S.Y. and Hsu, T.S., Vitamin C requirement of grass shrimp, *Penaeus monodon*, as determined with L-ascorbyl-2-monophosphate. *Aquaculture*, 122, 347–357, 1994.
3. Thomas, P., Bally, M.B., and Neff, J.M., Influence of some environmental variables on the ascorbic acid status of mullet, *Mugil cephalus* L., tissues. II. Seasonal fluctuations and biosynthetic ability. *J. Fish Biol.*, 27, 47–57, 1985.
4. Henrique, M.M.F., Morris, P.C., and Davies, S.J., Vitamin C status and physiological response of the gilthead seabream, *Sparus aurata* L., to stressors associated with aquaculture. *Aquacul. Res.*, 27, 405–412, 1996.
5. Thomas, P., Influence of some environmental variables on the ascorbic acid status of mullet, *Mugil cephalus* L., tissues. I. Effect of salinity, capture-stress, and temperature. *J. Fish Biol.*, 25, 711–720, 1984.
6. Parihar, M.S., Dubey, A.K., Javeri, T., and Prakash, P., Changes in lipid peroxidation, superoxide dismutase activity, ascorbic acid and phospholipid content in liver of freshwater catfish *Heteropneustes fossilis* exposed to elevated temperature. *J. Therm. Biol.*, 21 (5/6), 323–330, 1996.
7. Parihar, M.S. and Dubey, A.K., Lipid peroxidation and ascorbic acid status in respiratory organs of male and female freshwater catfish *Heteropneustes fossilis* exposed to temperature increase. *Comp. Biochem. Physiol.*, 112C (3), 309–313, 1995.
8. Thompson, I., White, A., Fletcher, T.C., Houlihan, D.F., and Secombes, C.J., The effect of stress on the immune response of Atlantic salmon (*Salmo salar* L.) fed diets containing different amounts of vitamin C. *Aquaculture*, 114, 1–18, 1993.
9. Wise, D.J., Tomasso, J.R., and Brandt, T.M., Ascorbic acid inhibition of nitrate-induced methemoglobinemia in channel catfish. *The Progressive Fish-Culturist*, 50, 77–80, 1988.
10. Blanco, O. and Meade, T., Effect of dietary ascorbic acid on the susceptibility of steelhead trout (*Salmo gairdneri*) to nitrite toxicity. *Rev. Biol. Trop.*, 28 (1), 91–107, 1980.
11. Andersson, T. and Förlin, L., Regulation of the cytochrome P450 enzyme system in fish. *Aq. Tox.*, 24, 1–20, 1992.
12. Lindesjöö, E., Fish diseases and pulp mill effluents; Epidemiological and histological studies. Ph.D. thesis. Uppsala University, ISBN 91-554-2765-0, 1992.
13. Murty, A.S., *Toxicity of Pesticides to Fish*. CRC Press, ISBN 0-8493-6059-5, 1986.

14. Mayer, F.L., Mehrle, P.M., and Crutcher, P.L., Interations of toxaphene and vitamin C in channel catfish. *Trans. Am. Fish Soc.,* 107 (2), 326–333, 1978.
15. Guha, G., Dutta, K., and Das, M., Vitamin C as antitoxic factor in DDT induced haematotoxicity in *Clarias batrachus. Proc. Zool. Soc.,* Calcutta, 46 (1), 11–15, 1993.
16. Agrawal, N.K., Juneja, C.J., and Mahajan C.L., Protective role of ascorbic acid in fishes exposed to organochlorine pollution. *Toxicology,* 11, 369–375, 1978.
17. Dunier, M., Vergnet, C., Siwicki, A.K., and Verlhac, V., Effect of lindane exposure on rainbow trout (*Oncorhynchus mykiss*) immunity; IV. Prevention of nonspecific and specific immunosuppression by dietary vitamin C (Ascorbate-2-polyphosphate). *Ecotoxicol. Environ. Saf.,* 30, 259–268, 1995.
18. Palace, V.P., Klaverkamp, J.F., Lockhart, W.L., Metner, D.A., Muir, D.C.G., and Brown, S.B., Mixed-function oxidase enzyme activity and oxidative stress in lake trout (*Salvelinus namaycush*) exposed to 3,3',4,4',5-pentachlorobiphenyl (PCB-126). *Environ. Toxicol. Chem.,* 15 (6), 955–960, 1996.
19. Thomas, P. and Wofford, H.W., Effects of cadmium and Aroclor 1254 on lipid peroxidation, glutathion peroxidase activity, and selected antioxidants in Atlantic croaker tissues. *Aquatic Toxicol.,* 27, 159–178, 1993.
20. Thomas, P. and Neff, J.M., Effects of a pollutant and other environmental variables on the ascorbic acid content of fish tissues. *Marine Environ. Res.,* 14, 489–491, 1984.
21. Thomas, P., Influence of some environmental variables on the ascorbic acid status of striped mullet, *Mugil cephalus* Linn., tissues. III. Effects of exposure to oil. *J. Fish Biol.,* 30, 485–494, 1987.
22. Hodson, P.V., Hilton, J.W., Blunt, B.R., and Slinger, S.J., Effects of dietary ascorbic acid on chronic lead toxicity to young rainbow trout (*Salmo gairdneri*). *Can. J. Fish Aquat. Sci.,* 37, 170–176, 1980.
23. Lanno, R.P., Slinger, S.J., and Hilton, J.W., Effect of ascorbic acid on dietary copper toxicity in rainbow trout (*Salmo gairdneri* Richardson). *Aquaculture,* 49, 269–287, 1985.
24. Yamamoto, Y., Ishii, T., Sato, M., and Ikeda, S., Effect of dietary ascorbic acid on the accumulation of copper in carp. *Bull. Jap. Soc. Sci. Fish.,* 43 (8), 989–993, 1977.
25. Yamamoto, Y., Hayama, K., and Ikeda, S., Effect of dietary ascorbic acid on the copper poisoning in rainbow trout. *Bull. Jap. Soc. Sci. Fish.,* 47 (8), 1085–1089, 1981.
26. Sastry, K.V. and Shukla, V., Influence of protective agents in the toxicity of cadmium to a freshwater fish (*Channa puntatus*). *Bull. Environ. Contam. Toxicol.,* 53, 711–717, 1994.
27. Kostic, M.M., Ognjanovic, B., Dimitrijevic, S., Zikic, R.V., Stajn, A., Rosic, G.L., and Zivkovic, R.V., Cadmium-induced changes of antioxidant and metabolic status in red blood cells of rats: in vivo effects. *Eur. J. Haematol.,* 51 (2), 86–92, 1993.
28. Ognjanovic, B., Zikic, R.V., Stajn, A., Saicic, Z.S., Kostic, M.M., and Petrovic, V.M., The effects of selenium on the antioxidant defense system in the liver of rats exposed to cadmium. *Physiol. Res.,* 44 (5), 293–300, 1995.
29. Zikic, R.V., Stajn, A., Saicic, Z.S., Spasic, M.B., Ziemnicki, K., and Petrovic, V.M., The activities of superoxide dismutase, catalase and ascorbic acid content in the liver of goldfish (*Carassius auratus gibelio* Bloch.) exposed to cadmium. *Physiol. Res.,* 45, 479–481, 1996.
30. Thomas, P., Bally, M., and Neff, J.M., Ascorbic acid status of mullet, *Mugil cephalus* Linn., exposed to cadmium. *J. Fish Biol.,* 20, 183–196, 1982.

31. Palace, V.P., Majewski, H.S., and Klaverkamp, J.F., Interactions among antioxidant defenses in liver of rainbow trout (*Oncorhynchus mykiss*) exposed to cadmium. *Can. J. Fish Aquat. Sci.*, 50, 156–162, 1993.
32. Rambeck, W.A., Bruckner, C., Meier, S., Zucker, H., and Kollmer, W.E., Cadmium bioavailability and the influence of feed components in chickens. In P. Brattler (Ed.) *Trace Element Analytical Chemistry in Medicine and Biology*. Walter de Gruyter, New York, 1988.
33. Sen, G., Behera, M.K., and Patel, P.N., Toxic effects of zinc on liver and brain of the fish *Channa punctatus* (Bloch). *Environ. Ecol.*, 10 (3), 742–744, 1992.
34. Bai, S.C. and Gatlin, D.M., Dietary rutin has limited synergistic effects on vitamin C nutrition of fingerling channel catfish (*Ictalurus punctatus*). *Fish Physiol. Biochem.*, 10 (3), 183–188, 1992.
35. Frischknecht, R., Wahli, T., and Meier, W., Comparision of pathological changes due to deficiency of vitamin C, vitamin E and combinations of vitamins C and E in rainbow trout, *Oncorhynchus mykiss* (Walbaum). *J. Fish Dis.*, 17, 31–45, 1994.
36. Wahli, T., Verlhac, V., Gabaudan, J., Schuep, W., and Meier, W., Influence of combined vitamins C and E on non-specific immunity and disease resistance of rainbow trout, *Oncorhynchus mykiss* (Walbaum) *J. Fish Dis.* 21, 127–137, 1998.
37. Ciereszko, A. and Dabrowski, K., Sperm quality and ascorbic acid concentration in rainbow trout semen are affected by dietary vitamin C: an across-season study. *Biol. Reprod.*, 52, 982–988, 1995.
38. Soilman, A.K., Jauncey, K., and Roberts, R.J., The effect of dietary ascorbic acid supplementation on hatchability, survival rate and fry performance in *Oreochromis mossambicus* (Peters). *Aquaculture*, 59, 197–208, 1986.
39. Sandnes, K., Ulgenes, Y., Braekkan, O.R., and Utne, F., The effect of ascorbic acid supplementation in broodstock feed on reproduction of rainbow trout (*Salmo gairdneri*). *Aquaculture*, 43, 167–177, 1984.
40. Dabrowski, K. and Blom, J.H., Ascorbic acid deposition in rainbow trout (*Oncorhynchus mykiss*) eggs and survival of embryos. *Comp. Biochem. Physiol.*, 108A (1), 129–135, 1994.
41. Blom, J.H. and Dabrowski, K., Ascorbic acid metabolism in fish: is there a maternal effect on the progeny? *Aquaculture*, 147, 215–224, 1996.
42. McDonald, G., Fitzsimons, J.D., and Honeyfield, D.C., Early life stage mortality syndrome in fishes of the Great Lakes and Baltic Sea. *American Fisheries Soc. Symposium 21*, ISBN 1-888569-08-5, 1998.
43. Fitzsimons, J.D., The effects of B-vitamins on a swim-up syndrome in Lake Ontario lake trout. *J. Great Lakes Res.*, 21, 286–289, 1985.
44. Amcoff, P., Börjeson, H., Eriksson, R., and Norrgren, L., Effects of thiamin treatments on survival of M74-affected feral Baltic salmon. *Am. Fish. Soc.*, 21, 31–40, 1998.
45. Norrgren, L., Andersson, T., Bergqvist, P.-A., and Björklund, I., Studies of adult feral Baltic salmon (*Salmo salar*) and yolksac fry suffering from abnormal mortality. *Env. Tox. Chem.*, 12, 2065–2075, 1993.
46. Vourinen, P.J., Paasivirta, J., Keinänen, M., Koistinen, J., Rantio, T., Hyötyläinen, T., and Welling, L., The M74 syndrome of Baltic salmon (*Salmo salar*) and organochlorine concentrations in the muscle of female salmon. *Chemosphere*, 34, 1151–1166, 1997.
47. Börjeson, H. and Norrgren, L., The M74 syndrome: A review of etiological factors. *Soc. Environ. Tox. Chem.* SETAC ISBN 1-880611-19-8, 1997.

48. Börjeson, H., Kallner, A., and Norrgren, L., The role of ascorbic acid in the etiology of M74. Centre for Reproductive Biology, Report 9, Uppsala, Sweden, ISSN 1403-0594, 1999.
49. Börjeson, H., Amcoff, P., Ragnarsson, B., and Norrgren, L., Reconditioning of sea-run Baltic salmon (*Salmo salar*) that have produced offspring than with the M74 syndrome. *AMBIO*, 28, 1, 30–36, 1999.

chapter twelve

The effect of ascorbic acid on the immune response in fish

Chhorn Lim, Craig A. Shoemaker, and Phillip H. Klesius

Contents

Abstract

Due to the lack of L-gulonolactone oxidase enzyme, most fish species are unable to synthesize vitamin C (ascorbic acid) in sufficient quantity to meet metabolic needs, and thus have a requirement for this vitamin. Deficiency symptoms reported are structural deformities, impaired collagen synthesis, decreased alkaline phosphatase activity, hemorrhages, anemia, low tissue levels of ascorbic acid, anorexia, growth retardation, poor feed efficiency, and delayed wound healing. The requirements of vitamin C by fish to prevent these deficiency signs ranged from 11 to about 100 mg/kg of diet. Vitamin C has also been shown to influence immunity and disease resistance in fish. The fish immune system consists of innate (natural) and acquired (specific) immune mechanisms as have been reported for terrestrial animals. Published information appears to indicate that a deficiency in ascorbic acid is immuno-suppressive, and animals fed ascorbic acid-deficient diets are more suscepti-ble to infectious diseases than those fed vitamin C-replete diets. However, evidence on the role of ascorbic acid in improving the immune response and disease resistance in fish is not consistent, although numerous studies have indicated that feeding fish ascorbic acid at levels higher than those required for normal growth and prevention of deficiency signs enhanced their immune response and resistance against bacterial infection. Without clear evidence on the beneficial effect of high levels of ascorbic acid on the immune response and disease resistance, feeding fish vitamin C at a level sufficient to meet their requirement for growth and prevention of deficiency signs is suggested.

12.1 Introduction

Ascorbic acid (AA) or vitamin C is known to participate in numerous physi-ological functions in animals including fish. Ancient fish such as gummy shark (*Mustelus manazo*), stingray (*Dasyatis akajei*), lamprey (*Lampetra japon-ica*) and African lungfish (*Protopterus aethiopicus*) can synthesize L-ascorbic acid from D-glucose but many species of teleost cannot.[1] Teleost fish such as coho salmon (*Oncorhynchus kisutch*), rainbow trout (*Oncorhynchus mykiss*), yellowtail (*Seriola quiquiradiata*), channel catfish (*Ictalurus punctatus*), blue tilapia (*Oreochromis aureus*), and common carp (*Cyprinus carpio*) are shown to have a dietary requirement for AA.[2-8] A lack of L-gulonolactone oxidase, an enzyme that catalyzes the last step of AA biosynthesis from D-glucose as the main precursor, or the inability to synthesize AA in sufficient quantity to meet metabolic needs leads to the exogenous requirement for this vitamin.[1,9-10] Thus, feeding fish diets devoid of AA will result in a number of deficiency signs such as structural deformities, impaired collagen synthesis, decreased alkaline phosphatase activity, hemorrhages, anorexia, growth retardation, poor feed utilization, and delayed wound healing.[11] The minimum dietary level of AA required for optimum growth and prevention of deficiency signs varies depending on the metabolic function, species, age, and size of fish.[10] The requirement values reported a range from 11 mg/kg of diet for channel catfish to about 100 mg for rainbow trout.[2, 12]

In addition to its requirement for normal growth and prevention of deficiency signs, AA has been shown to influence various parameters of the immune response and disease resistance in terrestrial animals and fish. However, evidence of the role of AA in fish immunity and disease resistance is not consistent. This chapter provides a review on the influence of AA on immune response and disease resistance of fish. To familiarize the readers with various immunological systems and terms, a brief description of fish immune system function is also included.

12.2 Immune system in fish

12.2.1 Innate (nonspecific) immunity

The fish immune system consists of innate (natural) and acquired (specific) immune mechanisms. Innate immunity consists of the nonspecific immune functions which fish use to combat common microbes and/or foreign agents in their environment. The skin and mucus are the first line of defense of the fish and will be considered as a component of the fish immune system. Skin and mucus create a barrier and do not allow entry of foreign agents into the fish. The mucus produced by goblet cells and secreted to the surface of the fish's skin possesses immune substances such as lysozyme, complement, and nonspecific immunoglobulin.[13] Fish also have "internal" natural resistance molecules and cells. Nonspecific immune molecules in the blood include lectins that recognize carbohydrates on the surface of pathogens or foreign agents. Lectins are also important in cellular communication by binding to cells and aiding in cell recognition.[14] Other components also include transferrins (iron binding molecules), acute-phase proteins, and components of the complement cascade. The nonspecific immune cells of fish include monocytes/macrophages, neutrophils (or granulocytes), and nonspecific cytotoxic cells (equivalent to natural killer cells in mammals). Monocytes and/or tissue macrophages are probably the most important cells involved in the immune response of fish.[15] Clem et al.[16] suggested their importance in the production of cytokines that help to regulate the immune system. Macrophages are the cells responsible for phagocytosis and killing of pathogens upon first recognition and subsequent infection.[17] The macrophage was also suggested by Vallejo et al.[18] to be responsible for presentation of antigen to B- and T-cells in fish. The macrophage is a link between the innate and acquired immune response. Neutrophils (granulocytes) are the cells of early inflammation in fish.[19] The function is probably in cytokine production and thus impacts recruitment of immune cells to the area of damage or infection. The third cell type is the non-specific cytotoxic cell. These cells are similar in function to mammalian natural killer cells. Evans and Jaso-Friedman[20] provide a review of these cells in fish. The function of these cells is destruction of target cells following receptor binding and signaling of the lytic pathway for destruction of the target. Nonspecific cytotoxic cells of fish have been suggested for immunity to parasites and viruses.[21, 22]

12.2.2 Acquired (specific) immunity

The acquired immune system consists of two arms, i.e., humoral and cell mediated, such as those seen in mammals. Fish have the ability to mount a humoral or antibody response. B-cells are responsible for production of antibody. The B-cells are covered with immunoglobulin which serves as the receptor for antigen recognition.[23] Once antigen binds the surface immunoglobulin, the B-cells proliferate and subsequent antibody production occurs. Antibody molecules of fish have been described as tetrameric immunoglobulin (IgM), i.e., structurally similar to the pentameric IgM of mammals. Kaattari[24] provides an excellent review of B-cell function and antibody response in fish. Specific immunoglobulin (i.e., antibody directed at an antigen) in fish functions in opsonization of bacteria, neutralization of viruses and toxins, and in activation of the complement cascade (classical pathway). The humoral immune response is responsible for defense against extracellular pathogens and/or toxins. The cell-mediated immune response is directed at intracellular pathogens. Cell-mediated immunity is dependent on accessory cell (such as macrophages) presentation of antigen to stimulate T-cells. Once stimulated, cascades of events occur and the activated T-cells produce cytokines (soluble factors which regulate the response) responsible for the activation of macrophages.[25] Activated macrophages have an increased capacity to kill intracellular pathogens, thereby inducing acquired immunity.[18, 26]

12.2.2.1 Immunological methods

Both *in vivo* and *in vitro* assays have been used in experiments to determine the influence of vitamin C on fish immune response. Assays include those to measure nonspecific immune and specific immune function. *In vitro* assays such as complement fixation (alternative pathway), chemotaxis, phagocytosis and killing by macrophages, determination of serum parameters (i.e., lysozyme), etc., have been used to determine the influence of nutrition on natural resistance or innate immunity. *In vivo* studies include the challenge of animals following feeding of varying levels of nutrients to determine if diets or dietary nutrients influence mortality following challenge with pathogen. Specific immune responses have been measured before and after immunization or challenge of fish and then *in vitro* monitoring of the antibody (B-cell function) response by agglutination or ELISA (enzyme linked immunosorbant assays). Specific *in vivo* tests would include challenge of fish following vaccination to monitor survival. Most assays used in fish nutrition and health research were adapted from mammalian immunology studies. An excellent source of the latest immunologic methods to measure immune function can be found in *Current Protocols in Immunology,* published by Greene Publishing Associates and Wiley Interscience.[27] Currently, many of the specialized reagents to determine immune function in fish are lacking. Thus, development of new reagents is necessary to accelerate and improve research on nutrition and fish health.

12.3   Ascorbic acid and immune response

Of all essential micronutrients, AA has generated the greatest interest concerning its interactions with host immune function and resistance to pathogens.[28] A great deal of information in this area has resulted from studies in humans and scurvy-prone terrestrial animals. AA is required for optimal functioning of cells, tissues, and organs. Bendich[29] suggested that the immune systems such as the thymus, spleen, and lymphatic system as well as the immune cells which secrete substances with bactericidal and viricidal actions are affected by the level of the intake of vitamin C. In biological systems, AA functions as a reducing agent as well as a cofactor in the hydroxylation reactions required for the formation of hydroxyproline.[30] The ability of AA to accelerate the formation of hydroxyproline is important because this compound is a major constituent of collagen, which is an essential component of a natural barrier, the skin. AA has also been found to play an important role in phagocytic cell function. Chemotactic response and phagocytosis by phagocytic cells (neutrophils) were stimulated by vitamin C supplementation.[28–30] It has been suggested that the influence of ascorbic acid on phagocyte mobility may be by its direct effects on the synthesis and assembly of microtubular structure of these cells.[28] High ascorbic acid levels in phagocytes have also been shown to protect them and surrounding tissues from oxidative damage caused by large amounts of highly reactive oxygen species produced during the oxidative burst when cells ingest and kill pathogenic organisms.[29, 30] In addition, vitamin C has been shown to be involved in the generation of energy required for the secretion of substances such as immunoglobulins, interferon, and other cytokines from cells.[29]

12.3.1   Channel catfish

Studies to evaluate the role of ascorbic acid on immune response and disease resistance of channel catfish have yielded contradictory results. Li and Lovell[31] showed that, under laboratory conditions, agglutinating antibody titers in response to *Edwardsiella ictaluri* antigen, serum hemolytic complement activity and phagocytic engulfment of *E. ictaluri* by peripheral phagocytes were impaired in small channel catfish fed the diet without supplemental AA. They showed that increasing ascorbic acid levels 100 times the amount required for growth and prevention of deficiency signs (3000 mg/kg diet) further enhanced antibody production and serum complement

activity but had no effect on phagocytosis. These immune capacities significantly increased but were similar among fish fed diets containing 30–300 mg
ascorbic acid/kg of diet. In contrast, Li et al.[32] found that increasing supplemental levels of vitamin C (L-ascorbyl-2-polyphosphate or AsPP) from 0 to
250 mg/kg diet had no effect on specific antibody levels of catfish 21 days
post-infection with *E. ictaluri*. However, in a field trial when large fingerlings
(averaging 21.8 g) were raised in ponds for 9 weeks then transferred to tanks
for 4 weeks, increasing the level of AA in the diet up to 4000 mg/kg did not
affect serum complement hemolytic activity and agglutinating antibody
titers against *E. ictaluri* antigen.[33] The reason for these differences may be
because natural food in ponds may have provided sufficient AA for fish[31] or
the effect of AA on humoral immune response might be regulated by a critical tissue level.[33]

Phagocytic engulfment of *E. ictaluri* by peripheral phagocytes of small
channel catfish raised in aquaria was not affected by increasing AA levels
from 30 to 3000 mg/kg diet.[31] Likewise, Johnson and Ainsworth[34] found that
for large catfish (320–420 g) raised in ponds, that increasing AA (AsPP) levels from 100 or 1000 mg/kg had no effect on percentage phagocytosis, phagocytic index, and bactericidal activities of the anterior kidney neutrophils. Lim
et al.[35] observed that mean number of macrophage migration decreased for
fish fed the vitamin C-deficient or replete diets but significantly increased for
channel catfish fed the diet containing 3000 mg AA (AsPP)kg. These results
indicate that macrophage migration is positively affected by high dietary levels of AA whereas their phagocytic activity is not.

12.3.2   *Salmonid*

An early study with rainbow trout indicated that fish fed a diet with 120 mg
AA/kg had significantly lower total iron binding capacity and phagocytic
index compared to those fed 1200 mg AA/kg.[36] In another study, Blazer and
Wolke[37] evaluated the immune response of rainbow trout maintained on
commercial and laboratory prepared diets and immunized with sheep red
blood cells or *Yersinia ruckeri*. Fish fed the laboratory prepared diet that contained higher than the requirement levels of vitamins C and E had increased
T and B cell responses to sheep red blood cells as well as enhanced serum
antibody titer against *Y. ruckeri*. Navarre and Halver[38] observed a progressive
increase of the serum agglutinating antibody titer at each incremental level of
AA supplementation (0, 100, 500, 1000, and 2000 mg/kg diet) in vaccinated
fish. They suggested that the increase in antibody titer in rainbow trout fed
high levels of AA may be the result of a stimulation of the lymphocyte activity. However, the effects of AA on humoral antibody production were
detected before weeks 6 and 8 post-vaccination.

Vitamin C has also been shown to have an impact on *in vitro* rainbow
trout lymphocyte responses.[39] Proliferation and macrophage activating factor

production from AA-depleted rainbow trout were increased by the presence of exogenous AA (sodium ascorbate and AsPP) in the culture medium. Parenteral addition of vitamin C also enhanced proliferation in leukocytes from vitamin C depleted trout. Supplementation of this vitamin in culture stimulated proliferation responses of leukocyte from trout fed a commercial diet but not when AA levels were already 3 to 4 times higher in the fish, as those injected with ascorbate. In contrast, Anggawati-Satyabudhy et al.[40] did not observe any differences in antibody production of rainbow trout fed diet containing different levels (20, 80, and 320 mg/kg) of AA (AsPP) and immunized with infectious hematopoietic necrosis virus. Similarly, Verlhac et al.[41] observed that antibody titer of rainbow trout after vaccination against enteric redmouth disease and serum hemolytic complement activity were not affected by dietary levels of AA (L-ascorbate-2-monophospate). However, AA deficiency decreased lymphocyte proliferation induced by concanavalin A and phagocytosis of latex beads by peripheral phagocytes. Another study by Verlhac and Gabaudan[42] showed increased leukocyte count, mitogen-induced proliferation of lymphocytes, and natural cytotoxicity in rainbow trout fed a diet containing 1000 mg AA/kg as compared to the group fed a diet containing 60 mg AA/kg.

In sockeye salmon (*Oncorhynchus nerka*), agglutinating antibody titer following immunization with *Aeromonas salmonicida* was not affected by dietary AA (Na-L-ascorbate-2-sulfate).[43] Lall et al.[44] found no conclusive evidence that AA affects the antibody production or the bactericidal activity of serum from Atlantic salmon (*Salmo salar*) immunized or nonimmunized with *Vibrio anguillarum* and *A. salmonicida*. The antibody response of Atlantic salmon against a soluble antigen (NIP_{11}-LPH) was slightly lower in fish deprived of AA, but there were no differences between the groups fed 500 or 5000 mg AA/kg suppled by either ascorbic acid or absorbyl-2-sulfate.[45] Hardie et al.[46] showed that some specific (agglutinating antibody against *A. salmonicida* or human gamma globulin, and lymphokine production) and non-specific immune responses (macrophage respiratory burst activity and erythrophagocytosis) of Atlantic salmon were unaffected by dietary AA at levels of 50, 310, or 2750 mg/kg. However, serum complement activity decreased in the AA-depleted fish, and increased in fish fed the diet containing 2750 mg/kg. With the same species, Waagbo et al.[47] observed that specific antibody against *V. salmonicida*, lysozyme activity in head kidney and serum complement activity in fish surviving the challenge with *A. salmonicida* were higher for the group fed a diet containing 4000 mg AA/kg. Increased antibody titer was also reported in Atlantic salmon fed high dietary L-absorbate-2-sulfate (4770 mg/kg) and immunized with *A. salmonicida* and *Yersina ruckeri*.[48] Mitogen-induced proliferation of lymphocytes and natural cytotoxicity were found to be higher in Atlantic salmon fed a diet containing 1000 mg AA/kg as compared to those fed a diet containing 60 mg AA/kg.[42]

12.3.3   Other species

The phagocytic index of pronephros cells of red sea bream (*Pagrus major*) increased with increasing levels of L-ascorbic acid (AA) or L-ascorbyl-2-phosphate Mg (APM) but fish fed 10,000 mg APM/kg diet had greater phagocytic index than those the same dietary level of AA. However, alternative pathway-complement activity was not influenced by dietary levels of AA or APM.[49] Roberts et al.[50] evaluated the effect of AA supplementation (0, 300, 1000, and 2000 mg/kg diet) on nonspecific immune response in juvenile turbot (*Scophthalmus maximus*). They found that serum lysozyme was significantly enhanced at the largest level of supplemental AA, while phagocytic indices of kidney and spleen phagocytes showed a positive correlation with dietary vitamin C levels. The total serum protein level, however, was not affected by dietary levels of AA. In contrast, Nitzan[51] reported that the serum antibody titer of pond-raised hybrid tilapia (*O. aureus* X *O. niloticus*) 16-day post-immunization with *Aeromonas hydrophila* was lower in fish fed the ascorbic acid polyphosphate-enriched diet (495 mg AA equivalent/kg) as compared to that of the unsupplemented control.

12.4   Ascorbic acid (AA) and disease resistance

12.4.1   Channel catfish

Earlier investigations have indicated that cage or pond-grown channel catfish fed diets deficient in ascorbic acid were more susceptible to bacterial infections than those fed ascorbic acid replete diets.[5, 52, 53] Durve and Lovell[54] demonstrated that, under laboratory conditions, 30 mg of supplemental AA/kg diet was sufficient for normal growth and prevention of deficiency signs as well as to improve the survival of channel catfish infected with *Edwardsiella tarda* (the bacterium causing emphysematous putrefactive disease) and held at the water temperature of 33°C (Table 12.1). At the lower water temperature of 23°C, an AA level of up to five times the normal requirement (150 mg/kg diet) was needed to increase resistance of channel catfish against *E. tarda* infection. The author suggested that the requirement for the resistance to infection was probably higher at lower temperature because of lower natural resistance of fish. In a subsequent study Li and Lovell[31] demonstrated that feeding high levels of AA (3000 mg/kg diet) enhanced resistance of channel catfish against *E. ictaluri*. Liu et al.[33] observed that when catfish were immunized with *E. ictaluri* and later challenged with virulent *E. ictaluri,* the LD_{50} of *E. ictaluri* increased with incremental concentrations of AA. The most dramatic increase occurred at AA levels between 100 and 1000 mg/kg diet. Vitamin C concentrations above 1000 mg/kg did not increase the LD_{50} appreciably. Li et al.[55] found high mortality of channel catfish fed vitamin C-free diet but increasing vitamin C (AsPP) concentration higher than the level required for growth did not improve disease resistance. In a more recent

Table 12.1 Effect of ascorbic acid on innate disease resistance

Fish species	Dose (mg/kg diet) and form of ascorbic acid	Challenge organism	Method of challenge	Effect[a]	Reference
Channel catfish	0 (L-ascorbic acid)	*Edwardsiella ictaluri*	Immersion	−	31
	30			+	
	60			+	
	150			+	
	300			+	
	3000			+++	
Channel catfish	0 (L-ascorbic acid)	*E. ictaluri*	IP injection	−	33
	100			+	
	500			+	
	1000			+++	
	2000			+++	
	3000			+++	
Channel catfish	0 (L-ascorbic acid)	*E. tarda*	IP injection	−	54
	30			+	
	60			+++	
	150			+++	
Channel catfish	0 (L-acorbyl-2-polyphosphate)	*E. ictaluri*	Immersion	−	55
	25			+	
	50			+	
	100			+	
	1000			+	
	2000			+	

Table 12.1   Continued

Fish species	Dose (mg/kg diet) and form of ascorbic acid	Challenge organism	Method of challenge	Effect[a]	Reference
Channel catfish	0 (L-ascorbyl-2-polyphosphate) 50 150 250	E. ictaluri	Immersion	− − − −	56
Rainbow trout	0 (L-ascorbic acid) 100 500 1000 2000	Vibrio anguillarum	IP injection or immersion	− + ++ ++ ++	38
Rainbow trout	0 (L-ascorbyl-2-polyphosphate) 80 320	IHN virus	Immersion	− + ++	40
Rainbow trout	0 (Silicone coated AA or L-ascorbyl-2-polyphosphate) 50 2000	Ichthyophthirius multifiliis	Adding theronts to cultured tanks	− − +	57
Sockeye salmon	0 (Na L-ascorbyl-2-sulfate) 10 100	Renibacteriun salmoninarum	IP injection	+ − −	43

Atlantic salmon	0 (L-ascorbic acid)	Aeromonas salmonicida or V. anguillarum	IP injection	−	44, 58
	50			−	
	100			−	
	200			−	
	500			−	
	1000			−	
	2000 or higher			−	
Atlantic salmon	90 (L-ascorbic acid or ascorbyl-2-sulphate)	V. salmonicida or Yersenia ruckeri	IP injection	−	48
	2980			−	
Atlantic salmon	50 (L-ascorbic acid)	A. salmonicida	Immersion	−	46
	310			+++	
	2750			+++	
Atlantic salmon	40 (Ca L-ascorbyl-2-monophosphate)	A. salmonicida	Cohabitation followed by IP injection	+	47
	400			+	
	2000			+	
	4000			+	

[a]Effect: None (−), little (+), slight (++), moderate (+++) and strong (++++).

study, however, Li et al.[56] showed that cumulative mortality of catfish 21 day post-challenge with *E. ictaluri* was lower in fish fed the practical-type basal diet (3 mg AA/kg) than those fed diets containing supplemental AA (AsPP) (50, 150, or 250 mg/kg).

12.4.2   Salmonid

Data on the interrelationship between AA and disease resistance in salmonid are also not consistent (Table 12.1). In rainbow trout, feeding AA 5 to 20 times the level required for growth (500–2000 mg/kg) increased the resistance of fish to *V. angillarum*.[38] Anggawati-Satyabudhy et al.[40] found that the survival of rainbow trout challenged with infectious hematopoietic necrosis virus was directly related to the dietary intake of AA (AsPP) and that fish fed the diet containing the highest AA (320 mg/kg) had the highest survival. High doses of AA (silicone-coated AA or AsPP) (2000 mg/kg) have also resulted in significant reduction of mortality in rainbow trout infected with *Ichthyophthirius multifiliis*. This reduction in mortality, however, is probably due to an increase of the general health status and ability to cope with different stressors.[57]

The resistance of sockeye salmon to experimentally induced bacterial kidney disease (caused by *Renibacterium salmoninarum*) was not affected by dietary levels of 10 or 100 mg of AA (Na L-ascorbyl-2-sulfate)/kg.[43] In Atlantic salmon, feeding large levels (2000 mg/kg or higher) of AA (L-ascorbic acid) did not improve the resistance of fish infected with *A. salmonicida* or *V. anguillarum*[44, 58] and *V. salmonicida* or *Y. ruckeri*.[48] In contrast, Hardie et al.[46] reported that Atlantic salmon fed the low vitamin C diet (50 mg/kg) exhibited an early onset of increased mortality when challenged with *A. salmonicida* compared to fish fed normal (310 mg/kg) and high (2750 mg/kg) levels of AA. Waagbo et al.[47] also found that Atlantic salmon fed high dietary levels of AA (4000 mg/kg) had higher survival than fish fed lower AA diets following exposure to *A. salmonicida*.

12.4.3   Other species

High dietary levels of AA have been reported to enhance the disease resistance of yellow tail (*Seriola lalandi*) to *Streptococcus* sp. infection;[59] in contrast, Nitzan[51] observed that supplementation of AA at a level of 495 mg/kg diets had no effect on disease resistance of pond-raised tilapia infected with *A. hydrophila*. With turbot (*Scophtalmus maximus*) larvae, although not significant, lower cumulative mortality and slower onset of mortality following challenge with *Vibrio anguillarum* for fish fed high AA diets.[50]

12.5   Interaction between ascorbic acid and other nutrients

Few studies have been conducted on the interactions between dietary ascorbic acid and other nutrients on immune response and disease resistance in fish. Duncan and Lovell[60] fed channel catfish diets containing a combination of folic acid (0, 0.4 an 4.0 mg/kg) and ascorbic acid (0, 20, and 200 mg/kg). They found that, after the fish were experimentally challenged with *E. ictaluri*, the highest survival and antibody titer were obtained with the groups fed the higher level of AA diets supplemented with either 0.4 or 4.0 mg folic acid/kg. At the lower level of AA, 4.0 mg folic acid/kg of diet was needed for significant improvement of survival. However, when fish were fed the lower dose of AA, addition of folic acid had no effect on antibody titer. Lim et al.[35] evaluated the interaction between dietary levels of AA (0, 50, and 3000 mg/kg) and iron (0, 30, and 300 mg/kg) on non-specific immune response and resistance of channel catfish to *E. ictaluri* challenge. Chemotactic response of macrophages in the presence or absence of exoantigen was affected by dietary levels of iron, AA, and their interactions. Mean macrophage migration was highest in fish fed the 30-mg iron diets. The effect of AA on this parameter was seen only when AA was added at the highest level. However, neither dietary levels of iron nor AA and their interaction influenced survival of channel catfish challenged with *E. ictaluri*. The influence of a combination of dietary yeast glucan and AA on non-specific and specific immune responses of rainbow trout were studied by Verlhac et al.[61] They observed that feeding rainbow trout the combination of glucan (level recommended by manufacturer) and high doses of AA for two weeks had a stimulatory effect on chemiluminescence response (oxidative burst) of head kidney neutrophil and macrophage, of complement activation (alternative pathway) and antibody response against *Y. ruckeri*. Wahli et al.[62] studied the combined effect of dietary levels of AA (0, 30, and 2000 mg/kg) and vitamin E (0, 30, and 800 mg/kg) on non-specific immune response and disease resistance of rainbow trout. The combination of high doses of AA and vitamin E significantly enhanced proliferation of lymphocytes by concanavalin A and macrophage oxidative burst activity (determined by chemiluminescence). Maximum survival of trout infected with hemorrhagic septicemia virus was also observed for diets containing high levels of both vitamins. A similar trend was observed in fish infected with a bacterial pathogen, *Y. ruckeri*.

12.6   Conclusions

Nutritional deficiencies have been shown, in general, to lead to increased susceptibility to infection. There is now substantial evidence that prolonged malnutrition is suppressive of the immune response through depression of cell-mediated immunity and humoral antibody production, as well as through diminished phagocytic activity and non-specific resistance factors. Nutritional excesses, on the other hand, do not necessarily result in increased resistance. In the case of ascorbic acid, there have been several reports demonstrating its efficacy in enhancing immune functions and disease resistance. Published information appears to indicate that a deficiency of AA is immunosuppressive and animals fed AA-deficient diets are more susceptible to infectious diseases than those fed vitamin C-replete diets. However, evidence on the efficacy of ascorbic acid in improving immune response and disease resistance in fish is not consistent, although numerous studies have indicated that feeding fish ascorbic acid at levels higher than those required for normal growth and prevention of deficiency signs enhanced their immune response and resistance against bacterial infection. The inconsistency between the results of these studies may be related to differences in fish species, strain, size, and/or nutritional status and chemical forms of AA. Other factors that may have influenced results are experimental design, feeding management and duration, culture facilities and/or conditions, pathogenicity of microorganisms, and the methods used or the dose of challenge. Thus, further studies on the effect of AA on various components of the immune system and disease resistance in fish should be conducted. However, in the absence of clear-cut information on the beneficial effect of high levels of AA on immune response and disease resistance, a level sufficient to meet the fish requirement for growth and prevention of deficiency signs is suggested. Li et al.[55] indicated that feeding elevated levels of AA to channel catfish to increase resistance to *E. ictaluri* infection should not be routinely recommended for commercial farmers. A level of 50 to 100 mg AA/kg commonly used in commercial feeds is probably adequate for growth and health of channel catfish.[33, 55, 56]

References

1. Touhata, K., Toyohara, H., Tomoaki, M., Kinoshita, M., Satou, M., and Sakaguchi, M., Distribution of L-Gulono-1, 4-Lactone oxidase among fishes, *Fisheries Science*, 61(4), 729, 1995.

2. Halver, J. E., Ashley, L. M., and Smith, R. R., Ascorbic acid requirements of coho salmon and rainbow trout, *Transactions of the American Fisheries Society*, 98, 762, 1969.

3. Kitamura, S., Suwa, T., Ohara, S., and Nakamura, K., Studies on vitamin requirements of rainbow trout, *Salmo gaidneri*, 1. On the ascorbic acid, *Bulletin of the Japanese Society Scientific Fisheries*, 33, 1120, 1965.

4. Sakaguchi, H., Takeda, G., and Tange, K., Studies on vitamin requirements by yellowtail, 1. Vitamin B_6 and vitamin C deficiency symptoms, *Bulletin of the Japanese Society Science Fisheries*, 35, 1201, 1969.

5. Lovell, R. T., Essentiality of vitamin C in feeds for intensively fed caged channel catfish, *Journal of Nutrition*, 103, 134, 1973.

6. Lim, C. and Lovell, R. T., Pathology of the vitamin C deficiency syndrome in channel catfish (*Ictalurus punctatus*), *Journal of Nutrition*, 108, 1137, 1978.

7. Stickney, R. R., McGeachin, R. B., Lewis, D. H., Marks, J., Riggs, A., Sis, R. F., Robinson, E. H., and Wurts, W., Response of *Tilapia aurea* to dietary vitamin C, *Journal of the World Mariculture Society*, 15, 179, 1984.

8. Dabrowski, K., Hinterleitner, S., Sturmbauer, C., El-Fiky, N., and Wieser, W., Do carp larvae require vitamin C?, *Aquaculture*, 72, 295, 1988.

9. Ikeda, S., Sato , M., and Kimura, R., Biochemical studies on L-ascorbic acid in aquatic animals, III. Biosynthesis of L-ascorbic acid by carp, *Bulletin of the Japanese Society Scientific Fisheries*, 30, 365, 1964.

10. Wilson, R. P. and Poe, W. E., Impaired collagen formation in the scorbutic channel catfish, *Journal of Nutrition*, 103, 1359, 1973.
11. NRC (National Research Council), *Nutrient Requirements of Fish*, National Research Council, National Academic Press, Washington, D.C., 1993, 114.
12. El Naggar, G. O. and Lovell, R. T., L-Ascorbyl-2-monophosphate has equal antiscorbutic activities as L-ascorbic acid but L-ascorbyl-2-sulfate is inferior to L-ascorbic acid for channel catfish, *Journal of Nutrition*, 121, 1622, 1991.
13. Zilberg, D. and Klesius, P. H., Quantification of immunoglobulin in the serum and mucus of channel catfish at different ages and following infection with *Edwardsiella ictaluri*, *Veterinary Immunology and Immunopathology*, 58, 171, 1997.
14. Sharon, N. and Lis, H., Carbohydrates in cell recognition, *Scientific American*, January, 1993, 82.
15. Shoemaker, C. A., Klesius, P. H., and Lim, C., Immunity and disease resistance in fish, in *Nutrition and Fish Health*, Lim, C. and Webster, C., Eds., Ha'p'orth Press, Inc. In press.
16. Clem, L. W., Sizemore, R. C., Alastair, C. F., and Miller, N. W., Monocytes as accessory cells in fish immune responses, *Developmental and Comparative Immunology*, 9, 803, 1985.
17. Shoemaker, C. A. and Klesius, P. H., Protective immunity against enteric septicemia in channel catfish, *Ictalurus punctatus* (Rafinesque), following controlled exposure to *Edwardsiella ictaluri*, *Journal of Fish Diseases*, 20, 361–368, 1997.
18. Vallejo, A. N., Miller, N. W., and Clem, L. W., Anitgen processing and presentation in teleost immune responses, in *Annual Review of Fish Diseases* vol. 2, Faisal, M. and Hedrick, F. M., Eds., Pergamon Press, New York, 1992, 73–89.
19. Manning, M. J., Fishes, in *Immunology: A Comparative Approach*, Turner, R. J., Ed., John Wiley & Sons, Chichester, Great Britain, 1994, 69.
20. Evans, D. L. and Jaso-Friedman, L., Nonspecific cytotoxic cells as effectors of immunity in fish, in *Annual Review of Fish Diseases* vol. 2, Faisal, M. and Hedrick, F. M., Eds., Pergamon Press, New York, 1992, 109.
21. Evans, D. L. and Gratzek, J. B., Immune defense mechanisms in fish to protozoan and helminth infections, *American Zoologist*, 29, 409, 1989.
22. Hogan, R. J., Stuge, T. B., Clem, L. W., Miller, N. W., and Chinchar, V. G., Antiviral cytotoxic cells in the channel catfish, *Developmental and Comparative Immunology*, 20, 115, 1996.
23. Janeway, C. A., Jr. and Travers, P., *Immunobiology: The Immune System in Health and Disease*, Current Biology Ltd., Garland Publishing, New York, 1994, chap. 3: 1–46.
24. Kaattari, S. L., Fish B lymphocytes: defining their form and function, in *Annual Review of Fish Diseases* vol. 2, Faisal, M. and Hedrick, F. M., Eds., Pergamon Press, New York, 1992, 161.
25. Lall, S. P. and Olivier, G., Role of micronutrients in immune response and disease resistance in fish, in *Fish Nutrition in Practice*, Kaushik, S. J. and Luquet P., Eds., INRA, Paris, France, 1993, 101.
26. Shoemaker, C. A., Klesius, P. H., and Plumb, J. A., Killing of *Edwardsiella ictaluri* by macrophages from channel catfish immune and susceptible to enteric septicemia of catfish, *Veterinary Immunology and Immunopathology*, 58, 181, 1997.
27. Coligan, J. E., Kruisbeek, A. M., Margulies, D. H., Shevach, E. M., and Strober, W., *Current Protocols in Immunology*, Greene Publishing and Associates and Wiley Interscience, John Wiley & Sons, New York, 1991.

28. Bessel, W. R., Single nutrients and immunity, *American Journal of Clinical Nutrition*, 35, 417, 1982.

29. Bendich, A., Vitamin C and immune responses, *Food Technology*, November, 112, 1987.

30. Muggli, R., Vitamin C and phagocytes, in *Nutrient Modulation of the Immune Response*, Cunningham-Rundles, S., Ed., Marcel Dekker, New York, 1993, 75.

31. Li, Y. and Lovell, R. T., Elevated levels of dietary ascorbic acid increase immune responses in channel catfish, *Journal of Nutrition*, 115, 123, 1985.

32. Li, M. H., Wise, D. J., and Robinson, E. H., Effect of dietary vitamin C on weight gain, tissue absorbate concentration, stress response, and disease resistance of channel catfish *Ictalurus punctatus*, *Journal of the World Mariculture Society*, 29, 1–8, 1998.

33. Liu, P. R., Plumb, J. A., Guerin, M., and Lovell, R. T., Effect of mega levels of dietary vitamin C on the immune response of channel catfish *Ictalurus punctatus* in ponds, *Diseases of Aquatic Organisms*, 7, 191, 1989.

34. Johnson, M. R. and Ainsworth, A. J., An elevated dietary level of ascorbic acid fails to influence the response of anterior kidney neutrophils to *Edwardsiella ictaluri* in channel catfish, *Journal of Aquatic Animal Health*, 3, 266, 1991.

35. Lim, C., Klesius, P. H., Li, M. H., and Robinson, E. H., Interaction between dietary levels of iron and vitamin C on growth, hematology, immune response and resistance of channel catfish (*Ictalurus punctatus*) to *Edwardsiella ictaluri* challenge, *Aquaculture*, 185, 313, 2000.

36. Blazer, V. S., *The Effect of Marginal Deficiencies of Ascorbic Aid and Alpha-tocopherol on the Natural Resistance and Immune Response of Rainbow Trout (Salmo gairdneri)*, Ph.D. dissertation, University of Rhode Island, Kingston, 1982, 113.

37. Blazer, V. S. and Wolke, R. E., Effect of diet on immune response of rainbow trout (*Salmo gairdneri*), *Canadian Journal of Fisheries and Aquatic Sciences*, 41, 1244, 1984.

38. Navarre, O. and Halver, J. E., Disease resistance and humoral antibody production in rainbow trout fed high levels of vitamin C, *Aquaculture*, 79, 207, 1989.

39. Haddie, L. J., Marsden, M. J., Fletcher, T. C., and Secombes, C. J., *In vitro* addition of vitamin C affects rainbow trout lymphocyte responses, *Fish & Shellfish Immunology*, 3, 207, 1993.

40. Anggawati-Satyabudhy, A. M. A., Grant, B. F., and Halver, J. E., Effect if L-ascorbyl-2-phosphates (AsPP) on growth and immunoresistance of rainbow trout to infectious hematopoietic necrosis (IHN) virus, *Proceedings, Third International Symposium on Feeding and Nutrition in Fish*, Toba, Japan, 1989, 411.

41. Verlhac, V., N' Doye, A., Gabaudan, J., Troutaud, D., and Deschaux, P., Vitamin Nutrition and Fish Immunity: Influence of Antioxidant Vitamins (C and E) on Immune Response of Rainbow Trout, *Fish Nutrition in Practice*, Kaushik, S.J. and Luquet, P., Eds., INRA, Paris, France, 1993, 167.

42. Verlhac, V. and Gabaudan, J., Influence of vitamin C on the immune system of salmonids, *Aquaculture and Fisheries Management*, 25, 21, 1994

43. Bell, G. R., Higgs, D. A., and Traxler, G. S., The effect of dietary absorbate, zinc, and manganese on the development of experimentally induced bacterial kidney disease in sockeye salmon, (*Onocorhynchus nerka*), *Aquaculture*, 36, 293, 1984.

44. Lall, S. P., Oliver, G., Weerakoon, D. E. M., and Hines, J. A., The effect of vitamin C deficiency and excess on immune response in Atlantic salmon (*Salmo salar* L.),

Proceedings, Third International Symposium on Feeding and Nutrition in Fish, Toba, Japan, 1989, 427.

45. Sandnes, K., Hansen, T., Killie, J. E. A., and Waagbo, R., Absorbate-2-sulfate as a dietary vitamin C source for Atlantic salmon (*Salmo salar*): I. Growth, bioactivity, hematology and humoral immune response, *Fish Physiology and Biochemistry,* 8, 419, 1990.

46. Hardie, L. H., Fletcher, T. C., and Secombes, C. J., The effect of dietary vitamin C on the immune response of the Atlantic salmon (*Salmo salar* L.), *Aquaculture,* 95, 201, 1991.

47. Waagbo, R., Glette, J., Raa-Nilsen, E., and Sandnes, K., Dietary vitamin C, immunity and disease resistance in Atlantic salmon (*Salmo salar*), *Fish Physiology and Biochemistry,* 12, 61, 1993.

48. Erdal, J. I., Evensen, O., Kaurstad, O. K., Lillehaug, A., Solbakken, R., and Thorud, K., Relationship between diet and immune response in Atlantic salmon (*Salmo salar* L.) after feeding various levels of ascorbic acid and omega-3 fatty acids, *Aquaculture,* 98, 363, 1991.

49. Yano, T., Furuichi, M., Nakao, M., and Ito, S., Effects of L-ascorbyl-2-phosphate Mg on the growth and nonspecific immune system of red sea bream *Pagrus major,* abstracts, World Aquaculture Society Meeting, June 10–14, 1990, Halifax, Canada, T29.12.

50. Roberts, M. L., Davies, S. J., and Pulsford, A. L., The influence of ascorbic acid (vitamin C) on non-specific immunity in the turbot (*Scophthalmus maximus* L.), *Fish & Shellfish Immunology,* 5, 27, 1995.

51. Nitzan, S., Angeoni, H., and Gur, N., Effects of ascorbic acid polyphosphate (AAPP) enrichment on growth, survival and disease resistance of hybrid tilapia, *The Israel Journal of Aquaculture-Bamidgeh,* 48, 133, 1996.

52. Lim, C., *Dietary Ascorbic Acid (Vitamin C) Requirements of Channel Catfish in Ponds and in a Controlled Environment,* Ph.D. dissertation, Auburn University, Auburn, AL, 1977, 76.

53. Lovell, R. T. and Lim, C., Vitamin C in pond diets for channel catfish, *Transactions of the American Fisheries Society,* 107, 321, 1978.

54. Durve, V. S. and Lovell, R. T., Vitamin C and disease resistance in channel catfish (*Ictalurus punctatus*), *Canadian Journal of Fisheries and Aquatic Sciences,* 39, 948, 1982.

55. Li, M. H., Johnson, M. R., and Robinson, E. H., Elevated dietary vitamin C concentrations did not improve resistance of channel catfish, *Ictalurus punctatus,* against *Edwarsiella ictaluri* infection, *Aquaculture,* 117, 303, 1993.

56. Li, M., Wise, D. J., and Robinson, E. H., Effect of dietary vitamin C on weight gain, tissue absorbate concentration, stress response, and disease resistance of channel catfish *Ictalurus punctatus, Journal of the World Aquaculture Society,* 29, 1, 1998.

57. Wahli, T., Frischknecht R., Schmill, M., and Gabaudan, J., A comparison of the effect of silicone coated ascorbic acid and ascorbyl phosphate on the sourse of itchthyophthiriosis in rainbow trout, *Oncorhynchus mykiss* (Walbaum), *Journal of Fish Diseases,* 18, 347, 1995.

58. Lall, S. P. and Oliver, G., Role of micronutrients in immune response and disease resistance in fish, in *Fish Nutrition in Practice,* Kaushik, S.J. and Luquet, P., Eds., INRA, Paris, France, 1993, 101.

59. Hosokawa, H., Teishima, H., Shimeno, S., and Takeda, M., Effect of dietary vitamin C for intensification of immunity on yellowtail, *Japanese Society of Fisheries Meeting,* October, 1990, Abstract #437, 1990.

60. Duncan, P. L. and Lovell, R. T., Influence of vitamin C on the folate requirement of channel catfish, *Ictalurus punctatus,* for growth, hematopoeisis, and resistance to *Edwardsiella ictaluri* infection, *Aquaculture,* 127, 233, 1994.

61. Verlhac, V., Gabaudan, J., Obach, A., Schuep, W., and Hole, R., Influence of dietary glucan and vitamin C on non-specific immune responses of rainbow trout (*Oncorhynchus mykiss*), *Aquaculture,* 143, 123, 1996.

62. Wahli, T., Verlhac, V., Gabaudan, J., Schuep, W., and Meier, W., Influence of combined vitamins C and E on non-specific immunity and disease resistance of rainbow trout, *Oncorhynchus mykiss* (Walbaum), *Journal of Fish Diseases,* 21, 127, 1998.

chapter thirteen

Critical review of the concentration, interactions with other nutrients, and transfer of ascorbic acid in algae, crustaceans, and fish

Malcolm Brown and Patrick Lavens

Contents

Abstract

While the ascorbic acid (AA) requirements of adult and juvenile marine fish
and shellfish are well known, less is understood about the AA requirements
of larvae. Most of the information available on larval requirements is based
on feeding trials conducted using live diets, such as microalgae or zooplankton (rotifers and *Artemia* sp.)

Nonenriched *Artemia* nauplii (0.5 to 0.7 mg AA g^{-1} dry weight) and rotifers
reared on yeast (0.1 to 0.6 mg AA g^{-1} dry weight) contain sufficient AA for the
normal growth and survival of most fish and crustacean larvae. Nevertheless,
zooplankton can be enriched with AA (e.g., 1.5 to 2.5 mg AA g^{-1}) by feeding
them on microalgae or commercial products. This may improve the physiological condition of larvae subsequently feeding on the zooplankton by
enhancing their AA tissue concentrations. While the commercial enrichments
may be more cost-effective, microalgae may have additional benefits such as
increasing the concentrations of other trace nutrients (e.g., other vitamins) in
zooplankton, which may further improve growth and/or survival of larvae.

For larvae feeding directly on microalgae (e.g., scallops, oysters,
prawns), more research is required to assess their AA requirements. Despite
differences in AA concentrations between microalgae, algal diets successfully
used as direct feeds contain 1 to 3 mg AA g^{-1} dry weight; AA requirements
are apparently met at these concentrations. Compared to microalgae, yeast
and bacteria appeared to be poor sources of AA (<0.3 mg g^{-1}). Their concentrations may be insufficient to meet dietary requirements and become a contributing factor to their low food value for bivalve molluscs.

13.1   Introduction

Ascorbic acid (AA) is recognized as an essential micronutrient for fish and
shellfish in aquaculture. While the quantitative requirements of adult and
juvenile animals are well documented,[1,2,3,4] less is known about the requirements of larvae. Their requirements may be higher than adults because of
their faster metabolism and growth; metabolic rate is thought to be the primary factor controlling AA requirement.[5] Also, high concentrations of AA in
fish eggs[6] suggest the larvae's requirement is high.

From the evidence, therefore, it is probably vital for rapidly developing
larvae to receive adequate concentrations of AA, which is logically through
their diet. As hatchery production for most species is dependent on live feeds,
it is important to define the AA concentrations of the feeds, and to understand
how their concentrations can be manipulated by different culturing.

This review covers several aspects relating to the transfer of AA through different trophic levels to larvae of cultured fish, crustaceans, and molluscs. We report on the AA concentration in microorganism diets used as direct feed for larval and juvenile animals, and as feed for zooplankton. AA concentrations in zooplankton fed these diets, and other formulated enrichment products are also compared. The nutritional requirements of larvae, deduced from feeding trials with AA-enriched live feeds, are also discussed.

13.2 AA concentrations of microorganism diets used in aquaculture

13.2.1 Live microalgae

In aquaculture, microalgae are fed to all growth stages of bivalves, and larval stages of some crustaceans and fish. They are also eaten by zooplankton (rotifers, brine shrimp, and copepods) cultured as food for late-larval and juvenile stages of most crustaceans and fish. Microalgae differ in their nutritional value, according to their shape, size, toxicity, digestibility, and biochemical composition.[7, 8, 9, 10] Attempts to correlate composition of nutrient fractions of microalgae with their nutritional value[8, 11, 12, 13, 14] have proved difficult because of differences in their digestibility and the variable, still poorly understood general composition of microalgae. However, the microalgal highly unsaturated fatty acids (HUFA) 20:5(n-3) and 22:6(n-3) (present in most species, except chlorophytes) are recognised as essential for growth of marine shellfish.[15] The importance of other trace nutrients in microalgae, including vitamins, is less well established.

Concentrations of AA in microalgae appear to vary more between species than most of the other vitamins.[16, 17] In a study of seven microalgae, Bayanova and Trubacheva[18] found a 50-fold difference in AA concentration between species. Similarly, Brown and Miller[16] found up to a 15-fold difference in AA concentrations between 11 microalgae commonly used in aquaculture. Generally, most microalgae were reported to contain between 1 and 3 mg AA g^{-1} dry weight (DW)* (Table 13.1). The highest AA concentrations were reported in *Chlorella vulgaris* (15 mg g^{-1}),[19] *Chaetoceros muelleri* (16 mg g^{-1} in log phase)[16] and *Chlorella pyrenoidosa* ($\approx$30 mg g^{-1}, a genetically modified strain grown heterotrophically).[20] Happette and Poulet[21] also examined the AA content (pg $cell^{-1}$) in five species of microalgae, but did not report concentrations on a dry weight basis.

Even the reported values for the same species can differ: for example, *Skeletonema costatum* has been reported to contain 0.06,[14] 1.5,[22] and 5.3 to 8.7 mg AA g^{-1}.[16] Some of these differences might be due, in part, to differences in

---

* Concentrations of AA are expressed in relation to the dry weight of the diet, unless otherwise indicated.

Table 13.1 Ascorbic acid (AA) concentrations (mg g^{-1} dry weight) in microalgae

Class/number of microalgae analysed	Growth conditions	Average AA (Range)	Reference
2 chlorophytes, 4 cyanophytes, 1 rhodophyte	Continuous light; T = 24 to 56°C; Late-log to stationary phase	1.6 (0.05–2.4)	18
3 chlorophytes, 1 cyanophyte	Continuous light; T = 15 to 25°C; stationary phase	5.3 (2–15)	19
1 prasinophyte, 2 prymnesiophytes, 2 diatoms	Continuous light; T = 18°C; harvest stage not specified	0.6 (0.06–0.8)	14
2 prymnesiophytes, 1 diatom	Continuous light; T = 18°C; late log phase	1.4 (0.7–1.8)	22
1 chlorophyte, 1 prymnesiophyte, 1 eustigmatophyte	Continuous light; T = 25°C; late log phase	2.3 (1.1–3.8)	24
2 chlorophytes, 1 prasinophyte, 4 diatoms, 1 prasinophyte, 1 cryptophyte, 1 eustigmatophyte	12:12 h L:D; T = 20°C; log phase	4.6 (1.3–16)	16
2 chlorophytes, 1 prasinophyte, 4 diatoms, 1 prasinophyte, 1 cryptophyte, 1 eustigmatophyte	12:12 h L:D; T = 20°C; stationary phase	3.1 (1.1–8.7)	16
16 chlorophytes, 3 diatoms, 2 cyanophytes, 1 chrysophyte, 1 euglenophyte	Dark (heterotrophic) growth; T = 35°C; late log (?) phase	0.9 (0.05–5.7)	20

culture conditions or methods of harvesting, extracting, and analysing the microalgae. For example, Happette and Poulet[23] found either acetonitrile or metaphosphoric acid + acetic acid to be three to four times more efficient than methanol or ethanol for extracting AA from plankton. The same authors found AA values from the extracted plankton were up to 10 times greater by HPLC analysis with reverse-phase columns than with anion-exchange columns. Samples should be analysed either fresh or after storage at −80°C or lower,[24] as freezing at temperatures such as −20°C[22] results in losses of up to 40% of AA in microalgae.[25] For the above reasons, the AA values from different studies are not easy to compare.

The broadest assessment, by Brown and Miller,[16] was of 11 microalgae (6 classes) in logarithmic and stationary growth phases (Figure 13.1). Concentrations ranged from 1.0 mg AA g^{-1} (*Thalassiosira pseudonana*, stationary phase) to 16 mg AA g^{-1} (*Chaetoceros muelleri*, log phase), but were unrelated to algal class. Neither were AA concentrations in logarithmic and stationary-phase cultures, which differed for most species. *Chaetoceros muelleri, T.*

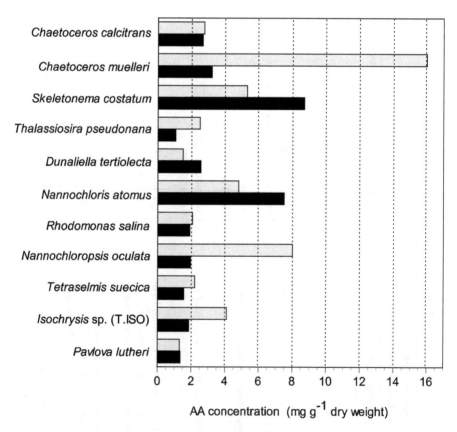

Figure 13.1 AA concentrations of various microalgae.

pseudonana, Nannochloropsis oculata and *Isochrysis* sp. (T.ISO) had higher AA concentrations during the logarithmic phase, whereas *Dunaliella tertiolecta* and *Nannochloris atomus* had higher during the stationary phase. There was no obvious relationship between the AA concentration of the microalgae, and their previously reported value as food: for example, *D. tertiolecta, Nannochloris atomus,* and *T. suecica* had similar or greater AA concentrations than most other species but had the poorest nutritional value for bivalve molluscs.[26]

Changes in AA concentration in microalgae going from logarithmic phase to stationary phase reflect the cells' physiological response to limitation of inorganic macronutrients such as nitrate or silicate. Concentrations of trace metals in culture media also affect the AA production of microalgae: high Fe (8 mg L^{-1}) and Cu (1.1 mg L^{-1}) concentrations reduced the AA production in heterotrophically grown *Chlorella pyrenoidosa*, with optimum production at 1.1 mg L^{-1} Fe and 0.11 mg L^{-1} Cu.[20]

The synthesis of ascorbate in microalgae is also affected by light intensity and photoperiod. In cultures of *T. pseudonana* grown under continuous light (24:0 h L:D), AA concentrations increased by 35% when the light intensity was increased from 50 to 100 μmol photon $m^{-2} s^{-1}$.[16] At a light intensity of 100 μmol photon $m^{-2} s^{-1}$ the same alga had 38% more AA when grown under 24:0 h L:D than under 12:12 h L:D.[16] In *Nannochloropsis* sp., the difference was 28%.[27] *Chlorella pyrenoidosa* previously grown heterotrophically in the dark, had a four-fold increase in AA concentration after exposure to 12 h light.[28] Further, average AA values in heterotrophically grown algae (0.9 mg g^{-1}; Table 13.1) are typically lower than most other reported values for phototrophically grown algae (Table 13.1). These limited data suggest that light has an important regulatory role in AA biosynthesis in microalgae, and more research is warranted.

Microalgae may release AA extracellularly, though this appears to be limited to a few species and it is not clear whether the vitamin is actively released, or released upon death or disintegration of the cell.[29] AA was not detected in the (extracellular) culture media from three green algae and one cyanobacteria.[19] It was detected in the media from stationary-phase cultures of *D. tertiolecta, Isochrysis* sp. (T.ISO) and *Rhodomonas salina* in concentrations ranging from 6 to 42% of the total (i.e., intracellular plus extracellular) AA detected,[16] which the authors suggest may have been an artefact caused by cell damage of these fragile species during harvesting by vacuum filtration.

13.2.2   Microalgal pastes and powders

For aquaculture hatcheries, off-the-shelf alternatives to live algal feeds may be less expensive and could enable them to more readily adjust to daily fluctuations in food requirements. Algal pastes and powders have been tested by hatcheries with varying success over the last decade.[30] Generally, they are considered useful as partial replacements for live microalgae, but not as complete diets because their nutritional values are significantly lower.[30]

Because AA is a readily oxidized vitamin, its measurement in paste or powder preparations could be a sensitive indicator of deterioration in nutritional value. Concentrations of AA in algal pastes of the diatom *Chaetoceros calcitrans*, prepared by centrifugation, decreased by 35% after 4 weeks' storage in the dark at 4°C.[25] However, more than 90% of the remaining AA was retained intracellularly. Studies with other pasted microalgae show that species that are easily damaged by centrifugation rapidly lose AA (e.g., *Isochrysis* sp. [T.ISO], which showed a complete loss after 2 weeks), whereas those with tough cell walls (*Nannochloropsis* sp., *Stichococcus* sp.) lost little or no AA over 6 months' storage.[31]

Frozen microalgal pastes have also been tested as alternatives to live microalgae, though their nutritional value is significantly lower.[32, 33] Frozen pastes of *Chaetoceros calcitrans* lost 39% of their AA after 1 week of storage at $-20°C$; when pastes were thawed and resuspended in seawater, 85% of the remaining AA was leached into the seawater after 10 min.[25]

Drying can also cause significant losses of AA in microalgae. *Scenedesmus acutus* contained 1.6 mg AA g^{-1} when freshly harvested, but lost 64% of its AA when freeze dried, and 90% when sun dried.[34] *Chaetoceros calcitrans* lost 34% of its AA through freeze drying, and 97% after drying at 60°C for 2 h.[25] When dried *C. calcitrans* was resuspended in seawater for 10 min, $\geq 95\%$ of the remaining AA leached out into the solution.[25]

13.2.3   Other microorganisms used in aquaculture

Other microorganisms tested as alternatives to live microalgae for aquaculture feeds include bacteria, yeast, and thraustochytrids.[30, 35, 36, 37] The rationale for their use is that they can be grown heterotrophically by fermentation, so they cost less to produce than phototrophically grown microalgae.[38]

Compared to microalgae, yeast and bacteria are deficient in AA. No AA (i.e., < 0.1 mg AA g^{-1}) was detected in 20 yeast strains.[20] *Saccharomyces cerevisiae* (baker's yeast), a popular food for rotifers, has been reported to contain from 0.1 to 0.3 mg AA g^{-1}.[24, 39] There is little published information on the AA concentration of bacteria, though some *Streptococcus* and *Lactobacillus* strains have been identified as synthesizing AA.[40] Two HUFA-containing bacteria—*Shewanella* sp. (ACAM 456)* and *Colwellia* sp. (ACAM 605)—contained 0.01 mg g^{-1} and 0.025 mg g^{-1} of AA, respectively.[41]

Thraustochytrids are marine eukaryotes with unclear taxonomy. RNA sequence analysis indicates they are a unique group distinct from the fungi, and related to the red and brown algae.[42] A spray-dried preparation of the thraustochytrid *Schizochytrium* sp. is sold commercially as AlgaMac-2000® (Aquafauna Biomarine, California) as a HUFA-enrichment for zooplankton.[36] AlgaMac-2000® is reported to contain 0.71 mg g^{-1} AA.[43] A freshly harvested thraustochytrid (unidentified species) contained 0.48 mg g^{-1} AA.[41]

* Code designation from the *Australian Collection of Antarctic Microorganisms*, Hobart, Australia.

13.2.4   Significance of AA concentrations in microorganisms to animals feeding on them

The animals fed cultured microalgae and other microorganisms in aquaculture include filter-feeding bivalve molluscs (oysters, scallops, mussels, clams), juvenile abalone, early shrimp larvae, and zooplankton (rotifers, *Artemia*, copepods). Unfortunately, there is little information on the AA requirements of these animals, and there is no evidence that they can synthesise AA. Therefore the AA concentrations in their diet are also a likely growth-determining factor.

Seguineau et al.[17, 22] found that, during development, the total AA concentration of scallop larvae increased by 20% of the AA they ingested from a mixed algae diet (containing $\approx$1 mg AA g^{-1}). It was not established whether the low efficiency was due to the AA uptake being saturated (i.e., reduced absorption),[44] or whether the rapidly growing larvae tissue had utilised the AA. However, because AA concentrations in the larvae were increasing with time, the authors inferred that the microalgae were adequately fulfilling the requirements for larval growth.[17]

Artificial diets for larval and juvenile bivalve molluscs are routinely supplemented by mixtures of vitamins, including AA at 10 mg g^{-1}, but quantitative requirements have not been established.[45, 46] The microalgae reported to have the lowest concentrations of AA (*Pavlova lutheri* and *T. pseudonana*; 1 mg g^{-1})[16] are considered to be excellent mollusc diets, either on their own or mixed with others.[26, 47, 48] Possibly, the variability of AA between microalgae may not be nutritionally significant for these filter feeders, and their requirements are fully met at dietary concentrations of 1 mg g^{-1}. Bacteria and yeast that have < 0.1 mg AA g^{-1} have not been successful as a complete diet for molluscs—though other factors, such as their poor digestibility[49, 50] and lack of essential polyunsaturated fatty acids[51] may be important contributing factors to their poor food value.

Prawn larvae are thought to require AA, though the quantity is unknown.[4, 52] Diatoms such as *Skeletonema* spp. and *Chaetoceros* spp. support excellent growth in the early larval stages of a range of *Penaeus* spp.[10, 53] Based on the AA concentration of these microalgae (Figure 13.1) it seems that AA above 2 mg g^{-1} dry weight are adequate for the diet of early prawn larvae.

There is no evidence that rotifers can synthesise AA, so presumably their requirements must be met through their diet. Rotifers appear to have a low AA requirement, as they can be reared through several generations on diets low in AA, such as baker's yeast.[54, 55] *Artemia* can also be grown up to reproductive stages at very high densities, using AA-deficient waste products from agricultural crops.[56] Adult *Artemia* deposits significant amounts of the vitamin C derivative, ascorbyl sulfate, in their eggs.[57] This derivatization of AA was not examined in other aquatic organisms. The transfer of AA from microorganisms to rotifers and other zooplankton has more nutritional significance to animals further up in the food chain that are fed such zooplankton (see Section 13.3).

13.3 Transfer of AA to zooplankton

13.3.1 Manipulation of AA content in live zooplanktonic organisms

As in microalgae, the biochemical composition (and hence nutritional value) of zooplankton can be quite variable. It is profoundly influenced by diet[58, 59, 60, 61, 62]—in particular, the essential fatty acids can be increased by feeding the zooplankter on microalgae or lipid emulsions rich in HUFA.[63, 64, 65] Recent research on methods for enriching AA in zooplankton has been stimulated by a better understanding of the need for adequate AA intake during larval fish and crustacean ontogeny, and a corresponding recognition that zooplankton can vary significantly in their AA, and therefore can potentially have suboptimal AA levels.

Happette and Poulet[21] found that the AA concentrations of zooplankton harvested from marine environments were related to their trophic position in the food chain. For example, carnivores had about a tenth of the AA of omnivorous and herbivorous calanoid species. Concentrations ranged from 0.016 to 0.44 mg AA g^{-1} wet weight[21] – or ≈0.1 to 3.5 mg AA g^{-1} DW based on estimate of 12% dry matter from copepod biomass[66] (Table 13.2). These authors subsequently assessed the incorporation and decrease of AA in copepods in the laboratory by starving and feeding experiments. After 8 days of starvation of adult calanoids, AA concentrations in female *Calanus helgolandicus* decreased by 54%. After resuming feeding with a microalga (*Thalassiosira weissflogii*) for 66 h, AA concentrations increased by 62%. The trends were similar in a trial with *Acartia clausi*. Happette and Poulet[21] estimated the transfer efficiency of AA (from alga to copepod) as between 40 and 80%. These results, taken with other data from this study and another,[67] suggest that copepods do not biosynthesise AA but derive it from their diet.

Rotifers are rapidly and efficiently enriched in AA when fed microalgae (Table 13.2). Furthermore, the concentrations in the rotifers reflect the AA concentrations in their diet.[24, 39, 68] Short-term enrichment in a laboratory-scale study increased AA concentrations in rotifers from 0.15 to 1.6 mg g^{-1} after 6 h feeding on *Isochrysis* sp. (T.ISO) (itself containing 3.8 mg AA g^{-1}).[24] A similar enrichment was found in rotifers (from 0.62 mg g^{-1} to 1.6 mg^{-1}) after 3 h feeding with the same alga (containing 4.2 mg AA g^{-1}).[39] With longer-term feeding, rotifers fed baker's yeast for 3 days contained 0.15 mg AA g^{-1}, whereas rotifers fed *Chlorella* sp. (containing 3.8 mg AA g^{-1}) for the same period contained 2.3 mg AA g^{-1}.[24] AA concentrations in rotifers (previously fed on yeast) increased from 0.37 mg g^{-1} to 1.5 mg g^{-1} after 3 d of feeding with *Isochrysis galbana* (containing 0.4 mg AA g^{-1}).[38] Brown et al.[39] fed three diets (baker's yeast, *Nannochloropsis oculata*, and *Isochrysis* sp. [T.ISO] containing 0.08, 2.6 and 4.2 mg AA g^{-1}, respectively) to rotifers; after 24 h the rotifers contained, respectively, 0.62, 1.7, and 2.5 mg AA g^{-1}. Between 56 and 76% of the AA from microalgae ingested by the rotifers was retained—though it was not established if the proportion of AA was truly assimilated, or how much was in nondigested (or partly-digested) microalgae in the rotifers' gut.[39]

Table 13.2 Ascorbic acid (AA) concentrations in cultured and natural zooplankton

Zooplankton/culture diet	AA (mg g^{-1} dry weight)	Reference
Rotifers		
BY ($\geq$ 3d)	0.13–0.37	24, 58, 68
Cultured on BY, + 30% AP for 24 h	1.5	24
Cultured on BY, + 30% AP for 72 h	2.0	24
Cultured on BY, + *Isochrysis galbana* for 23 h	1.3	68
3 d culture with BY + *Isochrysis* sp. (T.ISO) for 6 h	1.6	24
3 d culture with *Chlorella* sp.	2.3	24
7 d culture on CS®	0.14–0.25	24
7 d culture on CS® + PS for 6 h	0.21–0.28	24
7 d culture on CS® + PS vit C-boosted® for 6 h	0.94	24
7 d culture on CS® + *Chlorella* for 24 h	0.51	24
7 d culture on CS vit C-boosted®	0.33	24
7 d culture on CS vit C-boosted® + *Nannochloropsis* for 24 h	0.72	24
7 d culture on CS vit C-boosted® + *Isochrysis* sp. (T.ISO) for 24 h	1.6	24
7 d culture of CS vit C-boosted®	0.57	24
7 d culture of CS vit C-boosted® + PS vit C-boosted® for 6 h	1.3	24
7 d culture of CS vit C-boosted® + PS vit C-boosted® for 24 h	1.7	24
Cultured on 95% BY/5% mixed algae	0.62	39
Cultured on 95% BY/5% mixed algae + *Isochrysis* sp. (T.ISO) for 24 h	2.5	39
Cultured on 95% BY/5% mixed algae + *N. oculata* for 24 h	1.7	39
Artemia		
Non-enriched nauplii	0.31–0.66	24, 70
Enriched with 10% AP	1.1–1.6	70
Enriched with 20% AP	1.6–3.6	70
Copepods (natural)		
Acartia clausi	$\approx$ 0.2–2.1*	21
Temora longicornis	$\approx$ 1.0–3.0*	21
Calanus helgolandicus	$\approx$ 0.7–3.5*	21
Anomalocera pattersoni	$\approx$ 0.1*	21

* Original data was reported on a wet weight basis. These are estimates based on the assumption that copepod dry matter = 12.4% wet weight.[66] Abbreviations: BY = baker's yeast; CS = Culture Selco; PS = Protein Selco; AP = ascorbyl palmitate.

When starved rotifers rapidly ingest algae, most of the AA may be encapsulated in intact algal cells within the rotifers' guts and potentially unavailable to fish larvae which have a poorly developed digestive tract. This may be especially true of tough-walled algae such as *Nannochloropsis* spp. and *Chlorella* spp., which are commonly fed to rotifers. For example, milkfish fry that feed on microalgae can utilize *Isochrysis galbana*, but not *Chlorella* sp. because of its tough cell wall.[69] Generally though, rotifers are efficient at digesting microalgae, but at different rates for different species:[54] they can clear *Pavlova lutheri* from their gut within 40 min.[60] After prolonged feeding on algae (e.g., >3 h), a higher proportion of AA may be absorbed in the rotifer tissue and therefore be more readily available to feeding fish larvae. Further research is required to determine the availability of AA in algae-fed zooplankton.

Concentration of AA in rotifers reared in commercial hatcheries and fed various combinations of microalgae and/or commercial enrichment products have also been reported (Table 13.2).[24] AA was highest in rotifers fed *Isochrysis* sp. (T.ISO)—from 0.33 to 1.6 mg AA g^{-1} 24 h after feeding. Using a commercial enrichment product, Protein Selco vit C-boosted®,* the AA concentration in rotifers increased from 0.14 mg g^{-1} to 0.94 mg g^{-1} 6 h after feeding. In other laboratory-scale experiments, the AA concentrations in rotifers previously fed Culture Selco vit C-boosted®* for 7 d increased from 0.57 mg g^{-1} to 1.3 mg g^{-1} 6 h the addition of Protein Selco vit C-boosted®, and 1.7 mg g^{-1} 24 h later.[24]

Recently, technologies have been developed for AA enrichment of zooplankton by feeding them lipid emulsions containing the lipophylic ascorbyl palmitate (AP),[24] based on methods used for the HUFA-enrichment of zooplankton.[64] AP is a bioavailable form of AA within emulsions, rapidly converted to active AA when assimilated by the zooplankton.[24] Not only is it an effective method of boosting AA concentrations in zooplankton to potentially improve their nutritional value, but it also makes it possible to vary AA concentrations in live feed essentially independent of other nutrients. Hence, it is a useful technology for studying the effect of different AA concentrations on the growth, survival, and stress resistance of larvae (see next section).

To assess the enrichment of AA in rotifers, Merchie et al.[24] fed them with emulsions containing various proportions of AP (from 0 to 30%). The AA concentrations increased with time and were related to the percent of AP in their diet. The initial concentration was 0.13 mg AA g^{-1}, rising to 2.0 mg AA g^{-1} when provided in the 30% AP diet—a similar enrichment to that achieved with algae as feed in their other experiments (Table 13.2).[24]

AA in *Artemia* cysts occurs naturally as AA-sulfate, though this form is rapidly converted to free AA upon cyst hydration and hatching of the *Artemia* nauplii.[57] AA concentrations of newly-hatched *Artemia* nauplii are significantly higher than in yeast-fed rotifers and several-fold higher than in

---

* Registered trademark of INVE Aquaculture, N.V., Baasrode, Belgium.

wild freshwater zooplankton, though concentrations can vary with batch and strain, e.g., from 0.22 to 0.66 mg g^{-1}.[24, 57, 70] The most likely reason contributing to these concentration differences in the nauplii (and hence cysts) is adult nutrition during egg production.[24] This suggestion is supported by results of Poulet et al.,[67] who found that copepods collected from natural environments had seasonal differences in AA concentrations ranging from three- to ninefold, which the authors attributed to seasonal variation in the phytoplankton biomass on which the zooplankton were feeding.

Methods used for AA enrichment in rotifers have also been successfully used with *Artemia* (Table 13.2).[24, 57, 71, 72] After a 24 h enrichment with AP, concentration of AA in *Artemia* (initially 0.55 mg g^{-1}) increased to 0.85 mg g^{-1} when fed a 10% AP booster, to 2.0 mg g^{-1} with 20% AP and to 3.2 mg g^{-1} with 30% AP. When the 20% AP enrichment was provided in three daily rations instead of two, AA concentrations in *Artemia* increased to 3.4 mg g^{-1}.[24] Simultaneous measurement of AP in *Artemia* showed that after 24 h, about 90% had been converted to the active AA form.[24]

13.3.2   Retention of AA in zooplankton

Assimilated AA is efficiently retained in cultured zooplankton during non-feeding periods. During an 8 d starvation, AA concentrations in the copepod *Calanus helgolandicus* were reduced at a rate of 9% per day—five times lower than their rate of uptake during feeding.[21] AA concentrations in rotifers enriched by feeding with the microalgae *Isochrysis* sp. (T.ISO) and *N. oculata* did not change significantly over a 24 h non-feeding period at 22°C.[39] AA concentrations in rotifers enriched with AP did not change during a subsequent 24 h non-feeding period in fresh seawater at 25°C—in fact, concentrations showed an apparently small increase which was attributed to a further assimilation of AP from the rotifers' digestive tract.[24] Likewise, AA concentrations in *Artemia* enriched with AP did not change during a 24 h non-feeding period in fresh seawater at either 4°C or 28°C.[24] Observations across all these studies suggest that enriched zooplankton maintain their nutritional value with respect to AA during short-term storage or residence in larval fish culture tanks (i.e., for at least 24 h postenrichment).

13.4   AA transfer from zooplankton and requirements by larvae

The AA requirements of the juvenile and adult stages of fish and crustaceans have been determined by using semipurified diets with graded AA concentrations. Fish require 20 to 50 mg AA kg^{-1} diet[3, 73, 74] and shrimp 100 to 200 mg AA kg^{-1}.[2] Much less is known about the requirements of larval stages, partly because the output of larvae from hatcheries is variable (and low) and partly because artificial (i.e., of defined composition) diets that sustain the growth and survival of larvae are not generally available. Hatcheries therefore rely on live zooplankton diets of variable composition and (for some nutrients) unknown value. Live feeds also contain concentrations of AA that may be above those required by the larvae. Consequently, it has also been difficult to develop deficiency in fish or crustacean larvae and determine the AA requirements using live zooplankton diets.

There is strong evidence that larvae require more AA than do juveniles and adults, probably because of their higher rates of metabolism and growth. The AA requirement of fish appears to be related to their metabolic rate.[75] Catfish (*Ictalurus punctatus*) juveniles are more sensitive to dietary deficiency of AA than larger juveniles.[76] Further support is given by the high concentrations of AA in eggs, e.g., 0.28 to 0.32 mg AA g^{-1} wet weight (WW) for rainbow trout[6] coupled with a general decline in AA during fish ontogeny (e.g., from as high as 0.15 to 0.8 mg AA g^{-1} WW in larvae to 0.005 to 0.01 mg g^{-1} WW in several-month-old juveniles).[76] In turbot (*Scophthalmus maximus*) higher levels can be obtained in the eggs when an AA-rich maturation diet is provided (0.32 mg AA g^{-1} DW, up to tissue saturation levels of 1.15 mg AA g^{-1} DW);[77] these levels only decreased to 0.89 mg AA g^{-1} DW in metamorphosed juveniles when non-enriched live feed was offered.[83] In another study with first-feeding turbot larvae, Plañas et al.[78] found AA concentrations decreased in larvae when they were fed non-enriched rotifers (containing 0.36 mg AA g^{-1}). During the ontogeny of whitefish (*Coregonus lavaretus* L.) larvae, Dabrowski[79] noted varying responses to the level of dietary AA: (a) fish fed live food for 24 d showed a slow decrease in AA (expressed on a WW basis), (b) fish fed an AA-deficient diet showed a dramatic decrease in AA, and (c) fish fed a high level of AA were able to maintain body AA concentrations. These studies and others[80, 81] show that assimilated AA is readily catabolised by larvae fed AA-deficient diets, with a half-life of 4 to 10 d, i.e., at rates similar to scurvy-prone mammals.[82] AA appears therefore to be a critical nutrient in high demand during early larval development. Dabrowski suggests that the optimal dietary concentration of AA would maintain a steady-state tissue concentration.[79]

Other studies have shown the AA concentrations in some fish or crustacean larvae can increase when they are fed diets containing high concentrations of AA. Merchie et al. assessed the effects of feeding zooplankton enriched with AA (using AP) to several freshwater and marine fish (African catfish *Clarias gariepinus*,[72] European sea bass *Dicentrarchus labrax*,[72] and turbot *Scophthalmus maximus*[83]) as well as the giant freshwater prawn *Macrobrachium rosenbergii*.[71] In a review of these studies,[70] the authors reported that the AA concentrations of larvae fed the enriched *Artemia* diets increased significantly, indicating a net transfer of AA from diet to consuming larvae (Table 13.3). AA increases in larvae corresponded to the concentrations in the *Artemia*, though increasing dietary levels from 1.4 to 2.8 mg AA g^{-1} (i.e, from 10% and 20%-AP groups) did not significantly increase the AA levels in larval tissue. Consequently, the authors[70] suggested that larval tissue was saturated in AA when larvae where fed dietary levels of approximately 1.4 mg g^{-1} AA.

Table 13.3   Ascorbic acid (AA) concentration (mg g^{-1} dry weight) in the larvae of
crustaceans and fish before first feeding and after feeding rotifers and/or
Artemia sp. enriched with three ascorbyl palmitate (AP) concentrations

Experimental species	Before first feeding	Rotifer/*Artemia* enrichment level (mg AA g^{-1} DW)		
		0.60	1.40	2.80
Crustacean: giant freshwater prawn				
Experiment 1, larvae	0.15	0.37[a]	—	0.55[b]
Experiment 1, postlarvae		0.29[a]	—	0.33[a]
Experiment 2, larvae	0.29	0.35[a]	0.45[b]	0.51[b]
Experiment 2, postlarvae		0.26[a]	0.39[b]	0.43[b]
Fish: African catfish				
Experiment 1	—	0.47[a]	0.54[b]	0.55[b]
Experiment 2	0.29	0.41[a]	0.49[a,b]	0.55[b]
Experiment 3	0.24	0.43[a]	0.62[b]	0.75[b]
Experiment 4	0.46	0.51[a]	—	0.82[b]
Fish: European seabass	—	0.76[a]	1.60[b]	1.62[b]
Fish: Turbot				
Experiment 1	0.50	0.86[a]	1.22[b]	1.20[b]
Experiment 2	0.47	0.89[a]	1.12[b]	1.23[b]

Note: Experiments lasted approximately one month (except for the African catfish; 20 and 10 d
for experiment 1 and 2 to 4, respectively).

Numbers in the same row with common superscript are not significantly different. Average
dry weights (%) of the larvae samples were (before first feeding and at the end of the trial,
respectively): freshwater prawn: 15 and 20%, African catfish: 9 and 16%, European seabass: not
analyzed and 15%, turbot: 9 and 17%. (From Merchie, G., Lavens, P., and Sorgeloos, P.,
Aquaculture, 155, 165, 1997. With permission.)

According to Dabrowski,[79] the optimal dietary requirement may be that
which allows a steady-state AA concentration in larval and juvenile fish.
Therefore, unenriched *Artemia* (containing 0.56 mg g^{-1} AA) should have ful-
filled the requirement of the larvae assessed by Merchie et al. [70, 71, 72, 74, 83]—
since AA tissue concentration increased in all larvae fed on them. This
appeared to be the case, as additional enrichment with AP did not improve
the growth or survival of most fish larvae. The exception was tropical African
catfish, whose growth improved significantly in 3 out of 4 experiments[70, 72]
and was related to dietary AA over the range 0.1 to 2.6 mg AA g^{-1} diet.[70] This
difference may have been because the catfish grow faster, and eat more, than
the other species. Feed intake is possibly regulated by the amount of free radicals produced during digestion, and that high levels of AA, through efficient
scavenging of these radicals at the intestinal level, may prevent inhibition of
the feed intake process, allowing in this way a faster growth. This hypothesis
can be validated if ascorbyl phosphate is used instead of AP, the former compound being inert prior to uptake.

High levels of AA supplementation, however, had a positive effect on stress and/or disease resistance of the other larvae studied by Merchie et al.[70] Larvae of European sea bass,[72] African catfish,[72] turbot[83] and giant freshwater prawn[71] had greater stress resistance (salinity stress test) when fed diets supplemented with high AA concentrations. Also for turbot larvae, there was an apparent immunostimulation (less mortality after challenge with *Vibrio anguillarum*) and improvement in pigmentation from eating high-AA diets.[83] The enrichment of AA in zooplankton live feed—and its corresponding enhancement of AA in larval tissue up to a saturation point—may be valuable in conferring stress and disease resistance when culture conditions in hatcheries are suboptimal.[70]

Apart from stresses related to physical environment (e.g., handling, transportation, crowding, poor water quality), metamorphosis is a stressful, sensitive period where larvae undergo large morphological and biochemical changes, and may require additional AA.[70] In support of this, Merchie et al. reported[70] a significantly lower AA concentration in the juveniles of freshwater prawns than in larvae. These results and those of others[84] supports Dabrowski's hypothesis[86] that stress increases the AA requirement for larvae. Dabrowski suggests that body AA concentrations may be a useful indicator of physiological condition, and reflect potential for larval survival better than variation in growth rate.[86]

It is estimated that 5 to 20% of AA from non-enriched *Artemia* ingested by fish larvae is retained.[86] As adult fish apparently absorb 80 to 90% of dietary AA,[85] a low retention in larvae appears to be largely attributable to rapid catabolism during the fast-growing larval/juvenile stage.[86] However, degradation of AA in the digestive tract also contributes to differences between life stages and species in AA transfer. Declining AA concentrations during the development of fish larvae (e.g., roach *Rutilus rutilus*, whitefish *Coregonus lavaretus* and Arctic charr *Salvelinus alpinus*) corresponded to digestive tract physiology.[86] Loss of dietary AA was apparently highest in stomachless cyprinid fish where food was exposed to alkaline pH during its transit through the digestive tract.[86] Also, as AA absorption is a saturable process,[44] lower transfer efficiencies occur when higher dietary concentrations of AA (i.e., enriched *Artemia*) are presented to larvae.[70]

13.5 Interaction with other nutrients

The biological effects of nutrients are often studied in isolation, although they do not function as independent units but are related to, and interrelated with, other nutrients in terms of function and metabolism. Thus the requirement level of AA may be affected by the level of other nutrients in either the diet[87] or metabolically in the animal. More specifically, the role of AA as antioxidant by inactivating damaging free radicals produced through normal cellular activity and from various stressors may be influenced by other antioxidants such as vitamin E (α-tocopherol; α-T) and carotenoids (e.g., astaxanthin; ASX).[88] For example, several studies[89] demonstrated the ability of AA to reduce α-tocopheroxyl radicals and thereby regenerate them to α-T. It has been suggested that the antioxidant function of these micronutrients could interfere with the ingestion rate (see previous section, for catfish), the induction of several diseases (e.g., jaundice[90]), or could enhance the non-specific immunity by preserving the functional and structural integrity of immune cells. However, there are few studies in this field on aquaculture organisms, and almost none on larvae. The impact on larvae may be powerful, since high levels of oxidation-prone lipids (HUFA) are administered through the diet and the antioxidant enzyme system may not yet be fully developed.

Ruff et al.[91] recently investigated the interaction of AA, α-T, and HUFA in *Penaeus vannamei* postlarvae. No significant differences in survival, growth and anti-oxidative status could be detected for various combinations of the vitamins. Higher dietary levels of AA resulted in a significantly lower tissue concentration of α-T, which may be in conflict with the regeneration concept of α-T by AA. Merchie et al.[92] evaluated the interaction of AA and ASX on stress resistance of *Penaeus monodon* postlarvae. Three treatments were fed 230 mg ASX kg^{-1} diet combined with 100, 1700, or 3400 mg AA kg^{-1} diet, and two were fed 810 mg ASX kg^{-1} diet mixed with, respectively, 200 and 1700 mg AA kg^{-1} diet. For the high ASX treatments, the beneficial effect of extra dietary AA on stress resistance was not significant. This could be linked to the non-significant raise in AA incorporation in the shrimp tissues and may illustrate an impact on assimilation of both nutrients.

13.6   *Ecological significance of AA and its transfer between trophic levels*

This review has focused on the transfer of AA between trophic levels from an aquaculture viewpoint. It has been discussed that larvae may have a high requirement for AA, that their tissue concentrations can be rapidly depleted during environmental and nutritional stress, and that deficiencies make them more susceptible to mortality. From an ecological perspective, AA may have a regulatory role, so changes in its transfer in the food chain may cause significant changes in aquatic ecosystems.[86] Phytoplanktons are a rich source of AA[16] and a major contributor of AA in larval fish food chains, as zooplankton appear unable to synthesise this vitamin.[21] Hapette and co-workers[21, 67, 93] have found that concentrations of AA in natural zooplankton and fish larvae are related to their position within the phytoplankton–zooplankton–fish larvae food chain. They also speculate that the vitamin may have a role in inducing reproduction in copepods.[21] Dabrowski[86] suggested that AA could be monitored in aquatic systems to assess the quality of available food for larval fish and their nutritional status. He also proposed that AA concentrations may be a useful indicator of fish survival potential, and could be used as an early indicator of fish recruitment.[86]

13.7 Summary and conclusions

Non-enriched *Artemia* nauplii (0.5 to 0.7 mg AA g^{-1} dry weight) and rotifers reared on yeast (0.1 to 0.6 mg AA g^{-1} dry weight) apparently contained sufficient AA for the normal growth and survival of most fish and crustacean larvae. Nevertheless, zooplankton diets with enriched concentrations of AA (e.g., 1.5 to 2.5 mg AA g^{-1}) may improve the physiological condition of larvae, by enhancing their AA tissue concentrations. This may be particularly important in view of the likely protective role of AA in dealing with the stresses (sometimes unavoidable) in the hatchery environment. Moreover, in times of stress, the AA requirement of larvae would be increased because of an increased AA catabolism, although the direct evidence is missing. For some larvae at least, the extra cost of providing supplementary dietary AA would be more than offset by a more stable production output.

Enrichment of live zooplankton feeds in AA can now be done routinely. Enrichment with formulations containing AP is very effective, and has proved useful for assessing the effects of supplemental AA on larvae. However, such formulations would not be cost-effective commercially. Live microalgae and Protein Selco vit C-boosted® are both suitable for enriching AA in zooplankton to concentrations of ≥1 mg AA g^{-1}. Protein Selco vit C-boosted® may be a more attractive option, because it is an off-the-shelf product, and more cost-effective than continuous microalgal production. However, the cost-effectiveness of microalgae as a source of AA for enrichment might be significantly improved by (a) a better understanding of how AA production is regulated in microalgae so they can be grown under controlled conditions to maximise their AA, coupled with (b) adoption of new technology for mass cultivation of microalgae (photobioreactors, heterotrophic fermentors), which would reduce culture costs and make production more reliable. Enrichment of zooplankton with microalgae may have additional benefits. For some larvae, microalgae may provide increased concentrations of other trace nutrients (e.g., other vitamins) which may further improve growth and/or survival.

For aquaculture species feeding directly on microalgae, such as larvae and juvenile scallop or oysters, and early larval prawns, more research is required to assess their nutritional requirements for AA. Despite the differences in AA concentrations between microalgae (which can be greater than ten-fold), algal diets successfully used as direct feeds contain 1 to 3 mg AA g^{-1} dry weight; AA requirements are apparently met at these concentrations. Whether any additional benefit on stress resistance is conferred on these animals by feeding them high-AA diets (e.g., >1 mg AA g^{-1}) is not known. A reduced concentration and increased leaching rate of AA in processed microalgal feeds (pastes and powders) may contribute to their lower nutritional value. Compared to microalgae, yeast and bacteria appeared to be poor sources of AA (<0.3 mg g^{-1}). Their concentrations may be insufficient to meet dietary requirements and become a contributing factor to their low food value for bivalve molluscs.

References

1. Halver, J. E., Ashley, L. M., and Smith, R. R., Ascorbic acid requirements of coho salmon and rainbow trout, *Trans. Amer. Fish. Soc.*, 4, 762, 1969.
2. Conklin, D. E., Vitamins, in *Crustacean Nutrition: Advances in World Aquaculture,* D'Abramo, L. R., Conklin, D. E., and Akiyama, D. M., Eds., World Aquaculture Society, Baton Rouge, LA, 1997, 123.
3. Stickney, R. R., McGeachin, R. B., Lewis, D. H., Marks, J., Riggs, A., Sis, R. F., Robinson, E. H., and Wurts, W., Response of *Tilapia aurea* to dietary vitamin C., *J. World Maricult. Soc.*, 15, 179, 1984.
4. Kanazawa, A., Nutrition of penaeid prawns and shrimps, in *Proceedings of the First International Conference on the Culture of Penaeid Prawns/Shrimps*, Taki,Y., Primavera, J. H., and Llobrera, J. A., Eds., Iloilo City, Philippines, 123, 1985.
5. Dabrowski, K., Administration of gulonolactone does not evoke ascorbic acid synthesis in teleost fish, *Fish Physiol. Biochem.*, 9, 215, 1991.
6. Dabrowski, K. and Blom, H., Ascorbic acid deposition in rainbow trout (*Oncorhynchus mykiss*) eggs and survival of embryos, *Comp. Biochem. Physiol.*, 108A, 129, 1994.
7. Epifanio, C. E., Valenti, C. C., and Turk, C. L., A comparison of *Phaeodactylum tricornutum* and *Thalassiosira pseudonana* as foods for the oyster, *Crassostrea virginica, Aquaculture*, 23: 347, 1981.
8. Enright, C.T., Newkirk, G. F., Craigie, J. S. and Castell, J. D., Evaluation of phytoplankton as diets for juvenile *Ostrea edulis* L., *J. Exp. Mar. Biol. Ecol.*, 96,1, 1986.
9. Webb, K. L. and Chu, F. E., Phytoplankton as a food source for bivalve larvae, in *Proceedings of the 2nd International Conference of Aquaculture Nutrition, World Mariculture Society, Spec. Publ. No. 2,* Pruder, D. G., Langdon, C. J., and Conklin, D. E., Eds., Louisiana State University, Baton Rouge, 1982, 272.
10. Brown, M. R., Jeffrey, S. W., Volkman, J. K., and Dunstan, G. A., Nutritional properties of microalgae for mariculture, *Aquaculture,* 151, 315, 1997.
11. Whyte, J. N.C., Biochemical composition and energy content of six species of phytoplankton used in mariculture of bivalves, *Aquaculture*, 60, 231, 1987.
12. Thompson, P. A., Harrison, P. J., and Whyte, J. N. C., Influence of irradiance on the fatty acid composition of phytoplankton, *J. Phycol.*, 26, 278, 1990.
13. Brown, M. R., The amino-acid and sugar composition of 16 species of microalgae used in mariculture, *J. Exp. Mar. Biol. Ecol.*, 145, 79, 1991.
14. De Roeck-Holtzhauer, Y., Quere, I., and Claire, C., Vitamin analysis of five planktonic microalgae and one macroalga, *J. Appl. Phycol.*, 3, 259, 1991.
15. Langdon, C. J. and Waldock, M. J., The effect of algal and artificial diets on the growth and fatty acid composition of *Crassostrea gigas* spat, *J. Mar. Biol. Ass.*, *U.K.*, 61, 431, 1981.
16. Brown, M. R. and Miller, K. A., The ascorbic acid content of eleven species of microalgae used in mariculture, *J. Applied Phycol.*, 4, 205, 1999.
17. Seguineau, C, Laschi-Loquerie, A., Moal, J., and Samain, J. F., Vitamin requirements in great scallop larvae, *Aquacult. Int.*, 4, 315, 1996.
18. Bayanova, Y. I. and Trubachev, I. N., Comparative evaluation of the vitamin composition of some unicellular algae and higher plants grown under artificial conditions, *Appl. Biochem. Microbiol.*, 17, 292, 1981.
19. Aaronson, S., Dhawale, S. W., Patni, J., DeAngelis, B., Frank, O., and Baker, H., The cell content and secretion of water-soluble vitamins in several freshwater algae, *Arch. Microbiol.*, 112, 57,1977.

20. Running, J. A., Huss, R. J., and Olson, P. T., Heterotrophic production of ascorbic acid by microalgae, *J. Applied Phycology*, 6, 99, 1994.
21. Hapette, A. M. and Poulet, S. A., Variation of vitamin C in some common species of marine plankton, *Mar. Ecol. Prog. Ser.*, 64, 69, 1990.
22. Seguineau, C., Laschi-Loquerie, A., Leclercq, M., Samain, J. F., Moal, J., and Fayol, V., Vitamin transfer from algal diet to *Pecten maximum* larvae, *J. Mar. Biotechnol.*, 1, 67, 1993.
23. Hapette, A. M. and Poulet, S. A., Application of high-performance liquid chromatography to the determination of ascorbic acid in marine plankton, *J. Liquid Chromatogr.*, 13, 357, 1990.
24. Merchie, G., Lavens, P., Dhert, Ph., Dehasque, M., Nelis, H., De Leenheer, A., and Sorgeloos, P., Variation of ascorbic acid content in different live food organisms, *Aquaculture*, 134, 325, 1995.
25. Brown, M. R., Effects of storage and processing on the ascorbic acid content of concentrates prepared from *Chaetoceros calcitrans*, *J. Applied Phycology*, 7, 495, 1995.
26. Brown, M. R., Jeffrey, S. W., and Garland, C. D., Nutritional aspects of microalgae used in mariculture: a literature review, *CSIRO Mar. Lab. Rep.*, No. 205, Hobart, Tasmania, Australia, 1989.
27. Brown, M. R., Mular, M., Miller, I., Farmer, C., and Trenerry, C., The vitamin content of microalgae used in aquaculture, 11, 247, 1998.
28. Renstrøm, B., Grün, M., and Loewus, F. A., Biosynthesis of L-ascorbic acid in *Chlorella pyrenoidosa*, *Plant Sci. Lett.*, 28, 299, 1982/83.
29. Borowitzka, M., Vitamins and fine chemicals from microalgae, in *Micro-algal Biotechnology*, Borowitzka, M. A. and Borowitzka, L. J., Eds., Cambridge University Press, Cambridge, 1988, 153.
30. Robert, R. and Trintignac, P., Substitutes for live microalgae in mariculture: a review, *Aquatic Living Resour.*, 10, 315, 1997.
31. Knuckey, R., Australian Microalgae and Microalgal Concentrates for use as Aquaculture Feeds, Ph. D. thesis, University of Tasmania, Australia, 1998.
32. Brown, A., Jr., Experimental techniques for preserving diatoms used as food for larval *Penaeus aztecus*, *Proc. Natl. Shellfisheries Assn.*, 62, 21, 1972.
33. Lubzens, E, Gibson, O., Zmora, O., and Sukenik, A., Potential advantage of frozen algae (*Nannochloropsis* sp.) for rotifer (*Brachionus plicatilis*) culture, *Aquaculture*, 133, 295, 1995.
34. Venkataraman, L. V., Becker. H. E., and Shamala, T. R., Studies on the cultivation and utilisation of the alga *Scenedesmus acutus* as a single cell protein, *Life Sci.*, 20, 223, 1977.
35. Nell, J. A., MacLennan, D. G., Allan, G. L., Nearhos, S. P., and Frances, J., Evaluation of new microbial foods as partial substitutes for microalgae in a diet for Sydney rock oyster *Saccostrea commercialis* larvae and spat, in *New Microbial Foods for Aquaculture, Final Report to Fisheries Research Development Corporation (FRDC)*, Nell, J. A., MacLennan, D. G., Allan, G. L., Nearhos S. P., and Frances, J., Eds., NSW Fisheries, Brackish Water Fish Culture Research Station, Salamander Bay, New South Wales, 1994.
36. Barclay, W. and Zeller, S., Nutritional enhancement of n-3 and n-6 fatty acids in rotifers and *Artemia* by feeding spray-dried *Schizochytrium* sp., *J. World Aquacult. Soc.*, 27, 314, 1996.
37. Nichols, D. S., Hart, P., Nichols, P. D., and McMeekin, T. A., Enrichment of the rotifer *Brachionus plicatilis* fed an Antarctic bacterium containing polyunsaturated fatty acids, *Aquaculture*, 147, 115, 1996.

38. Glaudue, R. M. and Maxey, J. E., Microalgal feeds for aquaculture, *J. Applied Phycol.*, 6, 131, 1994.
39. Brown, M. R., Skabo, S., and Wilkinson, B., The enrichment and retention of ascorbic acid in rotifers fed microalgal diets, *Aquaculture Nutrition*, 4, 151, 1998.
40. Kalashyan, A. T. and Magakyan, A. T., Relationship between vitamin-synthesizing properties and acid production in thermophilic lactic acid bacteria, *Molochnaya Promyshlennost.*, 7, 18, 1975.
41. Brown, M. R. and Lewis, T. E., unpublished data, 1998.
42. Mannella, C. A., Frank, J., and Delihas, N., Interrelatedness of 5S RNA sequences investigated by correspondence analysis, *J. Mol. Evol.*, 24, 375, 1987.
43. Bio-Marine Inc. Aquafauna—Product information sheet for nutritional profile of AlgaMac—2000.
44. Gaby, S. K., Bendich, A., Singh, V. N., and Machlin, L. J., *Vitamin Intake and Health*, Marcel Dekker, New York, 1991.
45. Langdon, C. J. and Bolton, E. T., A microparticulate diet for a suspension-feeding bivalve mollusc, *Crassostrea virginica* (Gmelin), *J. Exp. Mar. Biol. Ecol.*, 82, 239, 1984.
46. Langdon, C. J. and Siegfried, C. A., Progress in the development of artificial diets for bivalve filter feeders, *Aquaculture*, 39, 135, 1984.
47. Ferreiro, M. J., Pérez-Camacho, A., Labarta, U., Beiras, R., Planas, M., and Férnandez-Reiriz, M. J., Changes in the biochemical composition of *Ostrea edulis* larvae fed on different food regimes, *Marine Biology*, 106, 395, 1990.
48. Thompson, P. A., Guo, M-X., and Harrison, P. J., Nutritional value of diets that vary in fatty acid composition for larval Pacific oysters (*Crassostrea gigas*), *Aquaculture*, 143, 379, 1996.
49. Epifanio, C. E., Comparison of yeast and algal diets for bivalve molluscs, *Aquaculture*, 16, 187, 1979.
50. Crosby, M. P., Newell, R. I. E., and Langdon, C. J., Bacterial mediation in the utilisation of carbon and nitrogen from detrital complexes by *Crassostrea virginica*, *Limnol. Oceanogr.*, 35, 625, 1990.
51. Brown, M. R., Barrett, S. M., Volkman, J. K., Nearhos, S. P., Nell, J. A., and Allan, G. L., Biochemical composition of new yeasts and bacteria evaluated as food for bivalve aquaculture, *Aquaculture*, 143, 341, 1996.
52. Jones, D. A., Yule, A. B., and Holland, D. L., Larval nutrition, in *Crustacean Nutrition: Advances in World Aquaculture*, D'Abramo, L. R., Conklin, D. E., and Akiyama, D. M., Eds., World Aquaculture Society, Baton Rouge, LA, 1997, 353.
53. Wilkenfeld, J. S., Lawrence, A. L., and Kuban, F. K., Survival, metamorphosis and growth of penaeid shrimp larvae on a variety of algal and animals foods, *J. World Aquacult. Soc.*, 15, 31, 1984.
54. Lubzens, E., Raising rotifers for use in aquaculture, *Hydrobiologia*, 147, 245, 1987.
55. Nagata, W. D. and Whyte, J. N. C., Effects of yeast and algal diets on the growth and biochemical composition of the rotifer *Brachionus plicatilis* (Muller) in culture, *Aquaculture and Fisheries Management*, 23, 13, 1992.
56. Lavens, P., De Meulemeester, A., and Sorgeloos, P., Evaluation of mono- and mixed diets as food for intensive *Artemia* culture, in *Artemia Research and its Application, Vol. 3. Ecology, Culturing, Use in Aquaculture*, Sorgeloos, P., Begtson, D.A., Decleir, W., and Jaspers, E., Eds., Universa Press, Wetteren, Belgium, 1987, 309.
57. Dabrowski, K., Some aspects of ascorbate metabolism in developing embryos of the brine shrimp (*Artemia salina*), *Can. J. Aquat. Sci.*, 48, 1905, 1991.

58. Sandnes, K., Lie, Ø., Haaland, H., and Olsen, Y., Vitamin content of the rotifer *Brachionus plicatilis, Fisk. Dir. Skr. Ernæring*, 6, 117, 1994.
59. Whyte, J. N. C. and Nagata, W. D., Carbohydrate and fatty acid composition of the rotifer, *Brachionus plicatilis*, fed monospecific diets of yeast or phytoplankton, *Aquaculture*, 89, 263, 1990.
60. Frolov, A. V., Pankov, S. L., Geradze, S. A., Pankova, S. A., and Spektorova, L. V., Influence of the biochemical composition of food on the biochemical composition of the rotifer *Brachionus plicatilis, Aquaculture*, 97, 181, 1991.
61. Dhont, J. and Lavens, P., Tank production and use of ongrown *Artemia*, in *Manual on the Production and Use of Live Food for Aquaculture*, Lavens, P. and Sorgeloos., P. Eds., FAO Fisheries Technical Paper No. 361, Rome, Italy, 1996, 161.
62. Fukusho, K., Arakawa, T., and Watanabe, T., Food value of a copepod, *Tigriopus japonicus*, cultured with ω-yeast for larvae and juveniles of mud dab *Limanda yokohamae, Bull. Jap. Soc. Scient. Fish.*, 46, 625, 1980.
63. Watanabe, T., Kitajima, C., and Fujita, S., Nutritional values of live organisms used in Japan for mass propagation of fish: a review, *Aquaculture*, 34: 115, 1983.
64. Léger, P., Bengtson, D. A., Simpson, K. L., and Sorgeloos, P., The use and nutritional value of *Artemia* as a food source, *Oceanogr. Mar. Biol. Ann. Rev.*, 24, 521, 1986.
65. Nichols, P. D., Holdsworth, D. G., Volkman, J. K., Daintith, M., and Allanson, S., High incorporation of essential fatty acids by the rotifer *Brachionus plicatilis* fed on the prymnesiophyte alga *Pavlova lutheri, Aust. J. Mar. Freshwater Res.*, 40, 645, 1989.
66. Dabrowski, K. and Rusiecki, M., Content of total and free amino acids in zooplanktonic food of fish larvae, *Aquaculture*, 30, 31, 1983.
67. Poulet, S. A., Hapette, A. M., Cole, R. B., and Tabet, J. C., Vitamin C in marine copepods, *Limnol. Oceanogr.* 34, 1325, 1989.
68. Lie, Ø., Haaland, H., Hemre, G.-I., Maage, A., Lied, E., Rosenlund, G., Sandnes, K., and Olsen, Y., Nutritional composition of rotifers following a change in diet from yeast and emulsified oil to microalgae, *Aquaculture International*, 5, 427, 1997.
69. Juario, J. V. and Storch, V., Biological evaluation of phytoplankton (*Chlorella* sp., *Tetraselmis* sp., and *Isochrysis galbana*) as food for milkfish (*Chanos chanos*) fry, *Aquaculture*, 40, 193, 1984.
70. Merchie, G., Lavens, P., and Sorgeloos, P., Optimization of dietary vitamin C in fish and crustacean larvae: a review, *Aquaculture*, 155, 165, 1997.
71. Merchie, G., Lavens, P., Radull, J., Nelis, H. Ollevier, F., De Leenheer, A., and Sorgeloos, P., Evaluation of vitamin C-enriched *Artemia nauplii* for larvae of the giant freshwater prawn, *Aquacult. Int.*, 3, 355, 1995.
72. Merchie, G., Lavens, P., Dhert, Ph., Pector, R., Mai Soni, A. F., Nelis, H. Ollevier, F., De Leenheer, A., and Sorgeloos, P., Live food mediated vitamin C transfer to *Dicentrarchus labrax* and *Clarias gariepinus, J. Appl. Ichthyol.*, 11, 336, 1995.
73. Tacon, A. G. J., Vitamin nutrition in shrimp and fish, in, *Proceedings of the Aquaculture Feed Processing and Nutrition Workshop, Thailand and Indonesia, September 1991*, Akiyama, D. M. and Tan, R. K. H., Eds., American Soybean Association, Singapore, 1991, 10.
74. Merchie, G., Lavens, P., Storch, V., Übel, U., Nelis, H., De Leenheer, A., and Sorgeloos, P., Influence of dietary vitamin C dosage on turbot (*Scophthalmus maximus*) and European sea bass (*Dicentrarchus labrax*) nursery stages, *Comp. Biochem. Physiol.* 114, 123, 1996.

75. Matusiewicz, M., Dabrowski, K., Volker, L., and Matusiewicz, K., Regulation of saturation and depletion of ascorbic acid in rainbow trout, *J. Nutr. Biochem.*, 5, 204, 1994.
76. Dabrowski, K., Moreau, R., and El-Saidy, D., Ontogenetic sensitivity of channel catfish to ascorbic acid deficiency, *J. Aquatic Animal Health*, 8, 22, 1996.
77. Lavens P., Lebegue, E., Jaunet, H., Brunel, A., Dhert, Ph., and Sorgeloos P., Effect of dietary essential fatty acids and vitamins on egg quality in turbot (*Scophthalmus maximus* L.) broodstocks. Submitted to *J. Exp. Mar. Biol. Ecol.*
78. Plañas, M., Carnero, D. G., Munilla, R., Merchie, G., and Lavens, P., Enrichment in ascorbic acid of the rotifer *Brachionus plicatilis* O. F. Müller for the rearing of marine fish larvae, in *Proceedings Fifth Spanish Aquacultural Congress, May 1995*, Castelló, F., Orvay, I., Calderer, A., and Reig, I., Eds., University of Barcelona, 1995, 137.
79. Dabrowski, K., Ascorbic acid status in the early life of whitefish (*Coregonus lavaretus* L.), *Aquaculture*, 84, 61, 1990.
80. Dabrowski, K., Hinterleitner, S., Sturmbauer, C., El-Fiky, N., and Weisner, W., Do carp larvae require vitamin C?, *Aquaculture*, 72, 295, 1988.
81. Dabrowski, K., Segner, H., Dallinger, R., Hinterleitner, S., Sturmbauer, C., and Wieser, W., Rearing of roach larvae; The vitamin C, minerals interrelationship and nutrition-related histology of the liver and intestine, *J. Animal Physiology and Animal Nutrition*, 62, 188, 1989.
82. Levine, M., New concepts in the biology and biochemistry of ascorbic acid, *New England J. Med.*, 314, 892, 1986.
83. Merchie, G., Lavens, P., Dhert, Ph., García Ulloa Gómez, M., Nelis, H., De Leenheer, A., and Sorgeloos, P., Dietary ascorbic acid requirements during the hatchery production of turbot larvae, *J. Fish Biol.*, 49, 573, 1996.
84. Ishibashi, Y., Kato, K., Ikeda, S., Murata, O., Nasu, T., and Kumai, H., Effect of dietary ascorbic acid on tolerance to intermittent hypoxic stress in Japanese parrot fish, *Nippon Suisan Gakkaishi*, 58, 2147, 1992.
85. Dabrowski, K. and Köck, G., Absorption and interaction with minerals of ascorbic acid and ascorbic sulphate in digestive tract of rainbow trout, *Canadian J. Fish. Aquat. Sci.*, 46, 1952, 1989.
86. Dabrowksi, K., Ascorbate concentration in fish ontogeny, *J. Fish Biol.*, 40, 273, 1992.
87. Hilton, J. W., The interaction of vitamins, minerals and diet composition in the diet of fish, *Aquaculture*, 79, 223, 1989.
88. Chew, B.P., Antioxidant vitamins affect food animal immunity and health, *J. Nutr.*, 124, 2033, 1995.
89. Lambelet P., Saucy F., and Löliger, J., Radical exchange reactions between vitamin E, vitamin C and phospholipids in autoxidizing polyunsaturated lipids, *Free Rad. Res.*, 20, 1, 1994.
90. Sakai, T., Murata, H., Endo, M., Yamauchi, K., Tabata, N., and Fukudome, M., 2-Thiobarbituric acid values and contents of α-tocopherol and bile pigments in the liver and muscle of jaundiced yellowtail, *Seriola aquiqueradiata*, *Agric. Biol. Chem.*, 53, 1739, 1989.
91. Ruff, N., Lavens, P., Huo, J.-Z., Sorgeloos, P., Nelis, H.J., and De Leenheer, A. Antioxidative effect of dietary tocopherol and ascorbic acid on production performance of *Penaeus vannamei* postlarvae. *Aquaculture International*, in press.

92. Merchie, G., Kontara, E., Lavens, P., Robles, R., Kurmaly, K., and Sorgeloos, P., Effect of vitamin C and astaxanthin on stress and disease resistance of postlarval tiger shrimp, *Penaeus monodon* (Fabricius), *Aquaculture Res.*, 29, 579, 1998.
93. Hapette, A. M., Coombs, S., Williams, R., and Poulet, S. A., Variation in vitamin C content of sprat larvae (*Sprattus sprattus*) in the Irish Sea, *Marine Biol.*, 108, 39, 1991.

chapter fourteen

Live food-mediated vitamin C transfer in sea bass (Dicentrarchus labrax, *L.*) during first feeding

Genciana Terova, Stefano Cecchini, Marco Saroglia,
Gaetano Caricato, and Zsigmond Jeney

Contents

14.1   Introduction

The vitamin C requirements of many of the species raised in the grow-out stage are now well known and vary from species to species, but also depend on the quality of the environment and the physiological state of the fish. By contrast, there is only a limited number of reports on the physiological needs of the very earliest stages of a fish's life, from the hatching embryo to the complete weaning with artificial feed. Signs of deficiency, such as serious spinal deformities and consequent retardation of growth, have also been reported in young fish.[1, 2]

In the larval stage, any dietary deficiency manifests itself in an accelerated fashion and becomes more severe than that in adult fish due to the much lower initial weight, the consequent lack of reserves, and the rapid growth.[3] This is also why symptoms of vitamin C deficiency become apparent more rapidly in larvae than in adult fish, and is also probably due to the higher vitamin C requirements during this critical stage of development.

In addition to its anti-scurvy activity, ascorbic acid is also one of the essential micronutrients necessary to reduce the negative impacts of stressing agents on the health of the fish and on their resistance to disease.[4, 5] This is why levels of vitamin C that are necessary for the growth of animals are lower than the doses necessary to overcome the damage caused by stress factors, and the former are often quoted in the literature. For example, it has been shown that the supplement of 30 mg of vitamin C per kg of feed is enough to guarantee normal growth and avoid signs of deficiency in catfish (*Ictalurus punctatus*), but doses five times higher are necessary to increase their resistance to *Edwardsiella tarda*.[6] Fish raised in intensive systems may be exposed to a stressful setting which causes an immune-depressive state. Diets lacking in vitamin C increase susceptibility to stress in fish. Scarano et al.[7] showed the anti-stress effects of vitamin C with respect to nitrites. The protective action against nitrites in sea bass takes effect with an addition of not less than 400 mg L-ascorbic acid (AA) /kg of feed, compared to a growth requirement of about 200 mg/kg of feed.

Vitamin C also seemed to play a protective role in respect to other common stresses produced by the intensive rearing environment. Studies on the effect of different dietary supplements of vitamin C in young gilthead sea bream suffering from oxygen depletion and thermal stress showed that a supplement of 900 mg AA/kg improved their physiological performance.[8]

One of the problems of supplementation of vitamin C in fish diets is that ascorbic acid is not stable, and so the fish may suffer from a lack of vitamin C even when it has been correctly added to the diet in sufficient quantities prior to processing. Among the more stable biologically active derivatives of ascorbic acid are ascorbate-2-monosulphate (AMS), ascorbate-2-monophosphate (AMP), ascorbate-2-polyphosphate (APP), and ascorbyl-6-palmitate (AP). The last mentioned, whose biological activity has been demonstrated in various studies, is less stable than the previously mentioned derivatives but is suitable as a vitamin supplement in live prey, brine shrimp (*Artemia salina*) and can be used during the rearing stage of larvae. By contrast, it is difficult or impossible for the fish to hydrolyze AMS.[9]

In salmonids, from the time of reabsorption of the yolk sac, an artificial diet can be used as the first feed. This is not possible for sea fish until the larvae develop a sufficiently complete enzymatic system.[10] There are currently many studies which show that it was not possible to abandon the zooplanktonic food as the first source of food. Later on, artificial feed can be integrated into the diet, depending on the species, the temperature of rearing, and technology of grow-out.

Among the studies carried out to show the vitamin C requirements in fish larvae which are still fed on live prey, the research of Merchie et al.[11] on sea bass and African catfish (*Clarias gariepinus*) is worth mentioning. The authors used rotifer, *Brachionus plicatilis* and brine shrimp nauplii as live food for the sea bass, brine shrimp alone for the catfish and AMP as the source of vitamin C. AMP was shown to be assimilated and metabolized by the zooplanktonic organisms, which convert it into a biologically active form.[12] The authors ascertained that the vitamin requirements are species-specific. While the vitamin concentrations in brine shrimp are enough in the case of sea bass, in African catfish there was an improvement in production performance (greater survival, larval length, and larval dry weight) proportional to the level of supplementation of vitamin C. Similar results were obtained when the larvae were subjected to salinity stress using the methods suggested by Dhert et al.[13] Survival after stress in catfish depended on vitamin C, while in sea bass there was not always a significant improvement in survival. The authors were persistent to conclude that while concentrations of vitamin C in brine shrimp (581 mg/g dry weight) were sufficient for the growth and resistance to stress in sea bass, in African catfish this was not the case. In African catfish this vitamin level was completely insufficient, and an improved performance was only obtained with vitamin levels four times greater.

It can be deduced that the vitamin concentrations in live prey rotifer and brine shrimp, and thus in the fish larvae fed on them, can increase depending on the strains used and, in the case of the rotifer, also on the strain of algae used to raise them.[11] Additionally, the chemical form of vitamin C in brine shrimp varies according to the life stage of the nauplii.[14] At the cyst stage, ascorbate is indeed present in biologically inactive sulphate form. Once the cysts are incubated in salt water at a suitable temperature, the ascorbyl-sulphate is hydrolyzed and becomes biologically available.

To verify if the regular live diet for the raising of the sea bass larvae contains enough ascorbate and if the larval performances could be improved by a further vitamin supplement, sea bass larvae were fed with an enriched brine shrimp, and production and physiological aspects of fish performances were monitored.

14.2   Materials and methods

A batch of 170,000 one-day-old sea bass embryos from the commercial hatchery of Nuova Acquazzurra (Civitavecchia, Roma) was disseminated in six cylindrical–conical fibreglass tanks with 28,000 in each at a density of c.100 embryo per liter in the Aquaculture Laboratory of the University of Basilicata. The tanks were connected to a single water recirculation and treatment plant. The eggs, kept at a temperature of 14°C ($\pm$0.5) and a salinity of 35 g/l, hatched about 72 hours after fertilization. Taking into consideration an embryo mortality rate of c. 15%, the density at hatching was about 85 larvae/l. The temperature was then maintained at 15–17°C for 30 d, after which it was raised progressively and reached the level of 20°C in the last period of the experiment. Larval feeding began on the eighth day after hatching with brine shrimp nauplii of the Bass entrée® type (INVE, Belgium). The size of hatched nauplii was 410 μm.

The first groups of sea bass larvae in three tanks were fed nauplii enriched with ascorbate palmitate (AP) with a fine emulsion in Super Selco® (S.S.) (INVE), the source of long chain polyunsaturated fatty acids (PUFA and HUFA), while the second group of three tanks received the same feed but enriched with S.S. as is the custom in commercial Mediterranean hatcheries. Until hatching and in the following days of the experiment, the temperature of the water was measured continuously to follow the development of the larvae based on the received heat (degree/day). Using the same feeding plan, from the 11th to the 24th day of life, the larvae were fed with brine shrimp nauplii AF480 (INVE), 480 μm in size. This was gradually replaced from the 24th day with *Artemia* EG® (also from INVE), of 600–700 μm in size, administered in the metanauplii stage after enrichment for 12 hours with Super Selco.®

Twenty days after hatching, the sea bass larvae were transferred into other tanks of different shapes and volumes. After the expected loss of fish occurring during the first 20 days and at the first transfer, the larvae in each treatment were reunited in a single cubic tank with a volume of 500 litres. The density of rearing in these tanks was 50 larvae per litre. In this second rearing phase, a quantity of feed greater than that suggested in the rearing plan was provided to allow the larvae to feed ad libitum. While the Bass Entrée® and AF 480® were distributed throughout all the water at feeding times, the

EG® was distributed to the larvae at precise points in the tanks by means of a plastic tube connected by a siphon to the appropriate 20 liter drum containing the brine shrimp metanauplii. The flow of the siphon was controlled and a continuous supply of metanauplii was provided.

The weaning period coincided with the administration of commercial scattered pelleted feed (Perla Marine,® Hendrix, Italy), and began on the 41st day after hatching with the distribution to the post-larvae of 6.0 type feed (<200 μm ∅). This was followed from the 48th day with 5.0 type (200–300 μm ∅). During the first days of weaning, the feed for the post-larvae was also accompanied by the distribution of brine shrimp, which was, however, progressively reduced. The weaning period finished on the 61st day after hatching. The post-larvae, or juveniles, were fed exclusively with Perla Marine type 4.0® (300–500 μm ∅), a commercial dry feed. Samples of the larvae, post-larvae, and juveniles fed with enriched *Artemia* in S.S.,® with or without the addition of AP, were taken 20, 30, 40, and 50 d after hatching. The total ascorbate and the presence of the collagen amino-acid indicators, hydroxyproline, and hydroxylysine and the relative relationship with the non-hydroxylated forms were assessed.

Samples of the fish were also taken for both the experimental group and the control group at 25, 35, 45, and 55 d after hatching to monitor their length, at fresh and dry weight.

In addition, to assess the anti-stress and immuno-stimulating effect of vitamin C at 40 and 50 d after hatching, samples from the two treatments were subjected to saline stress and infection tests.

14.2.1   Enrichment of brine shrimp

Different types of enrichment were adopted for the various types of *Artemia*. After unsatisfactory results with other forms of the vitamin, ascorbyl-6-palmitate was used in all experiments.

14.2.1.1   Enrichment of Artemia nauplii with Bass Entrée® and AF®
The *Artemia* cysts, after being disinfected and decapsulated, were immersed in two separate tanks for incubation and hatching. The nauplii hatched in the first tank were enriched with 120 mg/l of ascorbyl palmitate (AP) or 600 mg/l of S.S., a commercial product with a polyunsaturated fatty acid base (PUFA) with 400 mg/kg of ascorbic acid. The second tank nauplii were only enriched with 600 mg/l of S.S. In both cases, the enrichment took place 12 h before the distribution of the nauplii to the larvae. This means that as soon as the nauplii hatched they came into contact with the enrichment. The enriched nauplii were collected with a suitable plankton net and thoroughly rinsed in clean cold seawater. Before distribution to the two treatments of larvae, a sample of the enriched nauplii was weighed and conserved in a freezer at a temperature of −80°C to determine the concentration of vitamin C.

14.2.2   Enrichment of the Artemia EG® metanauplii

Artemia EG® nauplii, hatched at 25°C and at a salinity of 25 g/l, were placed at concentrations of 300,000/l into two tanks with water of 25 ppt salinity (25°C) and well aerated. One group of nauplii was enriched with 600 mg/l S.S. and 120 mg/l ascorbyl palmitate (AP), while a second group was enriched with S.S. alone. The enrichment was of 12 h duration before feeding the metanauplii to the larvae. The enriched metanauplii were collected with a plankton net and thoroughly rinsed in debacterized cold seawater. After rinsing a sample of the enriched metanauplii, it was weighed and frozen at −80°C to determine vitamin C.

14.2.2   Analysis of the vitamin C

To test for the AA concentrations of the artificial feed, three types of dry feed 6.0, 5.0, and 4.0 of Hendrix® (Trout Perla Marine) were analyzed.

The fish samples, taken at 10 d intervals, were immediately frozen on dry ice and stored at −80°C. The total ascorbate content was analyzed by HPLC, an inverse phase, with ionic couple, isocratic elution, and the use of UV or DAD (diode array) detectors and a single column C18.[5]

14.2.3   Determination of amino acids in the larvae

The method used was the Pico Tag® developed by Waters Co.®,[16] which employs phenylisothiocyanate (PITC), to derivatize amino acids with a simple one-step reaction. All reagents were then easily removed via vacuum and the resulting phenylthiocarbamyl amino acids (PTC-AA) rapidly separated on a highly efficient reverse phase column (Waters Pico•Tag™ column for free amino acids 3.9 × 30 cm). For hydrolysis, proteins or peptides were subjected to 6N HCl vapor in the presence of 0.5% phenol for 1 h at 150°C. Standards for collagen hydrolyzate, reagents, and amino acid analysis column were purchased from Waters Co.® External standards were added prior to hydrolysis to calculate the molar ratios of the amino acids of interest. Analyses were run in triplicate.

14.2.4   Larval biometry

14.2.4.1   Larvae sampling

The larvae of each feed treatment were sampled with a bucket lowered directly into the rearing tanks to reduce the stress to the larvae and to randomize. Post-larvae were sampled using a plankton net with 500 μm mesh.

The post-larvae were anesthetised by immersion in 100 mg/l solution of Finquell MS 222 (Sandoz). Once under deep sedation, the post-larvae were placed on a suitable slide designed to measure the total length of the body. Immediately after measuring the length, 50 larvae were blotted and transferred to a plastic basket to determine their wet weight, using a Sartorius® model BP 210 analytical balance. The group of 50 larvae previously sampled for length and wet weight was also used to determine the dry weight. The larvae were weighed in 5 sub-samples of 10 each. The samples were kept in an oven at 105°C for 24 h and then transferred to a silica gel dryer until cold.

14.2.5   Qualitative assessment test on the larvae

After 40 and 50 d groups of 10 post-larvae were taken from each experimental tank and subjected to saline stress tests.

The test consisted of evaluating the survival of the post-larvae when moving from 35 g/l to 42 g/l: from 35 to 50 g/l (as hyper-osmotic stress), and then from 35 to 1 g/l, from 35 to zero in deionized water (hypo-osmotic stress). Water from the recycling filtration plant, salted or diluted according to the level of salinity required, was used for the osmotic stress test, which was conducted three times in containers of 1 liter. The refractometer EUROMEX® mod. RF360 was used to measure the salinity. The cumulative mortality rates were measured at 15, 30, 45, 60, and 90 min.

14.2.5.1   Response to the infective stress test

A strain of *Vibrio anguillarum* "serotype 1," isolated from brown trout (*Salmo trutta*) in a fish farm near Centallo (CN, Italy) and supplied by the Zooprophylactic Institute of Piedmont and Liguria of Turin, was used for the infective test, carried out three times. The bacterial cultures were kept alive in BHIA (Brain Heart Infusion Agar) with 2% NaCl added, at an incubation temperature of 24°C (±1°C). The bacteria were collected in sterile conditions with swabs and then diluted in physiologically sterile solution. The bacterial concentration in the original culture, expressed in column forming units per ml (CFU/ml), was evaluated on the basis of the regression curve (absorption at 600 nm by a serial dilution of an original culture at known concentrations). The infective doses were then obtained by successive dilutions.

Three random samples of 10 juveniles were taken from each experimental group at 40 and 50 d. The animals were placed in 500 ml of infective solution for 15 min. The following infective doses were used: 101, 102, and 103 CSU/ml at 40 d and 101, 102, 103, 104, 105, and 106 CSU/ml at 50 d. After the infective bath, the animals were carefully collected and transferred to separate 1 liter containers containing fresh marine water. Cumulative mortality was monitored 1, 2, 3, 6, 12, and 24 h after infection.

14.2.5.2   Statistical analysis

All the data were presented as averages (d.s.) and were subjected to one-way variance analysis (ANOVA) ($P < 0.01$ and $P < 0.05$).

Table 14.1　Storage of vitamin C in *Artemia salina* metanauplii after different enrichment protocols (mean ± S.D.)

Sample	AA (μg/g wet weight)	DHA (μg/g wet weight)	TAA (μg/g wet weight)
Artemia EG® + S.S.® (600 mg/l) + AP (120 mg/l)	199.98 (5.57)	27.84 (18.61)	226.82 (18.11)
Artemia EG + S.S. (600 mg/l) + AP (120 mg/l) + AA (120 mg/l)	171.12 (67.24)	57.30 (42.16)	228.42 (41.05)
Artemia EG + S.S. (600 mg/l)	8.38 (6.82)	17.72 (15.29)	26.1 (9.25)
Artemia EG + S.S. (600 mg/l) + AA (120 mg/l)	19.81 (9.15)	22.34 (11.24)	38.93 (21.08)
Artemia BE®	2.29 (1.58)	39.72 (19.04)	41.91 (13.97)
Artemia AF®	16.01 (8.68)	25.13 (7.92)	38.73 (16.92)

AA = ascorbic acid; AP = ascorbyl-6-palmitate; S.S. = Super Selco®; TAA = total ascorbate; DHA dehydroascorbic acid; *Artemia* EG, BE, AF = different strains of *Artemia salina* from INVE.

14.3　Results

Table 14.1 shows the total ascorbic acid concentration (TAA) of brine shrimp nauplii enriched in S.S. and in AP and those enriched only in S.S. The vitamin C concentration in the artificial rearing feed utilised was 197.5 (2.8) μg/g.

Table 14.2 and Figure 14.1 show concentrations of TAA in the larvae subjected to two different feeding regimes. In all the stages monitored the difference in TAA per wet weight between the groups of larvae fed with

Table 14.2　Storage of vitamin C in sea bass (*Dicentrarchus labrax*, L) larvae, after 2 different feeding regimes (mean + S.D.)

Sample	Days from hatching	AA (μg/g wet weight)	DHA (μg/g wet weight)	TAA (μg/g wet weight)
Larvae fed	10	0	11.7 (1.1)	11.7 (1.1)
Artemia salina enriched	20	0	21.1 (1.8)	21.1 (1.8)
in S.S.® only	30	0	33.7 (1.0)	33.7 (1.0)
	40	27.9 (4.5)	16.8 (3.6)	46.3 (1.0)
	50	34.9 (1.3)	40.3 (1.0)	75.2 (0.6)
Larvae fed	10	2.2 (0.1)	21.9 (1.1)	24.1 (0.9)
Artemia salina enriched	20	22.1 (3.1)	18.9 (0.4)	42.6 (0.9)
in S.S. + AP	30	29.9 (3.5)	11.0 (1.4)	42.4 (2.0)
	40	42.6 (8.3)	16.8 (10.4)	61.1 (2.2)
	50	67.9 (4.3)	23.0 (6.3)	90.9 (2.7)

Note:　Significant differences between vitamin C supplemented and unsupplemented groups were found for every larval age. See Figure 14.1 for P values. S.S. = Super Selco®; AP = ascorbyl-6-palmitate; AA = L-ascorbic acid; TAA = total ascorbate; DHA = dehydroascorbic acid.

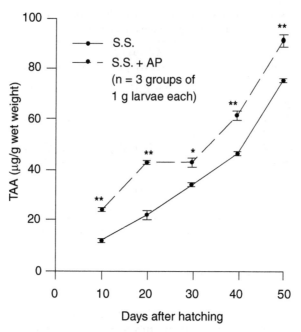

(*) (**) Significant differences between the two groups
at P < 0.05 and P < 0.01

Figure 14.1   Vitamin C concentrations in sea bass larvae fed brine shrimp enriched in Super Selco (S.S.), or in S.S. + ascorbyl-6-palmitate (AP).

Artemia nauplii enriched with S.S. and ascorbyl palmitate and the groups fed with *Artemia* nauplii enriched with S.S. alone were significant (P < 0.05, P < 0.01). The value of hydroxyproline (HYP %) in the group of sea bass larvae fed with brine shrimp supplemented with S.S. + AP (Table 14.3) was higher than values of the group fed with *Artemia salina* supplemented with S.S. alone.

Figure 14.1 shows the trend of the HYP/PRO (%) ratios in sea bass larvae which received two diets. The values for the larvae fed with *Artemia* enriched with S.S. + AP were significantly higher for all ages than for those fed on *Artemia* enriched only with S.S.

14.3.1   Larval biometry

For the group of larvae fed with both the enrichments, at 36, 46, and 56 d after hatching the average dry weight differences between the two groups of larvae were significant (P < 0.01) at all stages analyzed (Figure 14.2).

Table 14.3 Pattern of collagen indicating amino acids in sea bass larvae fed *Artemia* enriched or not in ascorbyl-palmitate (AP)

| | Amino acid | Days after hatching | | | | | | | | | |
| | | 10 | | 20 | | 30 | | 40 | | 50 | |
		M(%)	± S.D.(%)	M(%)	± S.D.(%)	M(%)	± S.D.(%)	M(%)	± S.D.(%)	M(%)	± S.D.(%)
Larvae fed *Artemia* not enriched in AP	Hydroxyproline	0.59	± 0.08b	1.00	± 0.05d	1.89	± 0.58e	5.21	± 0.12c	5.38	± 0.22f
	Proline	3.86	± 0.06	3.67	± 0.07	5.91	± 1.74	7.96	± 0.16	6.39	± 0.27
	Hydroxyproline /Proline Ratio	0.15	± 0.02 a	0.27	± 0.01c	0.32	± 0.01d	0.65	± 0.03a	0.84	± 0.04e
Larvae fed *Artemia* enriched in AP	Hydroxyproline	1.06	± 0.09c	2.83	± 0.23g	3.58	± 0.48d	4.26	± 0.05e	3.47	± 0.10b
	Proline	4.88	± 0.75	8.65	± 0.47	6.53	± 0.75	5.45	± 0.12	3.69	± 0.05
	Hydroxyproline /Proline Ratio	0.22	± 0.02 b	0.33	± 0.01f	0.55	± 0.01e	0.78	± 0.03c	0.94	± 0.02f

Note: For the same amino acid and for the ratio, values along a column with different letters are significantly different ($P < 0.01$ and $P < 0.05$).

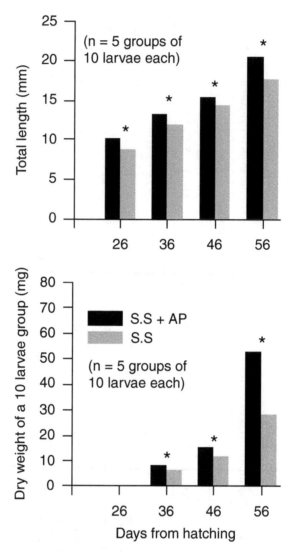

Figure 14.2 Growth of sea bass (*Dicentrarchus labrax* L.) larvae fed brine shrimp nauplii enriched with Super Selco (S.S.), or with S.S. + ascorbyl-6-palmitate (AP).

14.3.2   Osmotic stress

14.3.2.1   Tests at 40 days of life

In hyperosmotic conditions at 50 g/l (Figure 14.3 b), after 15 min of the test the mortality for the treatment fed vitamin supplements was 66.7%, whereas the group without vitamin supplements was 90% (10.0). The difference between the two treatments was not significant. At 30 min of the test, the cumulative mortality was 100%. In hyperosmotic conditions at 42 g/l

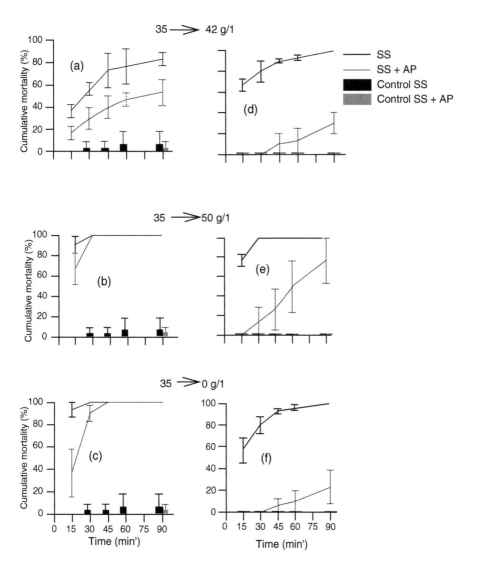

Figure 14.3 Resistance to salinity stress challenge in sea bass (*Dicentrarchus labrax*, L.) larvae 40 (a, b, c) and 50 (d, e, f) days of old fed brine shrimp nauplii enriched with Super Selco (S.S.), or with S.S. + ascorbyl-6-palmitate (AP).

salinity (Figure 14.3 a), after 15 min of the test, the mortality for the group fed with the enrichment in S.S. and AP was 16.7% (5.7), while the mortality for the post-larval group with the S.S. enrichment only was 36.7% (5.8). The difference between the two treatments was not significant. The differences in cumulative mortality rates after 60 and 90 min were significant ($P < 0.05$). The treatments subjected to stress in de-ionized water (Figure 14.3 c) with a diet enriched with AP had a mortality after 15 min of 36.7% (20.8), which was significantly lower (P < 0.05) than that for the post-larval group without supplements (mortality was 93.3%). After 30 min, the mortality rate for the group with vitamin supplements was 90% (17.3) and remained similar for the rest of the test. In treatments with juveniles without dietary supplements, the cumulative mortality was 100%.

The cumulative mortality in both groups in the saline test at 35 g/l, carried out as a control (Figure 14.3 a, b, c), was extremely modest and was caused by handling stress only.

14.3.2.2   Tests at 50 days of life
In hyperosmotic conditions, at 50 g/l of salinity (Figure 14.3 e), none of the fish in groups fed with vitamin supplements died. Mortality in groups without dietary supplements was 76.7% (5.8). The mortality in the group fed with vitamin supplements increased gradually from 26.7% (20.8) after 45 min, to 50.0% (26.5) after 60 min, and finally to 76.7% (23.1) after 90 min. The differences between the two groups were significant (P < 0.05).

In the test with de-ionized water (Figure 14.3 e) all of the larvae fed with vitamin supplements survived the first 30 min, while in the group without supplements the cumulative mortality was 56.6% (11.5) after 15 min and 80% (11.5) after 30 min. At 90 min, mortality in the group with vitamin supplements was 23.3% (15.3), whereas in the juveniles group without such supplements it was 100%. These results were significantly different (P < 0.05).

In the control at salinity of 35 g/l (Figure 14.3 d, e, f), all groups survived until the end of the tests.

14.3.3   Stress challenges (infections)

14.3.3.1   Tests at 40 days of life
Figure 14.4 shows the results of the infection tests with different concentrations (CFU/ml) of *Vibrio anguillarum* in sea bass larvae. In the two control groups, cumulative mortality due to handling was 13.3% (23.1) in the group fed the diet enriched with AP and 26.7% (5.8) in the group without AP.

None of the sea bass juveniles of the group fed with nauplii enriched with AP and S.S. and subjected to a bath infection of 10^1 (CFU/ml) died in the first 2 h of the test. Juveniles fed with nauplii enriched only with S.S. had mortality after 1 h of 13.3% (5.8), and that did not change in the following 3 h of the test.

14.3.3.2   Tests at 50 days of life
The sea bass juveniles fed diets with both enrichments and subjected to a bath infection test of 10^1 CFU/ml had mortality of 3.3% (5.8) after the first hour of the test, while there was no mortality in the juveniles fed with nauplii enriched only with S.S. While the results of the treatment fed with AP indicated that mortality remained similar throughout the test, in the treatment

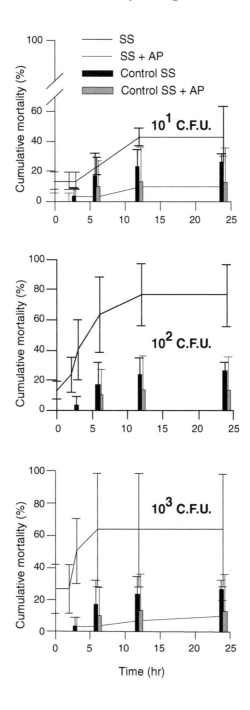

Figure 14.4 Resistance to a *Vibrio anguillarum* infective challenge in sea bass (*Dicentrarchus labrax* L.) larvae 40 days old, fed brine shrimp nauplii enriched with Super Selco (S.S.), or with S.S. + ascorbyl-6-palmitate (AP).

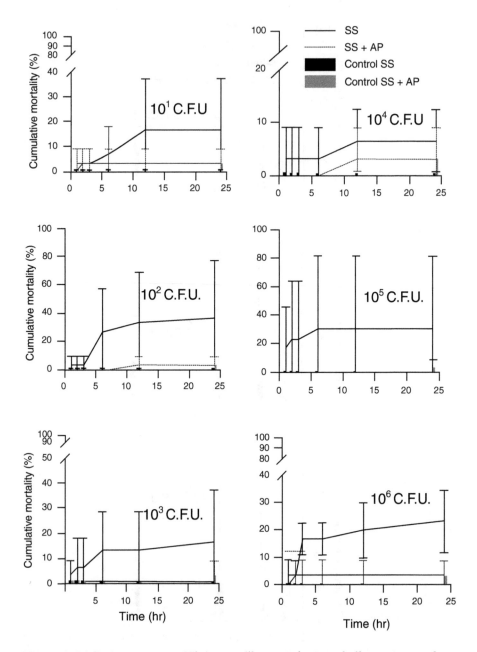

Figure 14.5 Resistance to a *Vibrio anguillarum* infective challenge in sea bass (*Dicentrarchus labrax* L.) larvae 50 days old, fed brine shrimp nauplii enriched with Super Selco (S.S.), or S.S. + ascorbyl-6-palmitate (AP).

without AP it rose to 3.3% (5.8) after the second and third hours and then rose to 6.7% (11.5) during the final stages of the test. However, the difference between the two groups was not significant.

14.4 Discussion

Ascorbic acid concentrations in the *Artemia* nauplii enriched with S.S. and AP were nine times higher than nauplii enriched only with S.S. Merchie et al.[11] obtained enrichment values of up to 3581 µg/g using a strain of *Artemia* which originally contained 581 µg/g (dry matter) of vitamin C. It has been shown that the vitamin concentrations in *Artemia* nauplii after hatching can vary enormously depending on the strain and the place of origin. Merchie et al.,[11] assessing the total ascorbate concentration in three different strains from Namibia, Brazil, and Vietnam, found from 310 µg/g to 581 µg/g. Dabrowski,[14] analysing the total ascorbate concentration strains of brine shrimp from North America, France, Australia and China, reported lower concentrations, even if of the same level of magnitude, between 162.4 and 428.5 µg/g wet weight.

A scarce vitamin supply in an extremely delicate stage of the early onto-genetic development could cause disturbances in larval development and irreversible damage in certain anatomical structures such as operculums and vertebrae.[11, 12] This type of damage is incompatible with the successive grow-out phase in the fish farms. "Disoperculated" fish are frequently found in euryhaline Mediterranean species in fish farms, particularly gilthead sea bream. Fish with operculum malformations are not appealing and therefore have a low market value. However, operculum malformation in gilthead sea bream and sea bass seem to be not solely vitamin C dependent (Saroglia, M., personal information). The consequences of the spine deformation (scoliosis, kyphosis, and lordosis) are more serious both from an aesthetic standpoint and in terms of the survival of sea bream. This is why commercial hatcheries select those juveniles which do not meet optimum requirements.

The lack of swim bladder inflation was a major malformation which until a few years ago was a problem for sea bream hatcheries. It has been proven that this cannot be associated with deficiency of vitamin C[11] but rather man-agement of the tanks in the first days of larvae rearing, and probably to a defi-ciency in polyunsaturated fatty acids (PUFA) in broodstock diets.[17]

The vitamin concentrations in larvae with AP integrated into their diet was always higher than the levels found in fish which did not have this vita-min supplement. This demonstrates the bioavailability of vitamin C in the *Artemia* nauplii when used as the first larval food.

The larvaes' ability to digest the esterified form of ascorbate is a relevant aspect. Sea bass and turbot larvae were analyzed after they had been fed with AMP and had lower body concentrations of total ascorbate than found in pre-sent work, which seems to demonstrate that the larvae were not capable of digesting the vitamin esther.[18] By contrast, Blom and Dabrowski[19] found in rainbow trout that the alevins obtained from females raised without vitamin C, or with a ration containing 360 mg/kg supplied as AMP in the feed, and themselves fed with 500 mg/kg ascorbic acid supplied as AMP, showed an increase in the body ascorbate concentrations from the first day of feeding. This may indicate that phosphatase enzymes are present in the intestines of the trout alevins perhaps also because of the different quantities of vitelline transferred with the egg. The presence of phosphates in the intestines of Mediterranean euryhaline species from the very first days of life has also been proven.[20]

Merchie et al.[12] obtained ascorbate concentrations greater than we achieved both in larvae without enrichment with AP (346–610 µg/g) (depending on the stage of life of the larvae) as well as those with AP integrated in the diet (1244–1624 µg/g dry weight). It must be emphasized that our experimental plan was different from that used by other authors. In brief, they kept the sea bass larvae at a density of 30 larvae/l in 60 l tanks at a temperature of 19°C, while in our case we used conditions that were similar to those found in commercial hatcheries, and the larvae were kept at 16°C, at a density of 100 larvae/l in 280 l tanks. The sea bass larvae were fed with *Artemia* nauplii from the 8th until the 60th d of life, with the weaning stage beginning on the 41st day.

In contrast, Merchie et al.[12] first fed the larvae with *Brachionus plicatilis* (4th–12th d of life), and then substituted it with *Artemia* nauplii (13th-46th d of life) until weaning was complete. This started on the 35th d, 6 days earlier, as a result of the higher temperatures, which is incidentally inappropriate in the early stage of the completion of the organogenesis.

Probably the greater larval metabolism caused by a temperature of 19°C increased the ingestion capacity of the larvae and thus created a greater vitamin demand. This may not, however, explain such a marked difference. It is difficult to reconcile the fact that the vitamin concentration in the larvae found by Merchie et al. was higher than the level of ascorbate in the diet, since there are no signs of bioaccumulation for the vitamin C, which is rather rapidly metabolized. Probably the other reasons for differences are the accuracy of analytical methods used, or the larger number of larvae representative in each of the samples. The difference in weight gain in the two groups subjected to different feeding regimes was of significance and showed that the supplementation of vitamin C improved the zoo-technical conditions. The growth in the control group (without supplements of vitamin C) is within the range reported in the literature.[21] This author indicated that the body length of 25 d old fish was 9.5 mm. In the present study the average body length of the 26 d old larvae without vitamin supplements was about 9 mm, while those subjected to AP enrichment were 10.2 mm. By contrast, Merchie et al.[12] did not find any differences in weight and growth between groups of larvae raised on diets with different quantities of vitamin C, while they did find such differences in the African catfish. They mentioned an average body length for sea bass larvae of c.15.5 mm after 35 days, while our experiments produced an average body length of 13.3 and 12.1 mm, for the groups of larvae with and without vitamin supplements, respectively, at 36 d after hatching. The reason for this difference can be found in the water temperature differential and the rate of increase in the daily temperatures. Indeed, from hatching of the eggs until the 38th d, temperatures in the present study were maintained at 16°C, whereas the experiment of Merchie was conducted at an average temperature of 19°C with a total, over 35 d, of 105°C per day.

It is worth mentioning that commercial hatcheries in the Mediterranean prefer to keep sea bass larvae at temperatures of 15–16°C because the quality of the juveniles is better.

The larvae fed nauplii with vitamin supplements were more homogenous in weight than those without. Disparity in growth in a fish population is one of the factors that can cause heavy losses due to cannibalism and the development of secondary infections. Thus, until selection by size can be carried out, it is advantageous that the size is uniform.

We observed that from the 46th to 56th d of life, the average weight of the fish increased from 15.6 and 12.2 mg to 53.3 and 28.6 mg, respectively, in treatments fed with and without vitamin supplements.

The supplement of vitamin C in the diet of the larvae has been shown to increase the resistance of the animals to saline stress. The results of the tests for cumulative mortality at 40 and 50 d of life were significantly higher in the groups of larvae fed without supplements of AP. The hypertonic stress test of 42 g/l conducted for 90 min at 40 d of life seems to be the most suitable for assessing the resistance capacity of the larvae, as it produced growing cumulative mortality in both groups from the beginning to the end of the experiment. The stress test conducted at 50 g/l and in deionized water, which resulted in the death of almost all the experimental fish, may be useful only in respect to salinity tolerance limits.

At 50 d of age, the salinity stress tests that provided the most information were conducted at 50 and 42 g/l and in deionized water. At 42 g/l, none of the larvae that had been fed the diet without vitamin supplements survived, while in the other groups the cumulative mortality was only 30% (± 10). The same trend was observed in the test carried out in the deionized water.

Merchie et al.[12] did not observe significant differences in survival in the larvae 18–35 days old subjected to salinity stress. Only in the test conducted with 27 d old were fish results of mortality significantly different. Nonetheless, from the presented data one can conclude that the mortality was higher in the fish that were not given vitamin C supplements, although not significantly so. The results of the stress test for African catfish larvae, however, were shown to be influenced by the quantity of vitamin C in the diet.[12]

The anti-stress role of vitamin C has been demonstrated in the larval stages, as it has already been for many fish species in the grow-out phase.[5–8, 22] The anti-stress action seemed to result in the inhibition of the cortico-steroid hormones, including hydrocortisone, through the peroxidization

of the unsaturated lipids, thus preventing their conversion into the choles-
terol components of hydrocortisone.[4]

The tests, aimed at assessing the influence of the supplementation of vitamins in the larval diet, did not result in significant differences, but a clear
trend of major resistance was identified in the larvae that had received vitamin supplements. The lack of statistical significance of the results was due to
high standard deviations. The cumulative mortality was not shown to be vitamin C dose dependent. This leads one to suspect that the observation period
should have been extended to one week after infection.

Acknowledgments

This work has been granted by the E.U. under the COPERNICUS CIPACT
930140 project, and by the Italian MURST.

References

1. Soliman, A. K., Jauncey, K., and Roberts, R. J. (1986a). The effect of varying
   forms of dietary ascorbic acid on the nutrition of juvenile tilapias (*Oreochromis
   niloticus*). *Aquaculture*, 52:1–10.
2. Dabrowski, K., Segner, H., Dallinger, R., Hinterleitner, S., Sturmbauer, C., and
   Wieser, W., (1989). Rearing of cyprinid fish larvae: the vitamin C-minerals interrelationship and nutrition-related histology of the liver and intestine of roach
   (*Rutilus rutilus* L.). *J. Anim. Physiol. Anim. Nutr.*, 62:188–202.
3. Dabrowski, K. (1984). The feeding of fish larvae: present "state of the art" and
   perspectives. *Reprod. Nutr. Develop.*, 24(6): 807–833.
4. Thompson, I., White A., Flecter, T. C., Houlihan, D. F., and Secombes, C. J.
   (1993). The effect of stress on the immune response of Atlantic salmon (*Salmo
   salar* L.) fed diets containing different amounts of vitamin C. *Aquaculture*. 114:
   1–18.
5. Mazik P. M., Brandt T. M., and Tomasso, J. R. (1987). Effects of dietary vitamin C
   on growth, caudal fin development, and tolerance of aquaculture-related stressors in channel catfish. *Prog. Fish. Cult.*, 49: 13–16.
6. Durve, V. S. and Lovell, R. T. (1982). Vitamin C and disease resistance in
   Channel Catfish (*Ictalurus punctatus*). *Can. J. Fish. Aquat. Sci.*, 39: 948–951.
7. Scarano, G., Saroglia, M., and Sciaraffia, F., (1991). Ruolo protettivo dell'acido
   ascorbico verso intossicazioni da nitriti in spigola (*D. labrax*, L.). *Riv. Ital.
   Acquacult.*, 26: 95–102.
8. Nunes G. D., Ninis, M. T., and Bucke, D. (1996). The response to environmental
   stress in juvenile *Sparus aurata* L. fed diets supplemented with high doses of
   vitamin C. *Bull. Eur. Ass. Fish Pathol.*, 16(6): 208.
9. Dabrowski, K., Matusiewicz, M., and Blom, J. H. (1994). Hydrolysis, absorption
   and bioavailability of ascorbic acid esters in fish. *Aquaculture*, 124: 169–192.
10. Dabrowski, K. and Glogowski, J. (1977). Studies on the role of exogenous prote-
    olytic enzymes in digestion processes in fish. *Hydrobiologia*, 54: 129–134.
11. Merchie, G., Lavens, P., Dhert, Ph., Pector, R., Mai Soni, A. F., Nelis, H., Ollevier,
    F., De Leenheer, A., and Sorgeloos, P. (1995a). Live food mediated vitamin C

transfer to *Dicentrarchus labrax* and *Clarias gariepinus*. *J. Appl. Ichthyol.,* 11:336–341.

12. Merchie, G., Lavens, P., Dhert, Ph., Dehasque, M., Nelis, H., De Leenheer, A., and Sorgeloos, P. (1995b). Variation of ascorbic acid content in different live food organisms. *Aquaculture,* 134: 325–337.

13. Dhert, P., Lavens, P., and Sorgeloos, P. (1992). Stress evaluation: a tool for quality control of hatchery-produced shrimp and fish fry. *Aquaculture Europe (EAS),* 17(2): 6–10.

14. Dabrowski, K. (1991). Some aspects of ascorbate metabolism in developing embryos of the brine shrimp (*Artemia salina*). *Can. J. Fish. Aquat. Sci.,* 48: 1905–1908.

15. Zs. Gy., Papp, Saroglia, M., and Terova, G., (1998). An improved method for assay of vitamin C in fish feed and tissues. *Chromatographia,* Vol. 48, No. 1/2, July : 43–47.

16. Cohen, S. A., Meys, M., and Tarvin, T. L. (1990) Section 1.4. Protocols for Unusual Samples; in: Waters, Milford, MD, 35-39.

17. Kitajima, C., Watanabe, T., Tsukashima, Y., and Fujita, S. (1994). Lordotic deformation and abnormal development of swim bladders in some hatchery-bred marine physoclistous fish in Japan. *J. World Aquacult. Soc.,* 25:64–77.

18. Merchie, G., Lavens, P., Storch, V., Ubel, U., Nelis, H., De Leenheer, A., and Sorgeloos, P. (1996). Influence of dietary vitamin C dosage on turbot (*Scophthalmus maximus*) and European sea bass (*Dicentrarchus labrax*) nursery stages. *Comp. Biochem. Physiol.,* Vol. 114A, No. 2: 123–133.

19. Blom, J. H. and Dabrowski, K. (1996). Ascorbic acid metabolism in fish: is there a maternal effect on the progeny? *Aquaculture,* 147:215–224.

20. Yufera, M., Fernàdez-Diaz, C., Pascual, E., Sarasquete, M. C., Moyano, F. J., Diàz, M., Alarcon, F. J., Garcia-Gallego, M., and Parra, G. (1998). Towards an inert diet for first-feeding gilthead sea bream (*Sparus aurata* L.) larvae, in *You can Bet on Aquaculture-Book of Abstracts.* World Aquaculture Society, pp 596.

21. Barnabé, G. (1990). Aquaculture. Vol. 2 Ellis Horwood, Chichester, West Sussex, pp. 1104.

22. Lim, C. and Lovell, R. T. (1978). Pathology of the vitamin C deficiency syndrome in channel catfish (*Ictalurus punctatus*). *J. Nutr.,* 108: 1137–1146.

chapter fifteen

In vitro *methods and results of ascorbic acid absorption in epithelial tissues of fish*

Michele Maffia, T. Verri, and C. Storelli

Contents

Abstract

L-ascorbic acid (AA) is an essential nutrient for several teleost species. It plays an important role in many enzymatic reactions to maintain prosthetic metal ions in their reduced form (for example, Fe^{2+}, Cu^+) and for scavenging free radicals to protect tissues from oxidative damage. In mammals, it has recently been shown that the facilitative sugar transporters of the GLUT type (GLUT1 and GLUT3) can transport the oxidized form of the vitamin dehydro-L-ascorbic acid. However, the bulk of the vitamin, which is present in the plasma essentially in its reduced form, is carried by the Na$^+$-dependent AA transporters SVCT1 (sodium-dependent vitamin C transporter 1) and SVCT2 (sodium-dependent vitamin C transporter 2), which have recently been functionally expressed in *Xenopus* oocytes, cloned and sequenced.[78] SVCT1 is mainly confined to epithelial tissues such as intestine, kidney, and liver. In fish, many results, as obtained by different *in vitro* techniques, have detailed the presence of an electrogenic Na$^+$-coupled AA transport mechanism at the brush border membrane of enterocytes, with kinetic characteristics similar to those found in mammals (apparent K_m ranging between 0.22 and 0.75 mM), when measured using the same experimental approaches. At the basolateral level of fish intestinal-absorbing epithelial cells, transport of dehydro-L-ascorbic acid (DHA) is mediated by Na$^+$-independent transport pathway(s), presumably belonging to the GLUT type family as found in mammals, although this has not been demonstrated so far. We report data on both the kinetic characteristics of vitamin C transport through biological membranes of epithelial cells and the experimental approaches used over the time to study AA absorption in fish. The possibility of getting new theoretical information on ascorbic acid absorption and metabolism in fish by using the *Xenopus laevis* expression system and the perspective to develop new biotechnological applications in aquaculture by gene transfer are also pointed out.

15.1 Introduction

AA plays a critical role in all living organisms. It is metabolically important as an antioxidant in biological systems and has a key role in hydroxylation of lysine and proline in collagen, thus preventing scurvy. It is involved in hydroxylation-mediated detoxification of xenobiotics, drugs, toxicants, and in metabolism of steroids.[1, 2]

AA is synthesized from L-gulonic acid, one of the intermediary metabolites of the D-glucuronic acid pathway, via L-gulono-γ-lactone in most animals. Although most vertebrates possess the enzymatic capability of synthesizing AA from D-glucose, certain species of mammals, including human, primates, guinea pigs and bats,[3–6] birds,[7] and fish[8] are unable to synthesize AA. In fact, they do not possess the microsomial enzyme L-gulono-γ-lactone oxidase (EC 1.1.3.8, GLO), which catalyzes the final step for the AA synthesis, and therefore must absorb the vitamin from the diet. Recently, the alteration of the GLO gene in humans and guinea pigs has been elucidated at the nucleotide level,[9, 10] and it has been concluded that in scurvy-prone species this alteration was not due to the absence of the gene but to mutations (which probably occurred several times during the evolution) in their nucleotide sequence. This has shed new light in the absence of a clear evolutionary lineage among taxonomic groups concerning the requirement of vitamin C.

AA homeostasis, absorption, and metabolism have been studied in mammals for five decades.[2, 11] On the contrary, related molecular processes in fish have not been fully addressed so far. AA is an essential nutrient for several teleost species such as rainbow trout,[12, 13] brook trout,[14] coho salmon,[15] catfish,[16] Indian major carps,[17] and tilapia.[18] As with scurvy-prone mammals, in these species AA appears to be mainly involved in collagen synthesis, since fish reared on AA-deficient diets develop signs traceable to impaired collagen biosynthesis (i.e., acute lordosis and scoliosis, vertebral dislocation, deformation of support cartilage, and delayed wound repair) and reduced growth rate, distortion of gill filament cartilage, short opercules, and hemorrhaging of fins, tail, muscle, and eyes.[19, 20] Interestingly, with respect to the above-mentioned species, claims were made that some cyprinids are able to synthesize AA in their livers and/or kidneys.[21] More ancient fishes, several species of *Chondrostei*, such as Siberian,[22] white and lake sturgeon, and paddlefish[23] do synthesize AA in the kidney.

In this chapter we focus on the *in vitro* methods used in fish to characterize transport phenomena across plasma membranes of epithelial cells of AA, oxidized form, DHA, and such of its derivatives as AA sulfate and phosphate esters. Data are available so far regarding these aspects of AA metabolism in lower vertebrates. In regard to the movement of AA and its derivatives across the membranes of intracellular organelles, to our knowledge, data on this subject have not been obtained in fish. Most recent aspects of AA transporters in mammals will be reviewed, as this will become critical to our understanding of intestinal mechanisms of vitamin transport in fish also. A perspective section will be provided in which we discuss recent advances based on the use of molecular biology techniques and transgenic technology, which might represent a starting point for future advances in the study of AA absorption, homeostasis, and metabolism in fish.

15.2   AA and its derivatives

AA (Figure 15.1) has a molecular weight of 176, ionizes at the hydroxyl C-2 (pK 4.17) or C-3 (pK 11.57) positions, and exists as a monovalent anion at physiological pH. Since it is highly water soluble, it does not diffuse in the lipid component of the membranes. Moreover, since it is too large to rapidly diffuse through aqueous membrane pores, its movement between different compartments of the animal body by simple diffusion is slow. Therefore, AA movement across membranes requires specific mechanisms, such as facilitated diffusion (e.g., carrier-mediated movement of substrate in the direction of the electrochemical gradient) and/or active transport (carrier-mediated movement against an electrochemical gradient). Because the prevalent form of AA carries a negative charge, the molecule has to move against the 30–90 mV electrical potential difference that typically exists across cell membranes

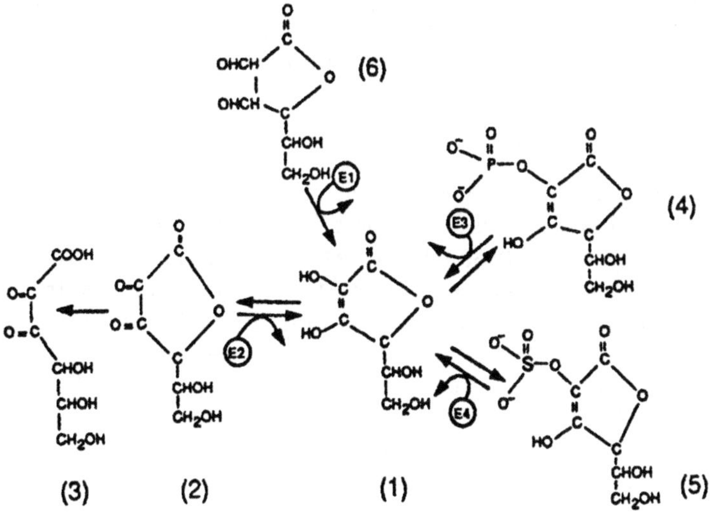

Figure 15.1   Ascorbic acid (1), its oxidation products, dehydroascorbic acid (2) and 2,3-diketogulonic acid (3), and its derivatives, ascorbyl-2-monophosphate (4) and ascorbyl-2-sulfate (5) are depicted. The precursor of ascorbic acid L-gulono-γ-lactone (6) is reported. Enzymes involved in single reactions are indicated: L-gulono-γ-lactone oxidase (E1), dehydroascorbate reductase (E2), alkaline phosphatase (E3), and ascorbyl-2-sulfate sulfatase (E4), respectively.

(cell interior negative); thus, an active transport process is required to raise the intracellular concentration equal to or above the level of the surrounding extracellular fluid.

Dehydro-L-ascorbic acid (DHA) is the product of AA oxidation (Figure 15.1). It lacks the dissociable protons at the C-2 and C-3 positions and is therefore electrically neutral under physiological conditions. In mammals, a small amount of total ascorbic acid is present under this oxidized form.[4] At higher concentrations DHA is toxic in intact animal,[24] erythrocytes,[25] leukocytes,[26] pancreatic islets *in vitro*,[27] and kidney.[28] It is a relatively unstable compound (with a half-life of a few minutes) and its stability decreases with increasing pH and temperature.[29] It decays through a biologically irreversible opening of the lactone ring to form 2,3-diketogulonic acid (Figure 15.1), which is lost in the urine, or degrades further to a variety of other compounds. The balance between reduction and delactonization of DHA is responsible for the conservation of ascorbic acid in the body. Most of the reports suggested that reduction of DHA to regenerate AA does not occur spontaneously, but requires a chemical reduction or an enzymatic process.[2] Only recently, spontaneous conversion of DHA to AA has been documented.[30]

With respect to AA, AA derivatives with sulfate (AS), phosphate (AP; Figure 15.1), glucose (AG) esters at the C-2 position, and palmitate at the C-6, in the lactone ring, are resistant to oxidation, although palmitate is not. Ascorbyl esters are added to commercial formulations, although they differ in the biological activity when supplied with the diet. The conversion of AS and AP to AA requires the hydrolysis of sulfate and phosphate by AS-sulfatase and alkaline phosphatase, respectively.[31]

15.3   AA homeostasis

As a general rule, vertebrates not able to synthesize AA need a membrane transport process allowing dietary ascorbic acid to be efficiently absorbed as chyme moves through the gut. Intricate mechanisms have evolved for extracting AA from food, conserving it in the kidney, and storing and transferring it into all body tissues in concert with existing needs. Once absorbed, AA has to be quickly distributed and eventually accumulated in different tissues of the body. In humans, AA is highly concentrated with respect to plasma in leukocytes, lungs, adrenals, pituitary, specific compartments of the eye and, to a lesser extent, in platelets, granulocytes, brain, kidney, and liver.[2] In fish, AA is also readily taken up by kidney, liver, intestine, spleen, and brain.[32] At the moment, evaluation of tissue AA concentration is the most reliable indicator of AA status of a fish.[33, 34] To maintain its homeostasis, AA content has to be preserved. Therefore, urinary loss of AA, which is freely filtered in the glomerulus of both species that synthesize it and species that have a dietary requirement, has to be avoided by the presence of mechanisms of reabsorption in the proximal convoluted tubule of the kidney. Thus, complete reabsorption of filtered AA may be normally achieved.

AA absorption, reabsorption, and interorgan transfer typically means that the vitamin must cross at least two cell membranes. For this reason, researchers have focused their attention on the characterization of translocation mechanisms for AA occurring both at epithelial (such as intestine and kidney) and nonepithelial (such as blood cells, adrenals, lung, etc.) tissues. In higher vertebrates, discovery of these mechanisms has required the development of *in vitro* methods and biochemical approaches. Results obtained have been widely reviewed.[2, 11, 35] Based on much of the information on intestinal absorption and renal reabsorption of AA in scurvy-prone mammals which require it in the diet, the following model has been proposed. Epithelial transport of the AA was described by Crane's model,[36] in which Na^+-dependent active transport accounts for uptake of the substrate across the brush border into intestinal and renal epithelial absorbing cells. The process is dependent on intact cellular metabolism. AA accumulates within the enterocytes or the proximal tubular reabsorbing cells so that a gradient develops that favors translocation movement toward the blood. Furthermore, dietary DHA is brought into the cell by facilitated transport[37] and rapidly reduced to AA by a process requiring a DHA reductase. AA exit proceeds by way of facilitated diffusion. By these processes, AA is effectively absorbed at the intestine and/or reabsorbed at the proximal tubular level in mammals and its plasma concentrations maintained within an adequate concentration. Once in the blood plasma, AA is distributed to different tissues, most probably by Na^+-independent transport mechanism(s) in non-epithelial tissues[38] and specific Na^+-dependent transport mechanisms in such epithelial tissues as retinal pigmented epithelium.[39] Mammalian species, such as rat and rabbit, that produce AA from D-glucose do not require it in the diet. These species showed a loss of the intestinal brush-border Na^+-dependent transporter for AA.[40] However, they retained other properties for processing it, such as the intracellular enzyme that reduces DHA and the transporter that takes up DHA across the serosal surface.

Concerning teleost fish, such a biochemical characterization of the transport mechanisms mediating ascorbic acid absorption at the membrane level is at its infancy. Experimental data obtained so far by biochemical methods are restricted to the intestine of some commercially important species; on the contrary, no information is available regarding either other epithelial tissues, such as kidney and retinal pigmented epithelia, or nonepithelial tissues.

15.4   AA absorption in fish intestine: in vitro *methods*

Absorption of vitamins by fish gastrointestinal tract has been under consideration by using methodologies which have evolved from *in vivo* to *in vitro*. *In vitro* methods, based on the use of different intestinal preparations, have allowed considerably more insight into the cellular mechanisms sustaining intestinal absorption of AA in vertebrates and an accurate analysis of cell membrane transporters and intracellular enzymes involved in the transport of this vitamin.

15.4.1   Isolated intestinal segments

Absorption and metabolism of AA and DHA in fish, such as rainbow trout, has been studied previously by using isolated loops of intestine.[41] This technique was first used to study and characterize guinea pig and rat small intestinal absorption processes.[42] Vitamin movement either in absorptive or in secretive directions has been determined by using everted and noneverted loops, respectively. Transepithelial fluxes of AA are measured by filling each loop with a known volume of a physiological buffer, containing an antioxidant agent such as thiourea or dithiothreitol to limit spontaneous oxidation of AA, and by incubating intestinal tissues (for at least 1 h) in a solution containing radiolabelled vitamin ([^{14}C]-AA). At the end of the incubation time, the loops are opened and the recovered fluid is assayed by high-performance liquid chromatography (HPLC) and scintillation spectrometry. In transport studies with [^{14}C]-DHA, exposure time is limited to no more than 20 minutes to prevent substrate degradation. Cellular uptake of vitamin C in isolated intestinal loops can be estimated by incubating tissues for at least 1 h with [^{14}C]-AA in both mucosal and serosal bathing solutions.[41] At the end of incubation time, the tissue is extracted in metaphosphoric acid and the amount and identity of ^{14}C label accumulated in the tissue evaluated.

15.4.2   Intestinal rings

In mammals, vitamin C absorption has been studied by using portions of intestine as rings (1–5 mm) cut either from noneverted or everted segments of intestine; the retraction of the circular and longitudinal muscle coats, which occurs after rings have been cut, causes a separation and spreading of the villi in a manner favorable for the access of oxygen and substrates. The advantage of cutting the rings from everted sacs of intestine is that villi do not lie in the center of the anulus; the disadvantage is that the extra manipulation required for the preliminary eversion involves an additional handling of the tissue. It was demonstrated that rings of scurvy-prone mammal guinea pig ileum absorb ascorbate against concentration gradient.[43] Uptake was inhibited by metabolic poisons, anaerobic conditions, or elimination of Na$^+$ from the bathing solution.

15.4.3   Schultz chambers

Information concerning ascorbic acid unidirectional fluxes across mucosal border of guinea pig and human ileum was mostly derived from the utilization of the Schultz chamber method.[44] This method has been successfully used also to study AA transport in fish intestine.[41] Such a technique consists of mounting intestinal segments (4–8 cm), with the mucosal surface up, in an influx chamber with a fixed number of ports (6–8), exposing a well-defined surface area (0.2–0.75 cm^2/port) that is bathed with oxygenated solutions (95% O_2–5% CO_2). After a defined incubation period with [^{14}C]-AA, and [^{14}C]-mannitol or [^{3}H]-inulin as nonpenetrating extracellular space markers, the exposed tissue is extracted in HNO_3, a process removing a majority of the radioactivity. Aliquots of the tissue extract are then analyzed for radioactivity and tissue uptake of AA are calculated after correction for [^{14}C]-mannitol or [^{3}H]-inulin space.

15.4.4   Membrane vesicles

Progress in the study of both transport and intracellular metabolism of AA has been achieved by using intestinal tissues disrupted by mechanical and osmotic stress. In particular, the advent and broad application of plasma membrane vesicle purification techniques furnished a method to accurately describe transport properties of substrates such as sugars, amino acids, or vitamins, across isolated apical and basolateral epithelial cell membranes. The earliest applications of brush border membrane vesicle (BBMV) purification technique to fish gastrointestinal epithelia employed Ca^{2+}[45] or Mg^{2+}[46] precipitation methods and have been applied to the study of nutrient transport in dogfish,[47] marine flounder,[48] European eel,[49] and tilapia.[50] Furthermore, methods used to purify intestinal basolateral membrane vesicles (BLMV) of mammalian epithelial cells involved density gradient centrifugation[51] and have been successfully applied to fish intestinal epithelia to examine transport mechanism of both nutrients[52] and vitamins.[53] When viewed under the electron microscope, vesicle preparations produced by different methods are composed by spherical, osmotically reactive fragments of the original membranes which are approximately 0.2 μm in diameter.[50] The final fraction containing BBMV is enriched in marker enzymes of the apical cell pole such as alkaline phosphatase, leucine aminopeptidase, and maltase, while final BLMV fraction is enriched in enzymes restricted to the serosal pole, including Na^+/K^+-ATPase and K^+-dependent phosphatase. Typical enzymatic properties of BBMV and BLMV fractions from the carnivorous eel are shown in Table 15.1 and are representative of results obtained when applying these preparative techniques to the intestinal epithelium of other teleosts. The main advantage in the use of these preparations is represented by the possibility of an accurate analysis of a transport system in terms of kinetic characteristics, effect of "driving forces" (i.e., ionic gradients, ATP dependence, electrical membrane potential, etc.), reaction mechanisms, selectivity, and specificity of the transport system. Analysis of experimental data obtained by this technique is enormously facilitated with respect to other *in vivo* and *in vitro* methods by the absence of cell organelles, cytosolic components and, consequently, of cell metabolism that complicates the understanding of transport phenomena. Generally, ionic gradients experimentally established are

Table 15.1 Enzymatic composition of isolated brush border membrane vesicles of eel intestine

Marker		Enrichment factor	Homogenate (mU/mg protein)	BBMV (mU/mg protein)	Enzyme recovery (% of homogenate activity)
Leucine aminopeptidase	Brush border membrane	15.6 ± 1.5	6.3 ± 0.6	97.9 ± 10.9	39.8 ± 6.6
Maltase	Brush border membrane	11.8 ± 1.1	82.6 ± 6.0	898.0 ± 93.5	25.2 ± 7.1
Alkaline phosphatase	Brush border membrane	12.8 ± 0.6	72.5 ± 2.5	927.3 ± 71.1	28.2 ± 7.0
Na^+-K^+-ATPase	Basolateral membrane	0.40 ± 0.02	12.8 ± 1.3	5.1 ± 0.6	1.3 ± 0.1
Succinate-cytochrome c oxidoreductase	Mitochondria	ND	9.0 ± 1.4	ND	ND
KCN-resistant NADH oxidoreductase	Endoplasmic reticulum	0.4 ± 0.1	278.3 ± 26.7	66.9 ± 9.6	4.8 ± 0.8

Enzymatic composition of isolated basolateral membrane vesicles of eel intestine

Marker		Enrichment factor	Homogenate (mU/mg protein)	BLMV (mU/mg protein)	Enzyme recovery (% of homogenate activity)
Na^+-K^+-ATPase	Basolateral membrane	14.9 ± 2.1	15.5 ± 0.6	296 ± 65	23.0 ± 7.1
K^+-stimulated phosphatase	Basolateral membrane	11.2 ± 1.5	1.2 ± 0.1	13.1 ± 1.6	37.8 ± 6.0
Alkaline phosphatase	Brush border membrane	1.1 ± 0.1	149 ± 22	157 ± 14	2.5 ± 0.9
Leucine aminopeptidase	Brush border membrane	1.2 ± 0.2	123 ± 19	126 ± 6	1.7 ± 0.6
Succinate-cytochrome c oxidoreductase	Mitochondria	1.1 ± 0.1	15.9 ± 2.6	13.1 ± 1.1	1.8 ± 0.1
KCN-resistant NADH oxidoreductase	Endoplasmic reticulum	1.6 ± 0.3	8.9 ± 0.5	14.6 ± 1.4	3.6 ± 1.5

Note: Values are means ± standard errors (SE). Enzymatic activities were measured at 25°C. ND, not detectable; BBMV, brush border membrane vesicles; BLMV, basolateral membrane vesicles. Enrichment factor is the ratio of specific activity in brush border or basolateral membrane fractions vs. specific activity in homogenate. Enzyme recovery represents the ratio between total enzymatic activity of BBMV vs. total activity of homogenate. ATPase, adenosinetriphosphatase.

dissipated by diffusion within 1–3 min. This is an important feature in the study of AA absorption, since short incubation times reduce problems related to substrate oxidation and degradation. However, it has to be taken into account that the amount of substrate transported within 20 s from the beginning of the reaction (i.e., the period of steepest ionic gradient occurrence) may sometimes not be sufficient to establish large intravesicular accumulation of a radiolabelled substrate against a concentration gradient (overshoot phenomenon). This effect, due to the low number of vitamin transporters in the plasma membrane, could limit information regarding the molecular processes involved in the transport of substrates, such as ascorbic acid and other water-soluble vitamins, such as myoinositol.[53, 54]

15.5   AA absorption in fish intestine: transport processes

In rainbow trout (*Oncorhynchus mykiss*), all gut organs exhibit AA absorption. In particular, sites of absorption are represented by stomach (20.7%), pyloric caeca region (23.4%), middle intestine (21.9%) and posterior intestine (20.1%).[55] In the case of stomachless common carp, which maintains an alkaline pH along the intestine, the major site of AA absorption is located in the anterior region.[56] Transport and metabolism of vitamin C have been evaluated in trout intestine by the *in vitro* technique using isolated intestinal segments. Transepithelial fluxes of [^{14}C]-AA were determined with the substrate present at 10 μM. The mucosa-to-serosa flux was four-fold greater than the serosa-to-mucosa with AA present in both bathing solutions. This net transcellular vitamin transfer was the result of an intracellular accumulation of AA in the enterocytes against the concentration gradient, which led to a tissue-to-medium ratio of 1.64.[41] However, this value is significantly lower than that observed in guinea pig intestinal mucosa, where uptake of [^{14}C]-AA proceeded to a tissue-to-medium ratio of 5.0.[57] In intestinal loops of rainbow trout, AA influx was not reduced due to the presence of the D- or L-stereoisomeric form of glucose, but was highly dependent on the presence of Na$^+$ in the bathing medium. These results suggested the occurrence of a secondary active transport phenomenon of AA at brush border membrane level of fish intestine, but the authors[41] did not examine in detail the main features of this putative transport system.

Information of the presence and kinetic characteristics of a transport system for AA in fish intestinal apical membranes has been provided by using BBMV isolated from the intestine of the euryhaline teleost *Anguilla anguilla*.[58] These studies showed that transport of [^{14}C]-AA by BBMV is specifically stimulated by an inwardly directed Na$^+$-gradient (Figure 15.2). However, in contrast with results obtained in guinea pig ileal BBMV,[59] the presence of a Na$^+$ gradient did not induce any intravesicular uptake of AA over the equilibrium value (Figure 15.2A). This finding could probably be related to the low J_{max} of the Na$^+$-dependent transport system for AA (0.33 ± 0.03 pmol/mg

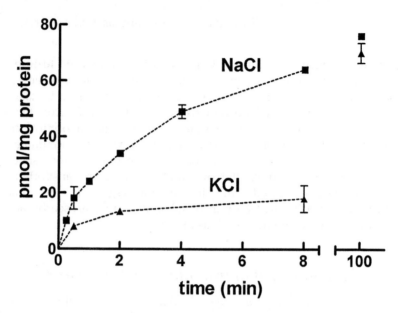

Figure 15.2A Effect of external cations on the time course of ascorbic acid uptake by eel intestinal BBMV. Vesicles were loaded with 300 mM mannitol and 20 mM HEPES adjusted to pH 7.4 with N-methyl-D-glucamine (NMDG) and were incubated in media containing 100 mM mannitol, 20 mM HEPES adjusted to pH 7.4 with NMDG, 0.075 mM ascorbic acid, 0.019 dithiothreitol, and either 100 mM NaCl or 100 mM KCl.

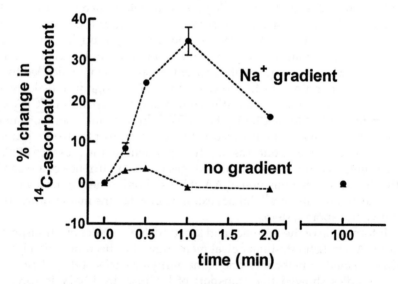

Figure 15.2B Effect of Na-gradient on the time course of ascorbic acid uptake by eel BBMV. Vesicles were preloaded for 1 h at 21°C with (in mM) 50 KCl, 100 mM mannitol, 0.1 L-[^{14}C]ascorbic acid, 0.025 dithiothreitol, 0.0244 valinomycin, 20 HEPES adjusted to pH 7.4 with KOH, and either 100 NaCl or 100 choline Cl. Vesicles were incubated in medium of the same pH containing (in mM) 100 NaCl, 100 mannitol, 0.1 L-[^{14}C]ascorbic acid, 0.025 dithiothreitol, and 50 KCl (short-circuited).

protein $\times$ min). To ascertain whether under physiological conditions Na^+ acts as an external activator or as a true driving force of AA transport, intravesicular uptake of $[^{14}C]$-AA was measured in eel intestinal BBMV preincubated for 1 h at 21°C with the radiolabelled substrate. Under these conditions a Na^+ gradient was able to transiently induce intravesicular accumulation of $[^{14}C]$-AA (Figure 15.2B), thus supporting the concept that the physiological intracellularly directed Na^+ gradient can energize the concentrative transport of AA across fish intestine apical membranes. These results are consistent with the presence of a secondary active transport mechanism[36] and are similar to those reported in guinea pig and human intestine,[40, 59] but in contrast with those reported for the intestine of the rat that does synthesize AA, where a net intestinal absorption of ascorbic acid did not occur.[60]

15.5.1 *Na^+-dependent AA transport: kinetic characteristics*

Na^+ dependent fluxes of AA, as measured in either isolated loops[41] or brush border membrane vesicles,[58] are saturable phenomena displaying apparent Michaelis-Menten constant (Km_{app}) values of 0.22 mM in rainbow trout and 0.75 mM in eel (at 23°C), respectively. These values, although obtained under different experimental conditions, are not significantly different from those reported in guinea pig intestine, where AA concentrations for a half-maximal influx were of 1 mM (at 37°C)[40] or of approximately 0.3 mM (at 20°C),[37, 59] depending on the experimental approach. This finding suggested that Km_{app} of Na^+/AA cotransport, which is a kinetic parameter inversely related to the enzyme-substrate affinity, was highly perturbed by physiological factors such as temperature, hydrostatic pressure, osmotic concentration, and pH, and was conserved among scurvy-prone vertebrates. A maintenance of substrate-carrier affinity could be important not only to retain the maximal responsiveness of the transporter to changes in intestinal concentrations of AA but also to control vitamin accumulation into the enterocyte. The assumption that the steric requirement of Na^+/AA cotransport is conserved among vertebrates is also confirmed by the finding that structural analogs, such as D-isoascorbate inhibited $[^{14}C]$-AA uptake in a fully competitive manner both in eel intestinal BBMV (Figure 15.3), with an inhibition constant (K_i) of 8.21 $\pm$ 0.63 mM, and in guinea pig, with an inhibition constant (K_i) of approx. 19.7 mM.[59]

15.5.2 *Na^+-dependent AA transport: electrogenicity*

An interesting feature displayed by Na^+/AA cotransport of eel intestine is its dependence on the application of a membrane electrical potential. AA uptake has been reported to increase significantly in the presence of a transient (inside negative) electrical potential across eel intestinal BBMV.[58] Since at pH 7.4 AA is an anionic molecule, such an electrical potential dependence of Na^+/AA cotransport could be explained by a Na^+ to AA flux stoichiometric

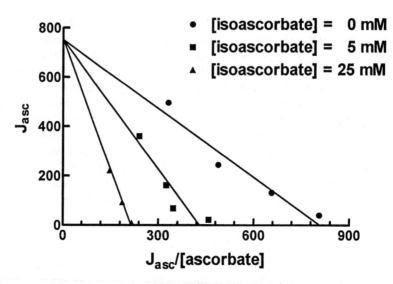

Figure 15.3 Inhibition of L-ascorbate influx by D-isoascorbic acid in eel BBMV. Vesicles were loaded with 300 mM mannitol, 50 mM KCl, and 20 mM HEPES adjusted to pH 7.4 with KOH and were incubated in a medium containing 100 mM mannitol, 100 mM NaCl, 22.4 μM valinomycin in ethanol, 20 mM HEPES adjusted to pH 7.4 with KOH, L-ascorbate concentrations ranging from 0.05 to 1.0 mM, and D-isoascorbic acid concentrations of 0, 5.0, or 25 mM. Plotted values represent the differences between influxes in NaCl medium and choline Cl medium for each ascorbic acid concentration. Best fit curves through each set of data were obtained using a computer regression program.

ratio greater than unity, as suggested in Figure 15.4. AA influx sigmoidally increased with increasing external Na^+ concentration; this behavior was described by a Hill equation for multisite cooperativity having a Hill coefficient greater than unity. This suggests that two or more Na^+ ions may move across the membrane for each AA molecule. In this regard, teleost intestinal AA transporter more closely resembles those of the mammalian kidney[61] and retinal pigmented epithelia,[39] which are also electrogenic and display apparent 2 Na^+ to 1 AA flux stoichiometric ratio, rather than the mammalian intestinal electroneutral process.[59]

By comparing the physiological capabilities of intestinal and renal AA transporters in mammals it was concluded that electrogenicity and a Na^+-to-AA flux ratio greater than unity in kidney aided in the complete withdrawal of the vitamin from the proximal tubule.[61] The increased driving force for AA accumulation in renal epithelial cells, brought about by the combination of the transmembrane chemical gradient for Na^+ and of the electrical potential, provides an efficient absorption mechanism for the vitamin during its brief passage along this part of the mammalian nephron. It has been argued that an electroneutral renal transporter exhibiting a 1 Na^+ to 1 AA flux

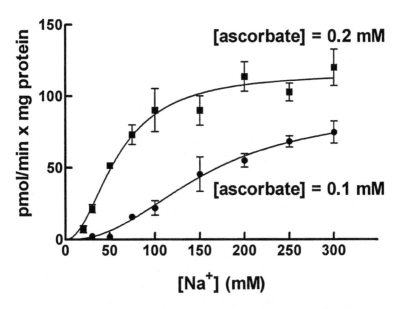

Figure 15.4 Effects of external Na^+ concentration on ascorbic acid influx measured over 30-s uptake interval. Vesicles were loaded with 400 mM choline Cl, 50 mM KCl, and 20 mM HEPES adjusted to pH 7.4 with KOH and were incubated in media containing 50 mM mannitol, 50 mM KCl, 22.4 μM valinomycin in ethanol, either 0.1 or 0.2 mM L-ascorbate, and concentrations of NaCl from 0 to 300 mM (choline Cl isosmotically replacing NaCl).

stoichiometric ratio would be too slow to recover a physiologically adequate amount of vitamin from the proximal tubule. On the other hand, the presence of an electroneutral AA transport system in mammalian small intestine, with a relatively slow transit time (compared with a flow of primary filtrate through the nephron) could be sufficient to ensure an adequate gastrointestinal absorption of the vitamin. The eel and other carnivorous teleosts possess a relatively short intestine[62] compared with omnivorous or herbivorous vertebrate species where the gut may be several times the body length. It may be that the electrogenic $2Na^+/1AA$ transporter exhibited by eel intestinal epithelium is an adaptation in carnivorous teleosts to ensure quick and efficient vitamin absorption during food passage through this short gastrointestinal tract. This has also been argued for vitamin reabsorption in the mammalian kidney. Electrogenicity of eel intestinal Na^+/AA cotransport strengthens an emerging concept considering the physiological electrical potential as one of the main driving forces involved in fish intestinal absorption of nutrients such as sugars, amino acids or peptides.[49, 63]

15.5.3 *DHA facilitated transport*

An important observation that emerges from *in vivo* studies is the significant presence of DHA, the oxidized form of vitamin C, in the digestive tract contents of stomachless and stomach-possessing fish.[31] The proportion of total vitamin C in food that is present in the reduced form depends on the length and conditions of storage. However, animals requiring ascorbic acid seemed to benefit also from absorbing dietary DHA, providing that in the mucosal cells there is an efficient way to absorb this substrate and reduce it to a biologically active form of vitamin C.

Evidence for the presence of a carrier mediated transport mechanism on brush border membranes isolated as vesicles from guinea pig intestine that serves to bring DHA into the enterocyte has been provided.[37] Main features of this transport system were to be independent of Na^+ gradient, saturable with a K_m of 0.17 mM and a maximal transport rate about 10 times lower than that of Na^+/AA cotransport. In fish, such direct experimental evidence is not available, although intracellular accumulation of ^{14}C-DHA has been shown in isolated loops of rainbow trout.[41] Once absorbed, fish enterocytes convert DHA to AA in a similar manner as mammals, by a DHA reductase using glutathione and probably NADPH.[11]

One additional transporter for vitamin C has been identified on the basolateral cell surface of guinea pig intestine and appears to be specific for DHA. Transport activity of this system, evaluated by using basolateral membrane vesicles, is regulated by features such as independence of Na^+ gradient, saturability (K_m = 0.228 mM), and inhibition by structural analogues similar to those displayed by the facilitated transport of DHA of brush border membranes. In fish intestine, the only investigation suggesting the presence of a carrier-mediated transport of DHA at basolateral level was performed on isolated segments of rainbow trout intestine.[41] No information regarding kinetic characteristics of this putative transporter of vitamin C is available in literature. The utility of this facilitated pathway located at the basolateral side of vertebrate intestinal epithelium is at the moment highly speculative. As a hypothesis, it might serve to regulate the whole-body redox state of vitamin C. An emerging concept is that ascorbic acid becomes oxidized in the body as a consequence of different functions involving it as an antioxidant. AA scavenges the reactive oxygen species potentially dangerous for cells and tissues.[2] The resulting metabolite, DHA, could be transported in the blood to several sites including the mucosal cells by the basolateral transporter, and reconverted to AA by the same enzyme that is active in handling dietary DHA. A summary of vitamin C transports in the intestine of teleosts compared to those described in higher vertebrates is reported in Table 15.2.

15.5.4 *Hydrolysis and absorption of AA derivatives*

Since native ascorbic acid is unstable and feed manufacturers must add more than the requirement to ensure adequate levels in fish diet,[64] there have

Table 15.2 Properties of vitamin C absorption in vertebrate small intestine

Vitamin	Species	Site of absorption	Process involved	Na+ dependence	Electrogenicity	K_m (mM)	Technique	Reference
L-ascorbic acid	Human	Ileum	Active transport	Yes		0.3	Intestinal segments	43
	Guinea pig	Ileum; mucosal border	Active transport	Yes	–	1	Schultz chambers	40
		Small intestine; mucosal border	Active transport	Yes	No	0.3	BBMV	59
		Ileum; mucosal border	Active transport	Yes	No	0.33	BBMV	37
		Ileum; basolateral border	Facilitated transport	No	No	0.22	BLMV	37
	Eel	Intestine; mucosal border	Active transport	Yes	Yes	0.75	BBMV	58
	Rainbow trout	Intestine; mucosal border	Active transport	Yes	–	0.22	Intestinal loops	41
Dehydro-L-ascorbic acid	Guinea pig	Ileum; mucosal border	Facilitated transport and intracellular reduction	No	No	0.17	BBMV	37
		Ileum; basolateral border	Facilitated transport intracellular reduction	No	No	0.28	BLMV	37
	Rainbow trout	Intestine; mucosal border	Facilitated transport intracellular reduction	No	No	–	Intestinal loops	41
		Intestine; basolateral border	Facilitated transport intracellular reduction	No	No	–	Intestinal loops	41

Note: BBMV, brush border membrane vesicles; BLMV, basolateral membrane vesicles.

recently been attempts to chemically stabilize AA, generally by complexing ascorbic acid with phosphate or other anionic groups (e.g., sulfate). It has been demonstrated that brush border membrane vesicles prepared from channel catfish intestine released phosphate from two monophosphate-stabilized forms of ascorbic acid ester, with this activity in proximal intestine exceeding that in distal intestine.[65] In this regard, BBMV of *Anguilla anguilla* (Table 15.1) and of other fish species, such as the carnivorous Antarctic teleost *Trematomus bernacchii*[66] and the herbivorous teleost (*Oreochromis mossambicus*)[67] display a significant activity of alkaline phosphatase. Apical absorption of AA by everted intestinal sleeves was greater in proximal intestine of catfish with maximal rates of absorption for the phosphate-stabilized forms at about 40% of those for native ascorbic acid.[65] This finding suggests that the bioavailability of monophosphate-stabilized forms of ascorbic acid depends on the enzymatic activity of alkaline phosphatase located on the brush border membrane. This enzyme is able to hydrolyze the phosphate group and release the native form of AA. Then released AA can be absorbed by the apical Na^+-dependent transport system. To the contrary, low rates of intestinal absorption, tissue accumulation, and marginal antiscorbutic potential of orally administered sulfate forms of ascorbic acid (such as ascorbyl-2-sulfate)[68] was most likely due to a low intestinal brush border sulfatase activity.[69] These results are in contrast with previous studies[1] suggesting that ascorbic acid sulfate may represent the major storage form of vitamin C in fish. Based on the above reported information, a model is reported in Figure 15.5.

15.6   AA reabsorption in teleost kidney

In mammals, ascorbic acid is reabsorbed at the proximal tubular level in the kidney. This reabsorption occurs by one Na^+-dependent transporter specific for ascorbic acid located in the brush-border of the reabsorptive cells[61] and one Na^+-independent transporter that is specific for DHA.[28] Reabsorbed ascorbic acid is maintained in the reduced state.[70] Furthermore, most DHA taken up into rat renal tubules is found in the reduced state. DHA-reductase-like activity has been found in the 55–70% ammonium sulfate fraction of rat kidney homogenates. The transport and metabolic properties for reduced and oxidized ascorbate are similar to those described previously for intestinal handling of ascorbic acid. In mammals, ascorbic acid is also secreted into the urine under some situations, perhaps by indirect Na^+-dependent transport across the basolateral membrane, this serving to buffer the vitamin concentration in plasma by increasing urinary loss when plasma levels become elevated.[71] To our knowledge, no data regarding the characterization of the membrane transport mechanisms involved in AA and DHA tubular reabsorption and secretion are available so far. However, the fact that teleost kidney is one of the organs in which high concentrations of ascorbic acid may accumulate, mostly after supplementation with diets containing ascorbic acid,[31] suggested that specific transport systems may be involved in ascorbic acid movement across renal apical and basolateral plasma membranes.

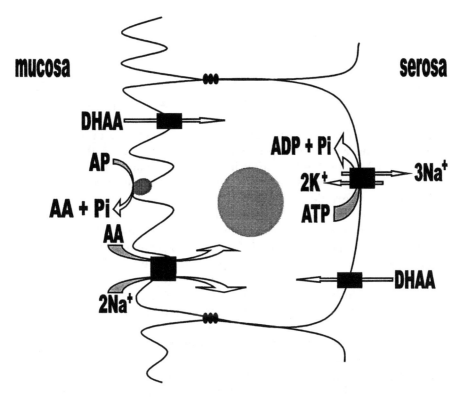

Figure 15.5 Model of known events in fish intestinal absorption of reduced, oxidized, and phosphate ester forms of vitamin C. L-ascorbic acid is taken up across brush-border membrane against a concentration gradient coupled with Na^+ diffusion down an electrochemical potential gradient entering the cell. Dehydroascorbic acid from either the lumen or interstitium probably enters the cell by facilitated diffusion and should be enzymatically reduced. Phosphate ester forms of ascorbic acid can be enzymatically hydrolized at brush border level to form the reduced vitamin C that can be translocated across apical membranes by a Na^+-coupled transport.

15.7   Perspectives in the study of AA absorption in fish: the cloning approach

15.7.1   Toward the molecular structure of plasma membrane transport system(s)

Plasma membrane proteins are generally expressed at a low level. This makes it difficult to isolate plasma membrane proteins by conventional protein purification techniques. As an alternative to the biochemical isolation, investigators have recently adopted a strategy of expression cloning using *Xenopus laevis* oocytes to gain molecular information on plasma membrane transport proteins. After the first cloning of the small intestinal

Na$^+$/D-glucose cotransporter,[72] this approach has been used to identify a number of membrane transporters and characterize their molecular structure and mechanisms.[73-76]

Figure 15.6 summarizes the essential steps in the expression cloning strategy, as based on the analysis of transport function in H$_2$O- or mRNA/cRNA-injected oocytes obtained from *Xenopus laevis*. Step 1 regards mRNA isolation, its injection into oocytes (up to 50 ng in a volume of up to 50 nl of water), and analysis of transport function in oocytes using radiolabelled substrates; the extent of expression being a function of the amount of injected mRNA and the time after injection was analyzed. Step 2 involves size fractionation of mRNA either on the basis of gel electrophoresis and/or sucrose density gradient centrifugation; again, the expression of transport activity is evaluated. At this step it is important to compare intrinsic activity (H$_2$O-injected oocytes) and expressed uptake activity (mRNA- minus H$_2$O-injected oocytes) to obtain information on whether the transport system of interest is indeed expressed or whether the possibility of stimulation/amplification of intrinsic activity has to be considered. For this purpose, the study of kinetic properties and/or cross-inhibition experiments (by substrates of other known membrane transporters) is required. On the basis of the above information, in step 3 size fractionation (library construction/screening) can be initiated. Reverse transcription (cDNA synthesis) of size-fractionated (functionally active) mRNA is followed again by the appropriate size selection (at least cutting off the lower length transcripts) and by the ligation of cDNA in the appropriate vector; directional ligation into a plasmid will permit the *in vitro* transcription of cRNA for oocyte injection after plasmid linearization. Step 3 contains the first screening steps (for transport expression); the library is separated (plated) before any amplification step in different pools (box), each containing approximately 1000 to 3000 independent colonies; cRNA is synthesized from plasmids grown separately from these pools. If the injection of one of these cRNAs leads to an increase of transport activity, the success of the cloning strategy is almost guaranteed. Further steps toward the isolation of a single clone involve so-called *sib selection* protocols, employing dilutions/replatings/replicas, as well as growth/isolation of plasmids and *in vitro* transcription of cRNA at all different steps.

15.7.2 *Expression of epithelial plasma membrane Na$^+$-dependent AA transporters*

Among the available information on the overall characteristics of AA intestinal absorption and renal reabsorption in vertebrates, nothing is known regarding the molecular features of the transport system(s) mediating AA transport across plasma membranes. Data regarding the molecular characteristics of putative intestinal Na$^+$-dependent mechanism(s) involved in AA absorption in vertebrates are not available so far. However, recently the expression of a

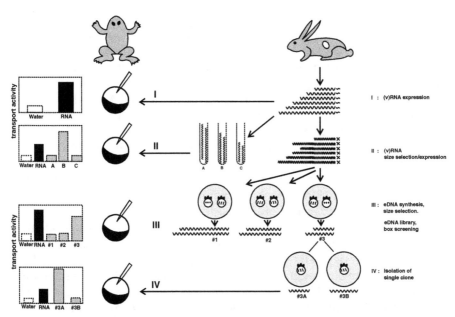

Figure 15.6 Strategy for expression cloning of transport systems. Four essential steps are outlined. Step 1 consists of mRNA isolation (e.g., intestine or kidney), its injection into oocytes from stages V or VI of *Xenopus laevis* toads, and analysis of transport by a tracer technique (e.g., $[^{32}P]$-PO_4). Step 2 consists of size fractionation of mRNA and injection of fractionated mRNA into oocytes followed by transport analysis. Step 3 consists of cDNA synthesis from size-fractionated (functionally active) mRNA, its size fractionation, and ligation (directional) into an appropriate plasmid vector; this step already contains screening for transport activity after the injection of *in vitro* transcribed cRNA from pools of plasmids grown and isolated from fractions of the library separated before an amplification step. Step 4 starts from a positive fraction (expressed transport activity, see step 3) and consists of the further fractionation (sib selection) of the library until a single clone-encoding transport activity is identified. Such a procedure has been used to identify membrane proteins suspected to be involved in membrane transport in several vertebrate species, including teleost fish (Werner et al., 1994). A, B, and C indicate different pools of size-fractionated mRNA species as obtained, e.g., from sucrose gradient centrifugation. #1, #2, and #3 stand for various pools of the library; #3A and #3B stand for single clones obtained in the sib selection procedure. This figure was kindly provided by A. Werner.

rabbit renal AA transporter in *Xenopus laevis* oocytes has been described.[77] Injection of poly(A)$^+$ RNA extracted from rabbit kidney cortex into oocytes produced over a five-fold increase in the uptake of [^{14}C]ascorbic acid (570 μM) compared to water-injected oocytes. Size fractionation of kidney cortex mRNA by sucrose gradient revealed that the mRNA species that induced AA transporter expression in oocytes is present in a fraction of about 2.0 kilobases (kb) (size range of 1.8–3.1 kb). Injection of the active fraction into oocytes produces a >40-fold increase in AA uptake compared with water-injected controls. Induced uptake of [^{14}C]ascorbic acid after injection of mRNA into oocytes is Na$^+$-dependent and significantly inhibited by unlabelled AA and by its structural analogue isoascorbic acid, but not by D-glucose. Saturation as a function of increasing AA concentration in the incubation medium (100–1000 μM) is evident, with an apparent K_m of 258 ± 72.5 μM and a maximum velocity of 29.6 ± 2.8 pmol × oocyte^{-1} × 2 h^{-1}. Because of these properties the expressed AA transporter seems to be that of the brush-border membrane of the renal reabsorptive cells, where it represents a major reabsorptive mechanism (as characterized by brush border membrane vesicle studies).[61]

Recent descriptions have revealed the isolation of two Na$^+$-dependent AA transporters, namely SVCT1 (sodium-dependent vitamin C transporter 1) and SVCT2 (sodium-dependent vitamin C transporter 2). SVCT1 has been cloned by screening a rat kidney cDNA library for Na$^+$-dependent L-[^{14}C]ascorbic acid transport activity in RNA-injected *Xenopus laevis* oocytes. Subsequent polymerase chain reaction (PCR)-based homology screening has yielded a related cDNA from rat brain coding for SVCT2, which shares with SVCT2 65% identity.[78] SVCT1 and SVCT2 have similar hydropathy profiles, each predicting 12 putative transmembrane spanning domains. These clones show homology to sequences of unknown function reported in data banks. Functional data suggest hyperbolic, high affinity, electrogenic, Na$^+$-dependent AA transport for both clones. Kinetic parameters for SVCT1 isoform, as determined from radiotracer assays, were apparent $K_{0.5}^{AA}$ = 18.7 ± 2.7 μM, V_{max}^{AA} = 3.5 ± 0.1 pmol × oocyte^{-1} × min^{-1}, $K_{0.5}^{Na}$ = 26.8 ± 3.8 mM, while kinetic parameters for SVCT2 were apparent $K_{0.5}^{AA}$ = 9.4 ± 1.9 μM, V_{max}^{AA} = 0.2 ± 0.01 pmol × oocyte^{-1} × min^{-1}, $K_{0.5}^{Na}$ = 10.4 ± 0.6 mM. Na$^+$/AA stoichiometric ratio was ~2 for both SVCT1 and SVCT2. Similar kinetics were obtained by voltage-clamp measurements.[78] SVCT1 is mainly confined to epithelial systems, such as intestine, kidney (straight segment S3 of the proximal tubule), and liver, whereas SVCT2 serves a host of metabolically active cells and specialized tissues in the brain, eye, and several other organs.[78] Taken together, these results indicate the ability of *Xenopus laevis* oocytes to be a suitable system to functionally express brush-border membrane AA transporter(s) theoretically at different tissues and levels.

15.7.3 *Na$^+$-independent transporters involved in DHA absorption in animal cells*

A molecular piece of the model depicted to comprehend how ascorbic acid is distributed in nonpolarized cells has been recently described.[38] Until 1993 it was not clear how AA is taken up into nonpolarized cells. The kinetics of cell and tissue accumulation of vitamin C *in vitro* indicated that the process is mediated by specific transporters at the cell membrane.[35] Some experimental observations had linked vitamin C transport with hexose transport systems in mammalian cells, although no clear information was available regarding the specific role(s) of these transporters, if any, in this process.[16, 37, 59, 61, 70, 79–83] Recently, it has been shown in the *Xenopus laevis* oocyte expression system that the mammalian facilitative hexose transporters, namely GLUT1,[84, 85] GLUT2 and the insulin-sensitive GLUT4[84–87] are efficient transporters of DHA.[38] These hexose transporters cover virtually most of the tissues in mammals, in that GLUT1 is widely distributed and, in particular, is localized in erythroid and brain, GLUT2 is localized in liver, β-cells, kidney and small intestine and GLUT4 is localized in fat and muscle.[88] Two transport pathways, one with low affinity (K_m value of about 3.5 mM and V_{max} of about 190 pmol $\times$ oocyte^{-1} $\times$ min^{-1}) and one with high affinity (K_m value of about 60 μM and V_{max} of about 4.8 pmol $\times$ oocyte^{-1} $\times$ min^{-1}) for DHA have been found in oocytes expressing the mammalian transporter GLUT1 and these oocytes accumulated vitamin C against a concentration gradient when supplied with DHA. In addition, these authors also attributed DHA transport activity to both GLUT2- and GLUT4-expressing oocytes.[38] However, more recent data obtained by *Xenopus laevis* oocytes and Chinese hamster ovary (CHO) cells overexpressing GLUT1–5 isoforms indicate that DHA is rapidly taken up by GLUT1 and GLUT3 and, with much less extent, by GLUT4.[89] Kinetic parameters for GLUT1, as determined by [^{14}C]DHA, were K_m value of about 1.1 mM and V_{max} of about 108 pmol $\times$ oocyte^{-1} $\times$ min^{-1}, while for GLUT3 were K_m value of about 1.7 mM and V_{max} of about 241 pmol $\times$ oocyte^{-1} $\times$ min^{-1}.[89] These observations indicate that mammalian facilitative hexose transporters are physiologically significant pathways for the uptake and accumulation of vitamin C by cells, and suggest that once inside the cell DHA would be reduced to AA, thus allowing AA accumulation against a concentration gradient. It has been well documented in fish that the proportion of DHA/AA is characteristic for each tissue and depends on vitamin C status.[90]

15.8 Perspectives in the study of ascorbic AA in fish: the transgenic approach

15.8.1 Methods for gene transfer in fish

Animals into which foreign genes have been artificially introduced are termed transgenic. In higher vertebrates such as mice, DNA can be injected into the pronuclei of fertilized eggs by a micropipette, and the injected embryos then incubated *in vitro* or implanted into the uterus of a pseudo-pregnant female for subsequent development. By this method, multiple copies of foreign genes may integrate at random locations into the genome of the transgenic animal in head-to-tail or head-to-head tandem arrays. If the foreign gene is introduced with a functional promoter, expression of the related protein is expected in some of the transgenic individuals. Eventually, the foreign gene is transmitted through the germ line to subsequent generations. Transgenic animals may show altered phenotypes. For instance, transgenic mice expressing human or rat growth hormone genes may possess elevated levels of growth hormone and consequently grow much faster than their control siblings.[91]

By microinjection, foreign genes have been introduced also into the fish genome. Transgenic common carp, catfish, goldfish, loach, medaka, salmon, tilapia, rainbow trout, and zebrafish have been generated.[92, 93] Since pronuclei from most fish species studied to date cannot be easily visualized, DNA is usually injected into the cytoplasm instead. Eggs and sperm from mature individuals are collected and fertilization is initiated by adding water and sperm to eggs and gently stirring. About 10^6–10^8 molecules of circular or linearized DNA in a volume of 0.2–20 nl are microinjected into one- to four-cell stage eggs. Since most of teleost fish undergo external fertilization, the injected embryos do not require the complex manipulations essential in mammalian systems, such as *in vitro* culturing of embryos and transferring of embryos into foster mothers. In addition, cytoplasmic injection in fish is less harmful to the embryos than injection into the nucleus, so the survival rate of injected fish embryos is much higher than that of mammalian embryos. Depending on the species, the survival rate of injected fish embryos ranged from 35 to 80%. However, the rates of DNA integration are variable ranging from 5 to 70%.[93] One exception to cytoplasmic injection is found in medaka oocytes in which the pronuclei are visible so that foreign DNA can be injected directly into the oocyte before fertilization.[94, 95] To obtain information about the stable transfer of foreign genes, DNA of presumptive transgenic animals is extracted from biopsy tissues and subjected to Southern blot hybridization, using the radiolabeled foreign gene as a probe, and/or to polymerase chain reaction (PCR) analysis, using specific oligonucleotides. Putative transgenic individuals can be subsequently raised for further studies and eventually used for producing transgenic stocks. Methods for foreign DNA insertion alternative to the tedious and time-consuming procedure of cytoplasmic microinjection, such as electroporation,[96] liposomes,[97] and infection by retroviral vectors,[98] are being developed mostly to face problems related to low efficiency of stable gene transfer and/or mass gene transfer.

15.8.2 *Transgenic expression of L-gulono-γ-lactone oxidase in the scurvy-prone medaka fish*

It is conceptually and practically possible to introduce useful genetic characteristics (as enhanced growth rates, disease resistance, cold resistance, etc.), to commercially important fish species, replacing or complementing the conventional genetic breeding approach. As stated above, in scurvy-prone species, lack of L-gulono-γ-lactone oxidase activity is not due to the absence of the gene but to mutations in the nucleotide sequence. Correction of this deficiency has recently been achieved by introduction of an active gene for L-gulono-γ-lactone oxidase into germ line of a scurvy-prone teleost species, namely medaka (*Oryzias latipes*). Rat liver GLO cDNA contained in the plasmid pSVL, a simian virus 40-bases eukaryotic expression vector, has been microinjected into the cytoplasm of fertilized one-cell stage eggs. Four male F_0 fish having the transgene in their germ line cells have come to maturity. F_1 progeny derived from one of the F_0 fish possessed L-gulono-γ-lactone oxidase activity, indicating that the transgene was functionally expressed in the fish. By genomic Southern blot analysis, the transgene has been found in both chromosome-integrated and extrachromosomal forms. This successful attempt is noteworthy. In fact, the application of transgenic technology in fisheries has been mainly confined to trials on growth promotion,[99, 100] but not to therapy of enzyme deficiencies. To our knowledge this is the first report demonstrating the gene transfer to fish of an enzyme that has been altered during evolution. This finding opens new and useful experimental designs to study gene therapy in fish species by a transgenic approach.

Acknowledgments

This work was supported as part of an international cooperative research program between the University of Lecce, Italy, and the University of Hawaii, U.S. Funding for this program was received from the Italian Ministry of University, Scientific, and Technological Research, the Programma Nazionale Ricerche in Antartide, the Ministry of Agriculture, Food, and Forestry (Fourth Plan on Fisheries and Aquaculture in Marine and Brackish Waters), and the Progetto Finalizzato Biotecnologie (Grant N. 99.00490.PF49) of the Italian Research National Council.

References

1. Tucker, B. W. and Halver, J. E., Ascorbate-2-sulfate metabolism in fish, *Nutrition Rev.*, 42, 173, 1984.
2. Rose, R. C. and Bode, A. M., Biology of free radical scavengers: an evaluation of ascorbate, *FASEB J.*, 7, 1135, 1993.
3. Chatterjee, I. B., Evolution and the biosynthesis of ascorbic acid, *Science*, 182: 1271, 1973.
4. Chatterjee, I. B., Majumder, A. K., Nandi, B. K., and Subramanian, N., Synthesis and some major functions of vitamin C in animals, *Ann. N.Y. Acad. Sci.*, 258, 24, 1975.
5. Nakajima, Y., Shanta, T. R., and Bourne, G. H., Histochemical detection of L-gulonolactone: phenazine methosulfate oxidoreductase activity in several mammals with special reference to synthesis of vitamin C in primates, *Histochemie*, 18, 293, 1969.

6. Birney, E. C., Jenness, R., and Ayaz, K. M., Inability of bats to synthesise L-ascorbic acid, *Nature*, 260, 626, 1976.
7. Chaudhuri, C. R. and Chatterjee, I. B., L-ascorbic acid synthesis in birds: phylogenetic trend, *Science*, 164, 435, 1969.
8. Touhata, K., Toyohara, H., Mitani, T., Kinoshita, M., Sato, M., and Sakaguchi, M., Distribution of L-gulono-1,4-lactone oxidase among fishes, *Fisheries Sci.*, 61, 729, 1995.
9. Nishikimi, M., Fukuyama, R., Minoshima, S., Shimizu, N., and Yagi, K., Cloning and chromosomal mapping of the human nonfunctional gene for L-gulono-γ-lactone oxidase, the enzyme for L-ascorbic acid biosynthesis missing in man, *J. Biol. Chem.*, 269, 13685, 1994.
10. Nishikimi, M., Kawai, T., and Yagi, K., Guinea pigs possess a highly mutated gene for L-gulono-γ-lactone oxidase, the key enzyme for L-ascorbic acid biosynthesis missing in this species, *J. Biol. Chem.*, 267, 21967, 1992.
11. Rose, R. C., Intestinal absorption of water soluble vitamins, *Proc. Soc. Exp. Biol. Med.*, 212, 191, 1996.
12. McLaren, B. A., Keller, E., O'Donnell, D. J., and Elvehjem, C. A., The nutrition of the rainbow trout. I. Studies of vitamin requirements, *Arch. Biochem. Biophys.*, 15, 169, 1947.
13. Kitamura, S., Suwa, T., Ohara, S., and Nakamura, K., Studies on vitamin requirements of the rainbow trout *Salmo gairdneri*. I. On the ascorbic acid, *Bull. Jpn. Soc. Sci. Fish.*, 33, 1120, 1965.
14. Poston, H. A., Effect of dietary L-ascorbic acid on immature brook trout, *Fish. Res. Bull.*, Cortland, New York, 30, 46, 1967.
15. Halver, J. E., Ashley, L. M., and Smith, R. R., Ascorbic acid requirements of coho salmon and rainbow trout, *Trans. Am. Fish. Soc.*, 90, 762, 1969.
16. Wilson, R. P. and Poe, W. E., Impaired collagen formation in the scorbutic channel catfish, *J. Nutr.*, 103, 1359, 1973.
17. Agrawal, N. K. and Mahajan, C. L., Nutritional deficiency disease in an Indian major carp, *Cirrhina mrigala* Hamilton, due to avitaminosis C during early growth, *J. Fish. Dis.*, 3, 231, 1980.
18. Soliman, A. K., Jauncey, K., and Roberts, R. H., The effect of dietary ascorbic acid supplementation on hatchability, survival rate and fry performance in *Oreochromis mossambicus* (Peters), *Aquaculture*, 59, 197, 1986.
19. Lim, C. and Lovell, R. T., Pathology of the vitamin C deficiency syndrome in channel catfish (*Ictalurus punctatus*), *J. Nutr.*, 108, 1137, 1978.
20. Soliman, A. K., Jauncey, K., and Roberts, R. H., The effect of varying forms of dietary ascorbic acid on the nutrition of juvenile tilapias (*Oreochromis niloticus*), *Aquaculture*, 52, 1, 1986.
21. Yamamoto, Y., Sato, M., and Ikeda, S., Existence of L-gulonolactone oxidase in some teleosts, *Bull. Jpn. Soc. Sci. Fish.*, 44, 775, 1978.
22. Moreau, R., Stockage tissulaire et utilisation de l'acide ascorbique. Evaluation des symptomes de carence en vitamine C chez l'esturgeon siberien (*Acipenser baeri*), Diplome d'Etudes Approfondies Biologie et Agronomie, Université de Rennes I, France, 1992.
23. Dabrowski, K., Primitive Actinopterigian fishes can synthesize ascorbic acid, *Experientia*, 50, 745, 1994.
24. Patterson, J. W. and Lazarow, A., Sulphydryl protection against dehydroascorbic acid diabetes, *J. Biol. Chem.*, 186, 141, 1950.

25. Bianchi, J. and Rose, R. C., Dehydroascorbic acid and cell membranes: possible disruptive effects, *Toxicology*, 40, 75, 1986.
26. Raghoebar, M., Huisman, J. A. M., van den Berg, W. B., and van Ginneken, C. A. M., Characteristics of the transport of ascorbic acid into leukocytes, *Life Sci.*, 40, 499, 1987.
27. Pillsbury, S., Watkins, D., and Cooperstein, S. J., Effect of dehydroascorbic acid on permeability of pancreatic islet tissue *in vitro*, *J. Pharmacol. Exp. Ther.*, 185, 713, 1973.
28. Bianchi, J. and Rose, R. C., Na$^+$-independent dehydro-L-ascorbic acid uptake in renal brush-border membrane vesicles, *Biochim. Biophys. Acta*, 819, 75, 1985.
29. Bode, A. M., Cunningham, L., and Rose, R. C., Spontaneous decay of oxidized ascorbic acid (dehydro-L-ascorbic acid) evaluated by high pressure liquid chromatography, *Clin. Chem.*, 36, 1807, 1990.
30. Jung, C. H. and Wells, W. W., Spontaneous conversion of L-dehydroascorbic acid to L-ascorbic acid and L-erythroascorbic acid, *Arch. Biochem. Biophys.*, 355, 9, 1998.
31. Dabrowski, K., Matusiewicz, M., and Blom, J. H., Hydrolysis, absorption and bioavailability of ascorbic acid esters in fish, *Aquaculture*, 124, 169, 1994.
32. Dabrowski, K., Lackner, R., and Doblander, C., Effect of dietary ascorbate on the concentration of tissue ascorbic acid, dehydroascorbic acid, ascorbic sulfate and activity of ascorbate sulfate sulfohydrolase in rainbow trout (*Oncorynchus mykiss*), *Can. J. Aquat. Sci.*, 47, 1518, 1990.
33. Hilton, J. W., Cho, C. Y., and Slinger, S. J., Effect of graded levels of supplemental ascorbic acid in practical diets fed to rainbow trout (*Salmo gairdneri*), *J. Fish. Res. Board Can.*, 35, 431, 1978.
34. Dabrowski, K., Hinterleitner, S., Sturmbauer, C., El-Fiky, N., and Wieser, W., Do carp larvae require vitamin C?, *Aquaculture*, 72, 295, 1988.
35. Rose, R. C., Transport of ascorbic acid and other water-soluble vitamins, *Biochim. Biophys. Acta*, 947, 335, 1988.
36. Crane, R. K., Hypothesis for mechanism of intestinal active transport of sugars, *Fed. Proc.*, 21, 891, 1962.
37. Bianchi, J., Wilson, F. A., and Rose, R. C., Dehydroascorbic acid and ascorbic acid transport systems in the guinea pig ileum, *Am. J. Physiol.*, 250, G461, 1986.
38. Vera, J. C., Rivas, C.I., Fischbarg, J., and Golde, D. W., Mammalian facilitative hexose transporters mediate the transport of dedydroascorbic acid, *Nature*, 364, 79, 1993.
39. Helbig, H., Korbmacher, C., Wohlfarth, J., Berweck, S., Kuhner, D., and Wiederholt, M., Electrogenic Na-ascorbate cotransport in cultured bovine pigmented ciliary epithelial cells, *Am. J. Physiol.*, 256, C44, 1989.
40. Mellors, A. J., Nahrwold, D. L., and Rose, R. C., Ascorbic acid flux across mucosal border of guinea pig and human ileum, *Am J. Physiol.*, 233, E374, 1977.
41. Rose, R. C. and Choi J. L., Intestinal absorption and metabolism of ascorbic acid in rainbow trout, *Am. J. Physiol.*, 258, R1238, 1990.
42. Rose, R. C., Li, J. K., and Koch, M. J., Intestinal transport and metabolism of oxidized ascorbic acid (dehydroascorbic acid), *Am. J. Physiol.*, 254, G824, 1988.
43. Stevenson, N. and Brush, M., Existence and characteristics of Na$^+$-dependent active transport of ascorbic acid in guinea pig, *Am. J. Clin. Nutr.*, 22, 318, 1969.
44. Schultz, S. G., Curran, P. F., Chez, R. A., and Fuisz, R. A., Alanine and sodium fluxes across the mucosal border of rabbit ileum, *J. Gen. Physiol.*, 50, 1241, 1967.

45. Schmitz, J., Preiser, H., Maestracci, D., Ghosh, B. K., Cerda, J., and Crane, R. K., Purification of the human intestinal brush border membrane, *Biochim. Biophys. Acta,* 323, 98, 1985.

46. Kessler, M., Acuto, O., Storelli, C., Murer, H., Muller, H., and Semenza, G., A modified procedure for the rapid preparation of efficiently transporting vesicles from small intestinal brush border. Their use in investigating some properties of D-glucose and choline transport systems, *Biochim. Biophys. Acta,* 506, 136, 1978.

47. Crane, R. K., Bogé, G., and Rigal, R., Isolation of brush border membranes in vesicular form from the intestinal spiral valve of the small dogfish (*Scyliorhinus canicula*), *Biochim. Biophys. Acta,* 554, 264, 1979.

48. Eveloff, J., Field, M., Kinne, R., and Murer, H., Sodium-cotransport systems in intestine and kidney of the winter flounder, *J. Comp. Physiol.,* 135, 175, 1980.

49. Storelli, C., Vilella, S., and Cassano, G., Na-dependent D-glucose and L-alanine transport in eel intestinal brush border membrane vesicles, *Am. J. Physiol.,* 251, R463, 1986.

50. Reshkin, S. J. and Ahearn, G. A., Effects of salinity adaptation on glucose transport by intestinal brush border membrane vesicles of a euryhaline teleost, *Am. J. Physiol.,* 252, R567, 1987.

51. Scalera, V., Storelli, C., Storelli-Joss, C., Haase, W., and Murer, H., A simple and fast method for the isolation of basolateral plasma membranes from rat small intestinal epithelial cells, *Biochem. J.,* 186, 177, 1980.

52. Reshkin, S. J., Vilella, S., Cassano, G., Ahearn, G. A., and Storelli, C., Basolateral amino acid and glucose transport by the intestine of the teleost *Anguilla anguilla, Comp. Biochem. Physiol.,* 91A, 779, 1988.

53. Reshkin, S. J., Vilella, S., Ahearn, G. A., and Storelli, C., Basolateral inositol transport by intestines of carnivorous and herbivorous teleosts, *Am. J. Physiol.,* 256, G509, 1989.

54. Vilella, S., Reshkin, S. J., Storelli, C., and Ahearn, G. A. Brush border inositol transport by intestines carnivorous and herbivorous teleosts, *Am. J. Physiol.,* 256, G501, 1989.

55. Dabrowski, K. and Kock, G., Absorption of ascorbic acid and ascorbic sulfate and their interaction with minerals in the digestive tract of rainbow trout (*Onchorhynchus mykiss*), *Can. J. Fish. Aqua. Sci.,* 46, 1952, 1989.

56. Dabrowski, K., Gastrointestinal circulation of ascorbic acid, *Comp. Biochem. Physiol.,* 95A, 481, 1990.

57. Patterson, L. T., Nahrwold, D. L., and Rose, R. C., Ascorbic acid uptake in guinea pig intestinal mucosa, *Life Sci.,* 31, 2783, 1982.

58. Maffia, M., Ahearn, G. A., Vilella, S., Zonno, V., and Storelli, C., Ascorbic acid transport by intestinal brush-border membrane vesicles of the teleost *Anguilla anguilla, Am. J. Physiol.,* 264, R1248, 1993.

59. Siliprandi, L., Vanni, P., Kessler, M., and Semenza, G., Na$^+$-dependent, electroneutral L-ascorbate transport across brush border membrane vesicles from guinea pig small intestine, *Biochem. Biophys. Acta,* 552, 129, 1979.

60. Hornig, D., Weber, F., and Wiss, O., Site of intestinal absorption of ascorbic acid in guinea pigs and rats, *Biochem. Biophys. Res. Commun.,* 52, 168, 1973.

61. Toggenburger, G., Hausermann, M., Mutsch, B., Genoni, G., Kessler, M., Weber, F., Hornig, D., O'Neill, B., and Semenza, G., Na$^+$-dependent, potential-sensitive L-ascorbate transport across brush-border membrane vesicles from kidney cortex, *Biochim. Biophys. Acta,* 646, 422, 1981.

62. Kapoor, B. G., Smit, H., and Verghira, I. A., The alimentary canal and digestion in teleosts, in *Advances in Marine Biology*, Russell, F.S. and Younge, C.H., Eds., Academic, London, 1975, vol. 13, 102.

63. Maffia, M., Verri, T., Danieli, A., Thamotharan, M., Pastore, M., Ahearn, G.A., and Storelli, C., H^+-glycyl-L-proline cotransport in brush-border membrane vesicles of eel (*Anguilla anguilla*) intestine, *Am. J. Physiol.*, 272, R217, 1997.

64. Robinson, E. H., *Vitamin C studies with catfish: requirement, biological activity, and stability*, Mississippi Agricultural and Forestry Experimental Station, Tech. Bull. 182, Mississippi State, MS, 1992, 8.

65. Buddington, R. K., Puchal, A. A., Houpe, K. L., and Diehl, W. J. III, Hydrolysis and absorption of two monophosphate derivatives of ascorbic acid by channel catfish *Ictalurus punctatus* intestine, *Aquaculture*, 114, 317, 1993.

66. Maffia, M., Acierno, R., Cillo, E., and Storelli C., Na^+-D-glucose cotransport by intestinal BBMVs of the Antarctic fish *Trematomus bernacchii*, *Am. J. Physiol.*, 271, R1576, 1996.

67. Ahearn, G. A. and Storelli, C., Use of vesicle techniques to characterize nutrient transport processes of the teleost gastrointestinal tract, in *Biochemistry and Molecular Biology of Fishes*, Hochachka, P.W. and Mommsen, T.P., Eds, Elsevier Science, New York, 1994, vol. 3, chap. 43, 513.

68. El Naggar, G. O. and Lovell, R. T., L-ascorbyl-2-monophosphate has equal anti-scorbutic activity as L-ascorbic acid but L-ascorbyl-2-sulfate is inferior to L-ascorbate for channel catfish, *J. Nutr.*, 121, 1622, 1991.

69. Dabrowski, K., Lackner, R., and Doblander, C., Ascorbate-2-sulfate sulfohydro-lase in fish and mammal. Comparative characterization and possible involve-ment in ascorbate metabolism, *Comp. Biochem. Physiol. B*, 104, 717, 1993.

70. Rose, R. C., Ascorbic acid transport in mammalian kidney, *Am. J. Physiol.*, 250, F627, 1986.

71. Friedman, G., Sherry, S., and Ralli, E., The mechanism of the excretion of vita-min C by the human kidney at low and normal plasma levels of ascorbic acid, *J. Clin. Invest.*, 19, 685, 1940.

72. Hediger, M. A., Coady, M. J., Ikeda T. S., and Wright, E. M., Expression cloning and cDNA sequencing of the Na^+/glucose cotransporter, *Nature*, 330, 379, 1987.

73. Hirayama, B. A., Wong, H. C., Smith, C. D., Hagenbuch B. A., Hediger, M. A., and Wright, E. M., Intestinal and renal Na^+/glucose cotransporters share com-mon structures, *Am. J. Physiol.*, 261, C296, 1991.

74. Hagenbuch, B., Stieger, B., Fogere, M., and Mejer, P. J., Functional expression cloning and characterization of the hepatocyte Na^+/bile acid cotransport sys-tem, *Proc. Natl. Acad. Sci. U.S.A.*, 88, 10629, 1991.

75. Kanai, Y. and Hediger, M. A., Primary structure and functional characterization of a high-affinity glutamate transporter, *Nature*, 360, 467, 1992.

76. Werner, A., Moore, M. L., Mantei, N., Biber, J., Semenza, G., and Murer, H., Cloning and expression of cDNA for a Na/P_i-cotransport system of kidney cor-tex, *Proc. Natl. Acad. Sci. U.S.A.*, 88, 9608, 1991.

77. Dyer, D. L., Kanai, Y., Hediger, M. A., Rubin, S. A., and Said, H. M., Expression of a rabbit renal ascorbic acid transporter in *Xenopus laevis* oocytes, *Am. J. Physiol.*, 267, C301, 1994.

78. Tsukaguchi, H., Tokui, T., MacKenzie, B., Berger, U. V., Chen, H. Z., Wang, Y., Brubaker, R. F., and Hediger, M. A., A family of mammalian Na^+-dependent L-ascorbic acid transporters, *Nature*, 399, 70, 1999.

79. Bigley, R., Wirth, M., Layman, D., Riddle, M., and Stankova, L., Interaction between glucose and dehydroascorbate transport in human neutrophils and fibroblasts, *Diabetes,* 32, 545, 1983.
80. Ingermann, R. L., Stankova, L., and Bigley, R. H., Role of monosaccharide transporter in vitamin C uptake by placental membrane vesicles, *Am. J. Physiol.,* 250, C637, 1986.
81. Padh, H. and Aleo, J. J., Characterization of the ascorbic acid transport by 3T6 fibroblasts, *Biochim. Biophys. Acta,* 901, 283, 1987.
82. McLennan, S., Yue, D. K., Fisher, E., Capogreco, C., Heffernan, S., Ross, G. R., and Turtle, J. R., Deficiency of ascorbic acid in experimental diabetes. Relationship with collagen and polyol pathway abnormalities, *Diabetes,* 37, 359, 1988.
83. Washko, P. and Levine, M., Inhibition of ascorbic acid transport in human neutrophils by glucose, *J. Biol. Chem.,* 267, 23568, 1992.
84. Birnbaum, M. J., Haspel, H. C., and Rosen, O. M., Cloning and characterization of a cDNA encoding the rat brain glucose-transporter protein, *Proc. Natl. Acad. Sci. U.S.A.,* 83, 5784, 1986.
85. Vera, J. C. and Rosen, O. M., Functional expression of mammalian glucose transporters in *Xenopus laevis* oocytes: evidence for cell-dependent insulin sensitivity, *Mol. "'l. Biol.,* 9, 4187, 1989.
86. Vera, J. C., and Rosen, O. M., Reconstitution of an insulin signaling pathway in *Xenopus laevis* oocytes: coexpression of a mammalian insulin receptor and three different mammalian hexose transporters, *Mol. Cell. Biol.,* 10, 743, 1990.
87. Thorens, B., Sarkar, H. K., Kaback, H. R., and Lodish, H. F., Cloning and functional expression in bacteria of a novel glucose transporter present in liver, intestine, kidney, and beta-pancreatic islet cells, *Cell,* 55, 281, 1988.
88. Hediger, M. A. and Rhodes, D. B., Molecular physiology of sodium-glucose cotransporters, *Physiol. Rev.,* 74, 993, 1994.
89. Rumsey, S. C., Kwon, O., Xu, G. W., Burant C. F., Simpson, I., and Levine, M., Glucose transporter isoforms GLUT1 and GLUT3 transport dehydroascorbic acid, *J. Biol. Chem.,* 272, 18982, 1997.
90. Dabrowski, K., Absorption of ascorbic acid and ascorbic sulfate and ascorbate metabolism in common carp (*Cyprinus carpio* L.), *J. Comp. Physiol. B,* 160, 549, 1990.
91. Palmiter, R. D., Brinster, R. L., Hammer, R. E., Trumbauer, M. E., Rosenfeld, M. G., Birnber, N. C., and Evans, R. M., Dramatic growth of mice that develop from eggs microinjected with metallothionein-growth hormone fusion genes, *Nature,* 300, 611, 1982.
92. Chen, T. T. and Powers D. A., Transgenic fish, *Trends Biotechnol.,* 8, 209, 1990.
93. Hew, C. L. and Fletcher, G. L., *Transgenic fish,* World Scientific, Singapore, 1992.
94. Ozato, K., Inoue, K., and Wakanatsu, Y., Transgenic fish: biological and technical problems, *Zool. Sci.,* 6, 445, 1989.
95. Ozato, K., Kondoh H., Inohara H., Iwamatsu T., Wakamatsu Y., and Okada T. S., Production of transgenic fish: introduction and expression of chicken delta-crystallin gene in medaka (*Oryzias latipes*) embryos, *Cell Differ.,* 19, 237, 1986.
96. Shigekawa, K. and Dower, W. J., Electroporation of eukaryotes and prokaryotes: a general approach to the introduction of macromolecules into cells, *BioTechniques,* 6, 742, 1988.

97. Felgner, P. L., Gadek, T. R., Holm, R., Chan, H. W., Wenz, M., Northrup, J. P., Ringold, G. M., and Danielsen, M., Lipofection: a highly efficient, lipid-mediated DNA-transfection procedure, *Proc. Natl. Acad. Sci. U.S.A.*, 84, 7413, 1987.
98. Varmas, H., *RNA Tumor Viruses*, Weiss, R. and Coffin, J., Eds., Cold Spring Harbor, 1982, 363.
99. Maclean, N., Penman D., and Zhu Z., Introduction of novel genes into fish, *Bio/Technology*, 5, 257, 1987.
100. Brem, G., Brenig, B., Horstgen-Schwark, G., and Winnacker, E.L., Gene transfer in tilapia (*Oreochromis niloticus*), *Aquaculture*, 68, 209, 1988.

chapter sixteen

Critical review of the effects of ascorbic acid on fish behavior

Shunsuke Koshio

Contents

16.1   Introduction

Large numbers of fish seedlings have been produced for a long time in Japan to provide fish for farmers and also to release in open waters (Table 16.1). Since the transition of artificially raised seedlings and juvenile fish from the hatchery to a wild habitat will be very critical, the seedlings should be expected to have

Table 16.1 Seedling production of major cultured fish species for restocking
and culture in Japan (1996)

Number of seedlings produced ($\times$ 1000)			
Common name	Scientific name	Restocking	Culture	Total
Red sea bream	*Pagrus major*	30330	88762	119092
Black sea beam	*Acanthopagrus schlegeli*	11630	649	11019
Striped knifejaw	*Oplegnathus fasciantus*	55	216	271
Japanese flounder	*Paralichtyus olivacesu*	33427	14240	47667
Striped jack	*Caranx delicatissimus*	605	2038	2643
Yellowtail	*Seriola quinqueradiata*	465	4	469

some characteristics such as normal morphology, low standard metabolism, quick acclimatization to the environmental changes, good health and resistance to disease, to have developed escape behavior from predators, and to have a strong association with species-specific normal behavior for maintaining the high survival and growth rates under open water habitat after release.[1]

Recently, several studies on the relationship between the seedlings' quality and behavioral pattern have been carried out to determine how to improve the quality of seedlings. These indicated that the restocking effectiveness could be greatly affected by the behavioral patterns of fish, rather than the general performance of seedlings.[2]

It has been reported that environmental factors such as temperature, feeding conditions, light intensity, water depth, circadian rhythms,[3] and fish density[4] controlled behavioral patterns of fish. In red sea breams, for example, higher survival, rates after restocking were obtained from those fish which showed the typical "frightened" response. This "tilting behavior" increased survival as long duration of this behavior or the high frequency of this behavior can result in avoidance of the predation.[5,6] Furthermore, Furuta[7] demonstrated that the ability of predator avoidance in Japanese flounder was higher in wild fish than in cultured individuals. Thus, it is very important to produce seedlings which have developed normal feeding and antipredator behaviors similar to those of their wild counterparts during the course of seedling production.

The brain is one of the most important organs that controls fish behavior, and many nutrients are responsible for maintaining normal behavior functions in animals. Micronutrients such as ascorbic acids (AA), docosahexaenoic acid, and phospholipids, are present in large amounts in the brain, as compared to other nutrients in fish.[8] However, there is a lack of information on the nutritional factors that control behavioral patterns of fish. Mammalian studies indicated that ascorbate acts as a neuromodulator and affects noradrenergic, dopaminergic, and cholonergic neurotransmission.[9–11] We have investigated the effect of two nutrients (AA and docosahexaenoic acid) on the changes of behavioral patterns of fish.

Our findings on ayu (*Plecoglossus altivelis*), yellowtail (*Seriola quinque-radiata*), and red sea bream (*Pagrus major*) are reviewed, and the implications and problems are discussed. Some applications are discussed concerning the improvement of seedling quality, which can be controlled by micronutrient intake.

16.2   Background

This chapter reviews the recent progress of the role of AA in relation to the behavior patterns of fish, which might affect the efficiency of stock enhancement. The survival rates and adaptability to the wild habitat after releasing depend on the formation of normal behavior of fish. The rearing trials were conducted to investigate whether dietary ascorbic acids change behavioral patters of ayu, yellowtail, and red sea bream.

In smaller ayu, the schooling rate and the distance to the nearest neighbor were highest in fish fed the diet containing 144 mg AA/kg. Aggressive behavior and distance to the nearest neighbor of larger ayu were greater in fish fed AA-supplemented diets than in those fed an AA-free diet. However, ayu fed an AA-free diet did not show any schooling behavior. Yellowtail fingerlings with the highest brain AA level expressed similar behavioral patterns to wild fish, and had a higher resistance against stresses. The red sea bream showed different tilting patterns under the varied AA intake, but it depended on the level of dietary docosahexaenoic acid. The tilting percentage increased with increased AA intake but this was not affected by AA intake when fed the diet containing 3.2% dietary docosahexaenoic acid (DHA). Dietary DHA levels did not affect the tilting duration, but it was lowest when fed the AA-free diet. Furthermore, after a 6-week trial, tilting duration was affected by the dietary AA level, in which it was higher in fish fed the diet containing more than 40 mg AA and 3.2% DHA.

16.3   Studies on the effects of ascorbic acid (AA) intake on fish behaviors

We used L-ascorbyl-2-monophosphate-Mg (APM) as a vitamin C source because of its bioavailability, convenience, and efficient application.

16.3.1   Young stage ayu, Plecoglossus altivelis

We conducted our behavior experiments using 10 g ayu to investigate the effects of AA.[12] We prepared four experimental diets containing APM as a vitamin C source with 0, 39, 327, and 1176 mg/kg diet (AA equivalent 0, 18, 150, and 541 mg/kg diet), respectively. In a feeding trial, 250 fish were placed in a 500 L polycarbonate round tank with a flow-through system of freshwater and fed test diets for 36 days. At the end of the trial, we randomly chose several fish for the behavior experiment and observed the schooling behavior using a video camera system with ten fish from each treatment group.

There was no correlation between jumping behavior,[13] air dive (survival rate after exposure to the air), and swimming activity[14] of ayu and vitamin C intake. However, ayu fed an AA-free diet did not show any schooling behavior. Aggressive behavior and distance to the nearest neighbor were greater in fish fed AA-supplemented diets than in those fed an AA-free diet. Particularly, fish fed a diet with 150 mg AA/kg showed the greatest aggressive behavior (Figure 16.1). Since more than 10 g of wild ayu show the aggressive behavior to protect their territory, feeding a high level of vitamin C to artificially raised ayu can demonstrate the similar aggressive behavior that was found in wild fish.

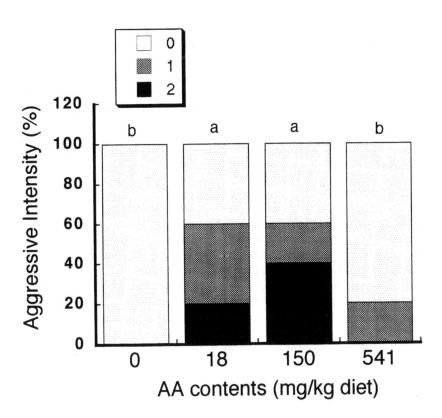

Figure 16.1 Aggressive behavior of ayu fed different contents of ascorbic acid for 36 days. The values with similar letters are not statistically significant (p > 0.05), 0: no aggression, 1: slight aggression, 2: strong aggression.

16.3.2 *Juvenile stage ayu*, Plecoglossus altivelis

We prepared four experimental diets containing 0, 15, 144, and 454 mg AA /kg diet.[15] Those diets were given to 3-g ayu juveniles for 30 days and 130 days, respectively. Filtered freshwater was filled in a 300 L roundtank and supplied at 2 L/min with the water temperature remaining constant at 12°C.

After a 30-day trial, we randomly chose 160 fish from each treatment group for the jumping behavior, 40 fish for the swimming activity test, 10 fish for the air dive test, and 10 fish for the schooling behavior observations, respectively. In the 30-day trial, ayu survived very well (more than 95% survival rate) and there were no vitamin C deficiency signs. The condition factor was the highest in fish fed the diet containing 144 mg AA, and both liver and brain AA concentrations increased with increased intake of AA. The schooling rate and the distance to the nearest neighbor were the highest in fish fed the diet containing 144 mg AA, whereas the swimming activity and jumping behavior were not affected by intake of vitamin C (Figure 16.2). In the 130-day trial, the aggressive behavior was higher in fish fed the diet containing 144 and 454 mg AA. Swimming activity and the distance to the nearest neighbor were observed as the highest in fish fed the diet containing 144 mg AA (Figure 16.3).

We also measured the plasma cortisol concentration to estimate stress conditions of fish fed test diets for 30 days and 130 days. Results demonstrated (Figure 16.4) that in the 30-day study the concentration was the lowest in fish fed 144 mg AA, and fish fed less than 40 mg AA showed higher concentrations. This indicated that fish with 144 mg AA were less stressed and can be linked with the better behavioral performance. In the 130-day trial, fish fed the diet containing 144 mg AA indicated the highest cortisol concentration. This higher value might be connected with the strong aggressive behavior, which was observed in this fish group. Furthermore, we performed the bacteria challenge test using *Vibrio anguillarum* on the fish fed test diets for 130 days. Ten fish were randomly chosen from each treatment group and subjected to a bacterial suspension for 5 min at 23.6°C and then transferred to the rearing tank for evaluating the survival. In the group of fish fed the diet containing no AA only 10 to 20% survived, whereas in those fed diets containing more than 144 mg AA almost 90% survived for 100 h.

16.3.3 *Fingerling stage yellowtail*, Seriola quinqueradiata

We conducted similar experiments to those described above on yellowtail fingerlings fed test diets containing ascorbic acid at the levels of 0.7, 157, 480, and 882 mg/kg diet.[16] Test diets were fed to 500 juveniles (mean TL: 3.6 cm, WT: 0.4 g) per treatment with duplicate tanks (500 L round type) for 20 days. The relatively short trial period was based on our preliminary observations, in which the effect of AA on behavior appeared in a short period of

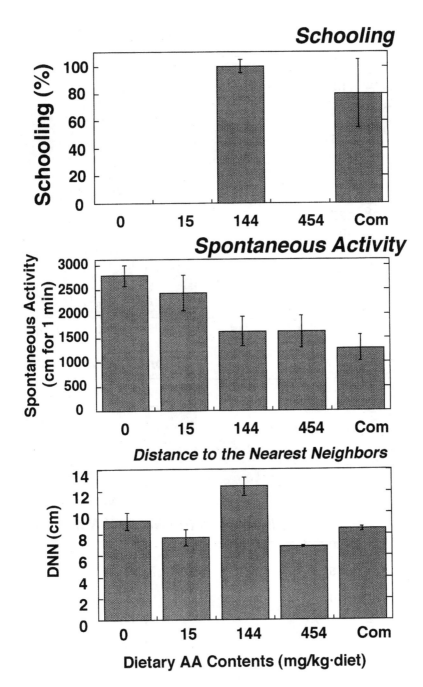

Figure 16.2   Schooling rate (%), spontaneous activity (cm/min), and DNN (cm) of
ayu fed diets containing different contents of AA for 30 days. Com indicates the commercial diet.

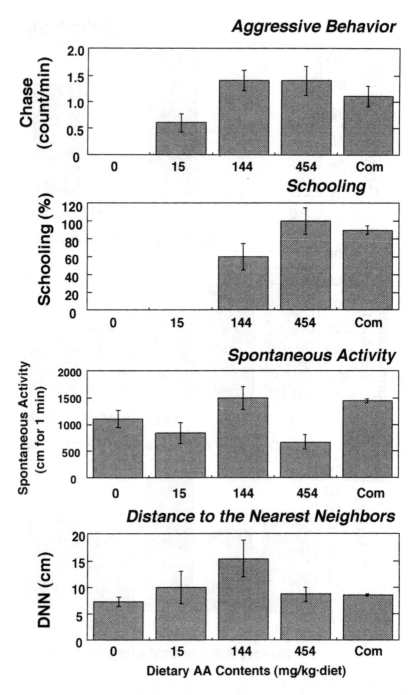

Figure 16.3   Aggressive behavior (chase, count/min), schooling rate (%), spontaneous activity (cm/min), and DNN (cm) of ayu fed diets containing different contents of AA for 130 days. Com indicates the commercial diet.

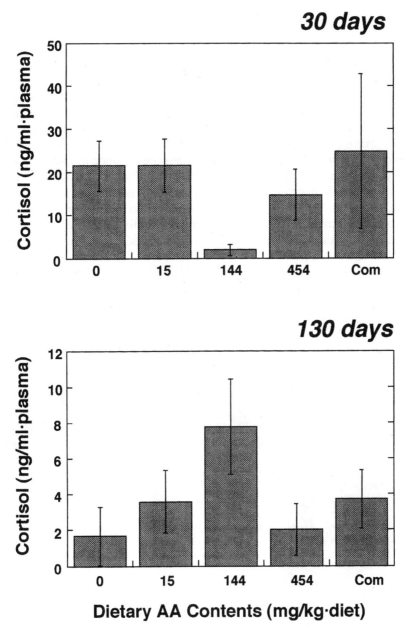

Figure 16.4 Plasma cortisol concentration (ng/ml) of ayu fed diets containing differ-ent contents of AA for 30 or 130 days.

time compared to the deficiency experiment. After the 20-day trial, we randomly sampled 10 to 20 juveniles from each treatment group for the test of "air dive" and schooling behavior. We conducted the "air dive" test to determine the degree of resistance to stress by counting the number of survivors after exposure to air. We recorded the schooling behavior such as schooling pattern, swimming activity, and individual distance using the video system, and the data obtained were analyzed by the computer graphic system.[16]

Survival was over 90% and there was no signs of AA deficiency in any group during the 20-day trial. The growth was not affected by different AA intake. The tolerance against air exposure seemed poor in fish fed the AA-free diet whereas fish fed diets containing more than 480 mg AA/kg showed the stronger tolerance against air exposure. Figures 16.5 and 16.6 showed the typical schooling and aggregation patterns which were observed in the yellowtail experiment. The schooling rate was significantly higher in fish fed test diets containing 480 or 882 mg AA (Figure 16.7). When we observed the AA concentrations in the liver and brain, the former increased with increased AA intake, whereas the latter leveled off at 480 mg AA. This study demonstrated that yellowtail juveniles with the highest brain AA level, by feeding the diets containing 480 mg AA/kg diet, expressed similar behavioral patterns to wild fish, and have a higher resistance against stresses. In other

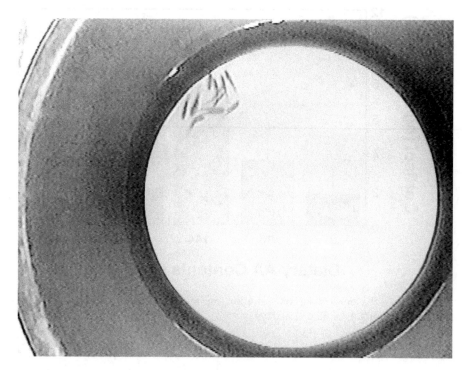

Figure 16.5   Aggregation pattern of yellowtail fed the diet containing no AA.

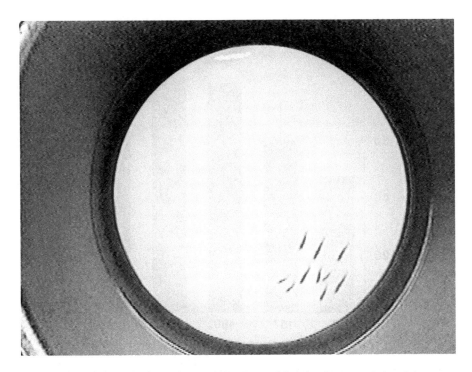

Figure 16.6 Schooling pattern of yellowtail fed the diet containing AA.

words, it is speculated based on this study that juvenile yellowtail fed the diet containing more than 480 mg AA would be able to show better growth and survival under natural conditions after release.

16.3.4 *Juvenile stage red sea bream,* Pagrus major

We observed the tilting behavior of red sea bream juveniles (2.6 g) when fed diets containing 0, 10, 40, 210, and 1000 mg AA/kg diet together with two levels of docosahexaenoic acids (1.6 and 3.2%).[17] To avoid the interference among fish in a tank, the individual rearing was applied in this study, and tilting frequency and duration were observed after a 14-day feeding trial.

The red sea bream showed different tilting patterns under the varied AA intake, but it depended on the level of dietary DHA. The tilting percentage, for example, increased with increased AA intake. The highest percentage was observed in fish fed the diet containing 1000 mg AA while the tilting percentage was not affected by AA intake when fed the diet containing 3.2% dietary docosahexaenoic acid. Dietary docosahexaenoic acid levels did not affect the tilting duration, but it was the lowest when fed the AA-free diet. Furthermore, after a 6-week trial, tilting duration was affected by the dietary AA level, in which it was higher in fish fed the diet containing more than 40 mg AA and 3.2% DHA.

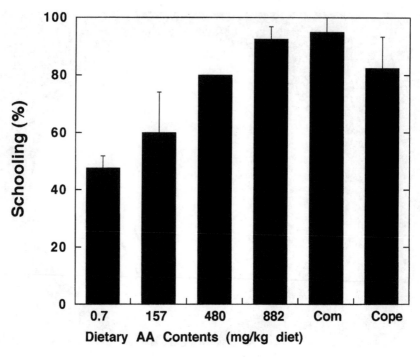

Figure 16.7 Schooling rate (%) of yellowtail fed diets containing different contest of AA for 20 days.

16.4   Conclusion

Since our studies demonstrated that the complex behavior such as schooling and tilting, being governed by the central nervous system, was affected by vitamin C and DHA intake, we suggest that both compounds might act as the modulator of neurotransmitters in fish. Insufficient intake of these nutrients might cause their abnormal distribution in the brain, which may lead to problems in the nervous system related to the control of behavior. We would also like to point out that by controlling dietary vitamin C or DHA contents it is possible for us to improve fish quality, and therefore the high survival and quick adaptation to a wild settling after restocking. However, there is still a limited number of studies on the relationship between nutrient consumption and behavior concerning fish and, therefore, further research is necessary to fully understand the mechanisms in this area of fish physiology.

Acknowledgment

The author would like to thank Dr. Y. Sakakura, Dr. K. Tsukamoto, Dr. J. Blom, Dr. K. Dabrowski, Mr. Y. Iida, Mr. T. Kida, and the students of the Aquatic Animal Nutrition Lab at the Faculty of Fisheries, Kagoshima University. Without their kind assistance this chapter could not have been written.

References

1. Nakano, H., Criteria for evaluating healthy fry for release, in *Healthy Fry for Release, and Their Production Techniques*, Vol. 93, Kitajima, C., Ed., Koseishakoseikaku, Tokyo, 1993, 9. (in Japanese)
2. Tsukamoto, K., Fry quality, in *Healthy Fry for Release, and Their Production Techniques*, Vol. 93, Kitajima, C., Ed., Koseishakouseikaku, Tokyo, 1993, 102. (in Japanese)
3. Uchida, K., Tsukamoto, K., and Kajihara, T., Effects of environmental factors on jumping behaviour of the juvenile Ayu Plecoglossus altivelis with special reference to their upstream migration, *Nippon Suisan Gakkaishi*, 56, 1393, 1990.
4. Tsukamoto, K., Uchida, K., Murakami, Y., Endo, M., and Kajihara, T., Density effect on jumping behavior and swimming upstream in the ayu juveniles, *Nippon Suisan Gakkaishi*, 51, 323, 1985.
5. Uchida, K., Kuwada, H., and Tsukamoto, K., Tilting behavior, a fear response to frightening stimuli, AA possible predictive index for stocking effectiveness in the juveniles of red sea bream *Pagrus major, Nippon Suisan Gakkaishi*, 59, 991, 1993.
6. Yamaoka, K., Yamamoto, E., and Taniguchi, N., Tilting behavior and its learning in juvenile Red Sea Bream, *Pagrus major, Bull. Mar. Sci. Fish. Kochi Univ.*, 14, 63, 1994.
7. Furuta, S., Predation on juvenile Japanese flounder (*Paralichthys olivaceus*) by diurnal piscivorous fish: Field observations and laboratory experiments, in *Survival Strategies in Early Life Stages of Marine Resources*, Watanabe, Y., Yamashita, Y., and Oozeki, Y., Eds., Balkema, Rotterdam, 1996, 285.
8. Ikeda, S., Sato, M., and Kimura, R., Biochemical studies on L-ascorbic acid in aquatic animals-II. Distribution in various parts of fish, *Bull. Jpn. Soc. Sci. Fish.*, 29, 765, 1963.
9. Kuo, C., Hata, F., Yoshida, H., Yamatodani, A., and Wada, H., Effect of ascorbic acid on release of acetylcholine from synaptic vesicles prepared from different species of animals and release of noradrenaline from synaptic vesicles of rat brain, *Life Science*, 24, 911, 1979.
10. Gardiner, T. W., Armstrong-James, M., Caan, A. W., Wightman, R. M., and Rebec, G. V., Modulation of neostriatal activity by iontophoresis of ascorbic acid, *Brain Research*, 344, 181, 1985.
11. Rebec, G. V. and Pierce, R. C., A vitamin as neuromodulator: ascorbate release into the extracellular fluid of the brain regulates dopaminergic and glutamatergic transmission, *Prog. Neurobio.*, 43, 537, 1994.
12. Koshio, S., Sakakura, Y., Iida, Y., Tsukamoto, K., Kida, T., and Dabrowski, K., The effect of vitamin C intake on schooling behavior of amphi-dromous fish, ayu (*Plecoglossus altivelis*), *Fish. Sci.*, 63, 619, 1998.
13. Tsukamoto, K. and Uchida, K., Spacing and jumping behaviour of the ayu, *Plecoglossus altivelis, Nippon Suisan Gakkaishi*, 56, 1383, 1990.
14. Tsukamoto, K., Masuda, S., Endo, M., and Otake, T., Behavioural characteristics of the ayu, *Plecoglossus altivelis*, as predictive indices for stocking effectiveness in rivers, *Nippon Suisan Gakkaishi*, 56, 1177, 1990.
15. Koshio, S., Sakakura, Y., Iida, Y., Tsukamoto, K., and Kida, T., The role of vitamin C on the improvement of fish quality for stock enhancement—The effect of dietary vitamin C on Ayu juveniles, Fall Meeting of the Japanese Society of Fisheries Science, Kyoto, Sep. 27–30, 1995 (Abstract in Japanese).

16. Sakakura, Y., Koshio, S., Iida, Y., Tsukamoto, K., Kida, T., and Blom, J. H., Dietary vitamin C improves the quality of yellowtail (*Seriola quinqueradiata*) seedlings, *Aquaculture*, 161, 427, 1998.
17. Koshio, S., Blom, J. H., Nomoto, S., Tsukamoto, K., Kida, T., Ishikawa, M., and Teshima, S., The role of vitamin C on the improvement of fish quality for stock enhancement—the effect of dietary vitamin C and squid liver oil on the tilting behavor of red sea bream juveniles, Fall Meeting of the Japanese Society of Fisheries Science, Hiroshima, Sep. 27–30, 1997 (Abstract in Japanese).

chapter seventeen

History, present, and future of ascorbic acid research in aquatic organisms

Konrad Dabrowski

Contents

Abstract

Barbara McLaren has to be credited for her 1947 discovery of ascorbic acid essentiality in teleost fish. This finding was a "scientific surprise" that was not uniformly accepted until 1969, when a study with salmonid fish demonstrated a decrease in tissue ascorbate concentration when fed a diet devoid of vitamin C. Many aspects of vitamin C metabolism in fish remain controversial today and this chapter focuses on some results obtained in salmonid fishes related to synthesis, deposition, hydrolysis, and availability of ascorbyl sulfate. Ascorbate secretion in gastric juice and reabsorption in the intestine are not unique to mammals although stomachless and stomach-possessing fish may handle ascorbate differently. The criteria for establishing ascorbate requirement in fish were reviewed and it was concluded that enormous plasticity in fish stemming from the nutritional specialization and environmental tolerance make it particularly difficult to come up with a uniform set of conditions for the purpose of comparative physiology. The mechanism of ascorbate action must include the hypothesis put forward by B. Peterkofsky and collaborators in 1991, according to which the insulin-like growth factor-binding proteins are elevated in scorbutic animals and binding to cellular receptors inhibits collagen, proteoglycan, and DNA synthesis. Studies should be continued to test a possibility of fish-to-fish (acipenserid to teleost) transfer of the gulonolactone oxidase gene through transgenic production technology.

17.1 Introduction

The purpose of this chapter is to summarize materials which seemed not to be covered by specific chapters but are of great value to the overall presentation of the status of ascorbate in aquatic organisms. Therefore, I concentrate on some aspects of the discovery of ascorbate essentiality to fish, areas concerned with the role of ascorbate in fish which are controversial, and finally I signal the areas where major progress can be made in understanding functions of vitamin C.

17.2 History

17.2.1 Discovery of ascorbic acid function in fish

Whether ascorbic acid is produced by renal or hepatic synthesis in most vertebrate animals or must be supplied in the diet of primates or teleost fish, it is transported to all peripheral tissues where it serves a vital role as an antioxidant, a cofactor in protein synthesis and amidation of peptides. Interest in the rapid depletion of this substance in humans, which resulted in severe diseases[1] and was known for several centuries, climaxed at the beginning of the twentieth century, first in discovering an animal model guinea pig,[2] essential for experimental studies, and later in the identification of the antiscorbutic property of hexuronic acid.[3] Independently, a chemical was isolated from lemons by King and Waugh.[4] Identification of the chemical nature of hexuronic acid was published simultaneously by Haworth et al.[5] following a description of experiments with guinea pigs carried out at the University of Szeged, Hungary, by Szent-Gyorgyi and his associates. In 1933, Szent-Gyorgyi[6] proposed the new name of the compound, ascorbic acid (AA). The discovery of AA sparked considerable public interest, and scientific controversies were not resolved until today, although most researchers agree[7] that the credit should go to both Szent-Gyorgyi and King. Vitamin C was synthesized chemically in 1933 by Reichstein et al.,[8] and in 1937 Szent-Gyorgyi, Haworth, and Reichstein were awarded the Nobel Prize.

The first observation of brook trout with deformed bodies (lordosis) and ventral sores was made by McCay and Tunison[9] when fish were fed for 44 weeks with meat preserved in 1% formalin. The other dietary ingredients were (one third each) cottonseed meal and dry skim milk. As Dr. G. Rumsey wrote (personal communication 1995)," . . . wonder when McCay and Tunison realized what they were working with?" It was not until McLaren et al.[10] (Figure 17.1) demonstrated with rainbow trout that lack of vitamin C in the diet resulted in increased mortality, decreased body gain, and hemorrhagic liver, kidney, and intestine. Using a modern semipurified diet consisting of 52% casein, 10% gelatin, 18% dextrin, corn and cod liver oil, minerals,

Figure 17.1 Dr. Barbara A. McLaren, discoverer of ascorbic acid essentiality in the diet of rainbow trout at the University of Wisconsin, Madison, in 1947. The picture was taken in Pullman, WA, before 1953. Later McLaren became the Chair of the Department of Household Science at the University of Toronto, Canada.

and vitamins, McLaren et al.[10] were able to show that 250 mg ascorbic acid per 1 kg diet was sufficient to satisfy the requirements of juvenile rainbow trout in 16 week-long studies.

Other studies failed to confirm ascorbic acid essentiality in salmonids[11] although a conclusive report by Kitamura et al.,[12] which has shown scoliosis and lordosis in three species of teleosts, rainbow trout, common carp, and guppy, when fed a vitamin C-devoid, a casein-gelatin diet, was the final proof that unlike many other animals, many species of fish are scurvy-prone. Poston's[13] work with brook trout and Halver et al.'s[14] work with coho salmon and rainbow trout further confirmed and expanded information on specific vitamin C-related pathologies and their roles in wound healing. Halver's work is also significant in vitamin C research in fish because it provides, for the first time, tissue concentrations of AA when fed graded levels of dietary vitamin C and points out interspecies differences (rainbow trout and coho salmon) in respect to the severity of vitamin deficiency. It is also characteristic that AA concentrations in the whole blood of coho salmon after 24 months of feeding a diet devoid of vitamin C, 22.3 ± 2.2 µg/ml, are enormously high for these conditions. The inaccuracy of the colorimetric method used is likely the cause of this erratic result.

Other reports appeared soon, in 1970, that further documented diseases in practical hatchery conditions (scoliosis and lordosis) which could be related to vitamin C deficiency.[15] Although, again, ascorbic acid concentrations were listed in diets and fish tissues, the levels were compromised by analytical inadequacies. Primbs and Sinnhuber[16] made a claim that ascorbic acid is not essential to rainbow trout and that pathologies such as scurvy symptoms in skeleton, and cartilage (gills) observed were due to hypervitaminosis A. In effect, the authors were half-correct; abnormality in skeletal formation due to a high level of dietary vitamin A was demonstrated many years later in marine fish, Japanese flounder.[17] No differences in growth of rainbow trout were noticed, although fish increased weight eight-fold. Primbs and Sinnhuber[16] used a colorimetric, dinitrophenylhydrazine (DNPH) method and have shown clear disparity in concentrations in several tissues between control (vitamin C 1200 mg/kg) and test fish. The authors interpreted those differences based on finding that vitamin A inhibits AA synthesis and no previous studies addressed this aspect of fish nutrition. The time proved that Primbs and Sinnhuber[16] were mistaken, because salmonids indeed do not express gulonlactone oxidase and require an exogenous source of vitamin C.

17.3   Present

17.3.1   Progress in AA analysis

We have reviewed AA analysis in fish tissues including sensitivity and possible interferences when colorimetric methods are used without the necessary controls and high pressure liquid chromatography (HPLC) methods are relying exclusively on retention time.[18] Interferences from uric acid, hypoxanthine, or xanthine in HPLC determinations are possible among other substances. Since then, several modifications were introduced that allow (1) simultaneous analysis of AA and its esters[19–21] and (2) AA and dehydroascorbate.[22–23] Simultaneous analysis of reduced AA and oxidized ascorbate (DHA) is particularly of value as controversies on the physiological relevance of AA/DHA ratio continue.[24] Moeslinger et al.[25] added a new method for physiological fluids analysis of DHA (or total AA following ascorbate oxidase reaction) where methanol in a phosphate buffer, pH 6, produces carbonyl groups in C2 and C3 positions of DHA, then monitored at 346 nm. According to the authors, by removing reactive iron with desferrioxamine solution, samples were stable up to 24 h at 4°C. Koshishi et al.[26] provided significant information to our understanding of DHA behavior *in vivo* in body fluids. These authors utilized a postcolumn derivatization with benzamidine and simultaneous detection with fluorescence spectrophotometry of AA, DHA, and 2, 3-diketogulonate. Bicarbonate present in rat blood plasma (20–40 mM) promotes DHA degradation and nearly 85% of intravenously injected DHA is recovered in urine after lacton ring cleavage.

17.3.2 *Presence and utilization of ascorbyl sulfate (AAS) in fish*

Ascorbyl sulfate is naturally occurring vitamin C ester in aquatic organisms, best known from cysts of brine shrimp (*Artemia salina*).[27] Oral administration of AAS to humans (6–7.5 g/man) resulted in 2.5% being excreted in the urine in 48 h, but most was excreted in feces.[28] The same authors did not find significant increase in urinary free AA and concluded that only 0.1 % was hydrolyzed. In monkey, Machlin et al.[29] used oral and intramuscular administration but AAS resulted in body weight loss and a decrease in blood ascorbic acid. Signs of scurvy (gingivitis, hemorrhage of skin and gums) developed in animals receiving AAS. In the review of 30 years' research on AAS activity of vitamin C, Tsujimura[30] stated that fish, in contrast to mammals, are capable of AAS hydrolysis. However, Tsujimura failed to consider evidence of high concentrations of AAS measured in feces of rainbow trout and its marginal absorbability,[31, 32] and concentrated on repeating data from oral administration of large doses of AAS. Figure 17.2 summarizes data on tissue concentrations of AA in rainbow trout fed diets with two ascorbyl esters along with controls groups supplemented with free AA or a diet devoid of an ascorbate source. We have reviewed the evidence contrary to earlier work on AAS in fish[23] and I will concentrate on results published after 1994.

Halver and Hardy,[33] in a 17-week-long study on Atlantic salmon fed semipurified diet (casein–gelatin–dextrin based) indicated that an equivalent of 50 mg/kg of AA in the form of AAS eliminated growth differences and vitamin C-related pathologies in comparison to fish fed vitamin C-devoid diet. This result would not be surprising in the light of Cho and Cowey's[34] conclusions that salmonid fish do not require more than 10 mg/kg diet of available form of vitamin C and growth depression was not evident after a

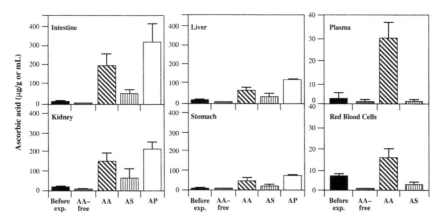

Figure 17.2   Ascorbic acid concentrations in tissues of rainbow trout fed diets devoid of vitamin C (AA-free) or diets supplemented with an equivalent of 500 mg AA/kg in the form of "unprotected" ascorbate (AA), ascorbyl sulphate (AS) or ascorbyl phosphate (AP).[31, 32]

30-fold body increase. In the same study, equivalent of 15–30 mg AA in the form of AAS caused severe mortality and a significant decrease in tissue AA concentrations. Going back to Halver and Hardy's[33] results, where different preparations of AAS were used over the years, some containing significant amounts of free AA; therefore, 50 mg/kg seemed to be a quite satisfactory dose in the case of 12-week duration experiments allowing for ten-fold body weight increase. We have also demonstrated that some limited hydrolysing capacity of AAS exists in the livers of salmonids and other fishes.[31]

Two other results from the work of Halver and Hardy[33] make a satisfactory explanation of why AAS bioavailability in salmon is unlikely. First, no increase in liver AA concentrations of salmon was found as dietary dose of AAS increased from 50 to 100, to 300 mg/kg, and deficient fish had higher AA level (17.1 μg/g) than those fed a diet with 100 mg/kg AAS (10.2 μg/g). The trend and absolute concentrations clearly indicated vitamin C deficiency. Second, AAS concentrations in carcasses of fish without a dietary source of vitamin C and those fed only dietary AA seemed to suggest again that there is a naturally occurring AAS, synthesized by the fish, stored in tissues. I can only refer to an earlier admission of the analytical laboratory providing these results, that "exact quantification (of AAS and AA) was not possible due to interfering co-eluted substances."[35]

Results of Felton et al.[36] on AAS utilization have a lot of weight in the controversy aimed at establishing the biological value of this substance in fish because they rekindle earlier claims that AAS is transformed in the fish body to AA and vice versa and because two techniques are used that should help to resolve the problem for good, "the latest high-performance liquid chromatography (HPLC) and electron ionization mass spectrometry

(EIMS)." AAS was force-fed to juvenile fish in pharmacological doses, 45 mg per 77 g, equivalent to 584 mg/kg. This would be 584 times higher than the requirement, estimated by Hilton et al.[37] at 1 mg/kg body weight in salmonids. It is difficult to reconcile that after this treatment concentration of AA varied in blood between 0 and 565 µg/g. In some treatments concentrations of AAS in the liver reached 2364 µg/g. What physiological meaning do those results have when in control fish concentrations of AAS in most tissues including muscle were not detectable? Consequently, the claim made by the authors that AAS can be absorbed through the gastric tissue and converted to AA has no basis, as the blood level of most of the fish force-fed with AAS has shown 0 concentration of AA. It raises even more questions about the analytical procedures of Felton et al. since we had never observed healthy salmonids without ascorbic acid in their blood plasma.[36]

Kittakoop et al.[19] and Shiau and Hsu's[39] studies with tiger prawn and tilapia hybrid, respectively, indicated ascorbyl sulfate sulfohydrolase activity in the liver of these two species. Utilization of AAS was inferior to equivalent amounts of ascorbyl monophosphate based on response measured from tissue AA concentrations. Purity of the AAS source was not examined. Shiau and Hsu[39] also indicated presence of AAS in liver and muscle tissues of tilapia fed a diet devoid of vitamin C, which are surprising results and contradict earlier findings in this species.

To summarize, it is evident that AAS utilization in fish is limited at best and large doses might provide some ascorbic acid via sulfohydrolase activity in the liver. However, in comparison to ascorbyl phosphates, where most hydrolysis takes place at the intestinal level and results in absorption of free AA into circulation,[38] a limited amount of AAS is transported via the concentration gradient as membrane carriers for AA and dehydroascorbic acid are highly specific[40] and ester form would be in competition with glucose for a common transporter. In the presence of physiological concentration of hexoses, only extremely high doses of AAS[36] may be transported into circulation.

17.3.3 Enzymatic synthesis of ascorbyl esters

Ascorbic acid-2-sulfate (AAS) was identified in cysts of brine shrimp, suggesting that ascorbic acid sulphotransferase must be involved in deposition of the stable ester of vitamin C.[41] After cyst hydration and the onset of their embryonic development, AAS is hydrolyzed to release free ascorbic acid (AA).[18]

The sulfation of ascorbic acid with ATP as an obligatory cofactor was analyzed in rat liver and intestine homogenates.[42] In this assay AAS formed was estimated in a supernatrent fraction ($12,000 \times g$) after the addition of $BaCl_2$ which precipitated Na_2SO_4 or 3-phosphoadenyl-sulfate (PAPS) (99.9%), as sulfate donors. In the presence of 5 mM $NA_2 SO_4$ or 0.25 mM PAPS and 1 mM AA the sulfation reaction was very efficient. We have reviewed information on the distribution of AAS in fish tissues and there is no direct evidence of synthesis or AA-sulfotransferase activity. Baker et al.[43] claimed AAS to be a urinary metabolite of vitamin C in humans and spawned interest in the biological activity of this component and parallel studies in fish; however, even the most deserving model of AAS synthesis in *Artamia* was not explored any further. Biosynthesis of the activated sulfate donor, PAPS, was investigated in the mammalian system and confirmed that the sequential action of two enzymes (I. ATP + sulfate and II. APS phosphorylation) resides on one bifunctional protein.[44]

Ascorbic acid-2-glucuronide was isolated in the urine and plasma from patients suffering from uremia.[45] The authors suggested that the compound where glucuronic acid coupled to the C_2 of ascorbic acid amounts to 0.5 mg/L. The role of this compound needs further investigation and *in vitro* glucuronidation with microsomal fraction of cells[46] seemed to be the reliable test for its physiological importance.

17.3.4 Gastric and enterointestinal ascorbate circulation

We have revised the evidence of basal secretion of ascorbic acid into fish stomach (salmonids) and intestine (stomachless cyprinids).[47] There is experimental evidence in humans that the concentration of AA in gastric secretion highly correlates with blood plasma concentration and furthermore, plasma vitamin C (35 μM/l).[48] One of the major findings suggesting a 20 to 50-fold increase of ascorbate concentrations in feces in case of gastrointestinal disorders is rarely brought up, although it has enormous implications. Rathbone et al.[49] calculated that taking into account average gastric juice secretion, approximately 60 mg AA is secreted daily. Dabrowski[47] concluded that in both scurvy-prone human and teleost fish, the amount secreted daily is almost twice the recommended daily allowance, implying that gastrointestinal circulation and reabsorption are critical to maintain physiologically required AA concentration in the body.

Fish fed ascorbate-free diets maintained detectable levels of ascorbate in the gastrointestinal lumen. More precisely, most vitamin C in the digestive tract was in the oxidized form.[18] The function of DHA as a chemopreventive agent and its role in nutrient absorption and/or availability needs further investigation. Naito et al.[50] demonstrated that a mixture of physiological concentrations of ascorbic acid (40–4000 μM) and ferrous sulphate injected into gastric mucosa resulted in gastric ulcers in a dose-dependent manner, whereas each of the compounds, injected alone, did not form gastric ulcers. Superoxide radicals were part of the mechanism of ulcer formation as the superoxide dismutase (SOD) reduced gastric injuries. Clearly, the finding of presence of dehydroascorbate in the alkaline intestinal environment would support the hypothesis of noninvolvement of vitamin C in oxygen radical-mediated lipid peroxidation.

Ascorbate acts as an important extracellular scavenger in the digestive tract, protecting gastric mucosa and reducing impact after injuries such as bleeding.[51] A dose of 1 mM ascorbate in blood plasma was effective in reducing the impact of hemorrhage shock in rats. Therefore, ascorbate pretreatment of steady dietary dose, resulting in maintaining high concentrations in circulation (normal level in blood plasma of fish would be 0.01–0.05 mM) will prevent possible pathogenic situations.

17.3.5   Interaction of ascorbate and flavonoids

Most *in vitro* evidence in scurvy-prone animals associates uptake of ascorbate in the intestine with Na-dependent absorption against a concentration gradient.[52] The large proportion of total ascorbate in formulated diets in the form of ascorbyl ester for fish will make the process more elaborate, involving brush border enzymes, alkaline phosphatase (Fig. 17.3). The released ascorbic acid will face immediate oxidation and possibly the mechanism

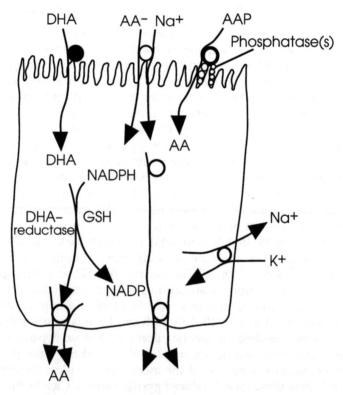

Figure 17.3   A working model describing absorption of ascorbic acid and its derivatives in the epithelial cell of digestive tract of teleost fish. Mucosal transport involves active, AA-Na dependent; DHA-transporter facilitated, Na-independent, and brush-border located phosphatase(s) activity involved in hydrolysis and transport of AA. Serosal transport involves (only) AA-Na dependent movement of ascorbate. The role of GSH (glutathione) and DHA-cytosole reductase is shown.

involved in transport of DHA into the enterocytes is quantitatively more important than any other. GLUT-1 is a primary DHA transporter expressed in the intestine, however, its transport is inhibited by D-glucose, while fructose and L-glucose showed no inhibition.[53] The apparent Km for the DHA transporter is 1.5 mM. The intestinal epithelium is also a site of ascorbate transporter, SVCT-1,[40] whereas SVCT-2 was found in gastric glands and might be involved in basolateral uptake of ascorbate for secretion into the gastric juice.

These findings are important because by characterization of ascorbate/DHA transporter systems, an interaction with other substances might be better understood. Plant-derived isoflavonoids are common ingredients of fish diets supplemented with soybean, rapeseed, or cottonseed meals.[54] Some isoflavonoids in the aglycone form, such as genistein, were recovered in bile that suggests effective hydrolysis and absorption. Vera et al.[55] demonstrated with cells expressing the GLUT-1 transporter that 50% inhibition of DHA was realized at 10–15 µM concentration of genistein within 1–10 min of preincubation. The authors have also shown that genistein inhibition of DHA glucose transporters comes from binding into transporter sites in a competitive manner. Therefore, if released into the gut lumen several isoflavonoids may effectively block transporter systems for vitamins. Park and Levine[56] used myricetin as the isoflavonid of choice and at concentrations of 18–22 µM, both DHA and AA transporter systems were inhibited, although different mechanisms must be involved. Na-dependent AA uptake was blocked by noncompetitive inhibition.

The most abundant flavonol in plants, quercetin was also an effective inhibitor of ascorbic acid accumulation in human cancer cells and this flavonoid was more effective than genistein.[57] The degree of inhibition of AA transport into cancer cells was associated with antiproliterative property and can be linked to AA deprivation for cell growth.

In summary, the above data from *in vitro* studies are directly relevant to *in vivo* situations where considerable amounts of flavonoids in fish diets, of plant–protein origin, are used and may affect the bioavailability of many nutrients including vitamin C.

17.3.6   *Requirement of ascorbate in fish*

The criteria for establishing scientifically sound levels of recommended vitamin C in the diets of fish and other aquatic organisms will evolve as conditions of culture, diet formulations, and species of interest change, and new associations between nutrition and health, reproduction, or environmental conditions (oxygen supersaturation, ultraviolet-B light) become proven in laboratory and/or epidemiological observations. Criteria of the nutritional adequacy of vitamin C should include basal needs for practically healthy fish, and saturation kinetics of vitamin C in vital tissues (such as lymphocytes, kidney) as a function of daily dose, although a direct link must be established between such indicator and production (growth, survival, resistance) benefits.[58]

Many of the experiments claiming essentiality of AA in teleost fish should come under justified criticism because of the inadequecy of the duration of study or accuracy of biochemical assays. Most recently, Reddy and Ramesh[59] reported that common carp did not show a decrease in growth of tissue ascorbate concentration after 70 days of feeding a vitamin C-devoid diet in comparison to control. The second most common artifactual evidence comes from experiments using practical ingredients in diet formulations. For example, Phromkunthong et al.[60] used 46% of plant ingredients in diet formulation for zebra fish (Brachydanio renio), whereas Martins[61] and Martins et al.[62] used diets containing 74% plant ingredients and argued that these were vitamin C-free diets. Calculation of the requirements based on such erratic data has a rather negative effect, and comparisons to study executed with more caution bear little relevance. Martins et al.[62] studied frugivorous Amazonian fish[63] where ascorbate concentrations in the food may be the highest found in terrestrial plants. Unfortunately, no data was presented on ascorbate concentration of blood plasma or other tissues. This aspect of interspecies comparison of vitamin C requirements in fish which occupy environment almost constantly at $-1.9°C$ or $+40°C$ water temperatures could be an asset for understanding the mechanisms of vitamin C action. To the contrary Gieseg et al.[64] examined concentration of ascorbate in Antarctic notothenoid fish blood and concluded that the levels (10–54 μM) were significantly higher only in one species than in temperate water fish.

In another Amazonian fish, *Astronotus ocellatus*, the diet of which also includes fruits, growth depression was observed in case of vitamin-C-free, semipurified diets at an extraordinarily low increase of body weight equal to 100%[65] although the cumulative period of deficiency was over 37 weeks. In most temperate water fish species, five to ten-fold body weight increase is needed to demonstrate scorbutic abnormalities in skeleton and gills. In oscars, deformations of jaw and gills filaments resembled those rarely observed in other fish species. Saroglia and Scarano[66] were probably the first who noticed abnormalities or complete lack of a urohyal bone and detachment of the splanchuric cranium in the marine teleost, *Dicentrarchus labrax*. Clinical signs of vitamin C deficiency are well manifested in advanced avitaminosis, however, an earlier detection criteria, such as white blood cells ascorbate concentrations[67] or specific enzyme activity[68] will require analytical work.

In conclusion, the purpose of this discussion was to highlight a conceptual basis for determining AA requirements in fish, acknowledging the enormous diversity of species and their environmental requirements and nutritional histories as well as to avoid most common mistakes in laboratory experiments which lead to unsubstantiated claims or may underestimate the true requirement. Nutrient requirements estimates have implications for fish feed formulations and may affect the practical applications and success of aquaculture industry.

17.4   Future

17.4.1   Mechanism of ascorbate action

Throughout the literature linking ascorbate deficiency with scurvy in fish, the idea has been expressed that collagen "abnormality" and delayed wound healing have been functionally related to poor hydroxylation of proline and lysine and consequently to the formation of unstable molecules of triple-helical procollagen. Concentrations of hydroxyproline residues in fish tissues were correlated to vitamin C-deficiency.[69] Homogenous preparations of the enzyme prolyl-hydroxylase from tissues of birds, mammals, and cultured cells[70] indicated that these enzymes required ferrous iron and that ascorbate acted as a reductant. The same authors reviewed the variability and contradictions related to the effect of ascorbate on (1) collagen synthesis and (2) proline and lysine hydroxylation. They also discussed possible reasons for the confusing results and listed a number of factors which were unmonitored and uncontrolled. Chojkier et al.[71] showed for the first time that calvaria cells from scorbutic guinea pigs maintained in culture displayed a decrease in the rate of collagen synthesis and this was unrelated to a decrease of proline residue hydroxylation. Guinea pigs losing weight on restricted feeding but provided with ascorbate showed a decline in collagen synthesis. At that point, investigators concluded that collagen mRNA levels were down-regulated in starving and ascorbate-deficient animals. There is now evidence that the mode of action of ascorbate associated with synthesis of collagen and collagen metabolism is complex, including gene transcription, cell proliferation, stabilization of procollagen mRNA, post-tranlational changes, and collagen secretion.

The idea that a humoral factor other than ascorbate is involved in collagen and proteoglycan synthesis in scurvy-prone animals was proposed by Beverly Peterkofsky in 1991.[72] Earlier experiments demonstrated that nutritionally inflicted deficiency of ascorbate in scorbutic guinea pigs could be transmitted via a medium containing sera to cell cultures, and would result in defects in the synthesis of collagen and proteoglycan. This clearly indicated that inhibition in collagen synthesis was not due to a missing factor (ascorbate) but to the presence of an inhibitor. This inhibition could be reversed by insulin-like growth factor-I (IGF-I). Peterkofsky and collaborators[72] have shown that a 10-fold increase in sera from scobutic guinea pigs of two proteins is capable of inhibiting the binding of IGF (IGF-BP) to its cellular receptors (Figure 17.4).

Kipp et al.[73] provided evidence of multiple phases in vitamin C deficiency development where the initial step involved a decrease in concentrations of procollagens mRNA by 26 to 40% when there was no weight loss or induction of IGF-BP observed. The decline of collagen synthesis is observed already when proline undergoes hydroxylation at a normal rate. Defective blood vessel formation and excessive angiogenesis also preceeds inhibition of proline hydroxylation. This time line of events related to ascorbate deficiency not only explains some discrepancies in earlier observations in different experiments, it also suggests that *in vivo* hydroxylation of proline and lysine will be supported at extremely low levels of ascorbate, and that earlier observations of reduced levels of hydroxyproline due to vitamin C deficiency were probably due to a decrease of collagen synthesis.

Synthesis and secretion of different collagens is affected by ascorbate and/or chelation of iron depending on the position and extent of proline hydroxylation. Type IV collagen, a major protein constituent of cellular basement membranes, has a characteristically high content of 4-hydroxyproline.[74] In cells producing exclusively type IV collagen, Fe-chelator (α, α-dipyridyl) and ascorbate absence completely inhibited collagen secretion. Interestingly, collagen synthesis was not stimulated by ascorbate "recovery" even after two days of treatment. Because type IV collagen contains segments of non-collagenous sequences, the unhydroxylated chain is the substrate to protease digestion resulting in the degradation of cellular (and vascular) structures.

Further confirmation of a decrease in type IV collagen mRNA concentration in serum came from experiments with scorbutic guinea pigs.[75] However, severely scorbutic animals with a weight loss of approximately 20% had significantly decreased levels of collagen mRNA (57% of normal) and elastin mRNA (3% of normal) in blood vessels. In food-deprived guinea pigs, elastin mRNA decreased to 34.5% of the normal value, indicating high specificity of this biochemical defect.

In conclusion, by explaining the complexity of vitamin C action and addressing the mechanism(s) of its function beyond antioxidant and enzyme co-activator roles, we will be able to provide a more specific and sensitive functional index of vitamin C status in animals.

17.4.2 Molecular (gene expression) regulation by ascorbate

Ascorbate deficiency impairs extracellular bone matrix synthesis (secretion), resulting in abnormal skeletal formation, but also evidently affects cell differentiation. Ragab et al.[76] reasoned that if precursor cells obtained from spleen or bone marrow require ascorbic acid in the medium as an essential component of acquiring certain functions (such as specific acid phosphatase activity), this could be a strong indication that control of osteoclast differentiation will also impact bone resorption-osteoporosis. Indeed, they found that relatively high (physiological) concentrations of ascorbate are required (12.5-25 μg/mL) for osteoclast differentiation, but vitamins also strongly increase the life span of precursor cells and matured osteoclasts. Very appropriately, new *in vitro* studies increasingly use ascorbyl phosphate as a vitamin C source,[77] avoiding ascorbate degradation in alkaline culture mediums prior to cellular absorption. Ascorbyl phosphate has been reported to increase the proliferation of human fibroblasts *in vitro*, whereas rat dermal fibroblasts (animal synthesizing ascorbate) responded to the vitamin C source by elevated DNA synthesis, but no enhanced proliferation or differentiation (secre-

A   **B**

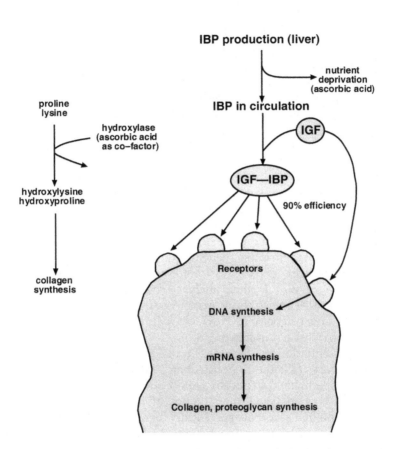

Figure 17.4   The mechanism of ascorbate function(s) at cellular level. A. Ascorbate as co-factor in hydroxylase activity leading to formation of hydroxylated amino acids. This leads to a significant decrease in collagen and cartilage proteoglycan synthesis, although effect of vitamin C deficiency and fasting cannot be separated. B. Illustration of Beverly Peterkofsky[69] hypothesis of ascorbate-deficiency affecting insulin-like growth factor-I (IGF) inhibitory binding proteins (IBP) synthesis and their level in circulation. In scorbutic guinea pigs IBP activity increased 10-fold in comparison to normal animals. IBP inhibited IGF binding to its cellular receptors and consequently downregulated synthesis of collagen(s).

tory activity) was observed.[78] Furthermore, 17.6 and 35 μg/mL of free ascorbic acid inhibited human fibroblast proliferation at 50 and 95%, respectively. The author associated this with ascorbic acid cytotoxicity, as equimolar concentrations of ascorbyl phosphate made no difference to cell growth. This is a puzzling result, as ascorbate degradation during the nine day cell culture or the accumulation of dehydroascorbate might have been a cause of the "toxicity".

Several of those models will most likely apply to fish cell cultures. Lee et al.[79] used goldfish skin cell lines to demonstrate the essentiality of ascorbic acid (50 μg/mL) (see above) to induce collagen secretion, whereas at lower concentration even after 21 days of culture, no fiber formation was detected with histological methods. Interestingly, two other cell lines from rainbow trout tissues were unresponsive to ascorbate presence in terms of collagen formation.

In vitro cell culture requires several days for the induction of a specific gene. Highly responsive cells such as osteoblasts require ascorbate to express osteocalcin-calcium binding protein. Xiao et al.[80] transfected mouse pre-osteoblasts with osteocalcin gene promoter driving expression of firefly luciferase. Ascorbate stimulated luciferase activity 20-fold, and this response was stereospecific. These investigators identified a specific region in the nucleotide sequence responsible for ascorbate-dependent induction, and demonstrated that there is a sequence requiring collagen matrix synthesis followed by osteocalcin gene expression for ascorbate-dependent induction to occur. Otsuka et al.[81] further analyzed ascorbate action specificity on osteoblasts and was able to show an increase in mRNA expression of alkaline phosphatase and osteocalcin. However, expression of one morphogenetic protein mRNA, was not induced.

Cell differentiation and the role of transforming growth factor (TGF) may in some cell types take the right direction, and in human dermal fibroblasts TGF and ascorbate have an additive effect on collagen synthesis.[82] However, when mRNA was analyzed in a situation of TGF and ascorbate addition it became evident that ascorbate specifically increases collagen type I synthesis and promotes post-transcriptional changes, whereas TGF increases overall protein synthesis.

Mahmoodian et al.[83] indicated that "scurvy effect" differed in many facets from "fasting effect" in guinea pigs subjected to vitamin C deficiency. Their results indicated a fall of mRNA concentrations for osteocalcin, type I collagen, and alkaline phosphatase. Furthermore, these authors found that bone alkaline phosphatase activity decreased to 31% of control when liver enzymes increased to 110% of control following a minimal loss (7%) of body weight. Investigators associated this with the specific response of alkaline phosphatase isoenzymes in different tissues. Independently, Matusiewicz and Dabrowski[38] found that a similar mechanism operated in rainbow trout, where vitamin C deficiency resulted in a liver/bone alkaline phosphatase activity ratio of 2.27 in comparison to 0.21 in fish on high dietary ascorbate status.

Experiments performed with rats carrying hereditary defect in GLO expression are particularly illuminating examples of the effect of ascorbate deficiency on gene expression not directly related to collagen formation and bone structure. Ikeda et al.[84] suggested that the expression of genes located in hepatic tissues in vitamin C-deficient rats bears similarity to the acute inflammation stage. The serum concentrations of haptoglobin and alfa-acid glycoprotein mRNA as well as the activity of interleucin (4-fold increase) and concentrations of corticosterone (5-fold increase) were significantly elevated in scorbutic animals. The time needed to downregulate expression of these genes was 14 days, whereas low hepatic ascorbate levels were reached on day 7 of dietary vitamin C withdrawal, suggesting that this effect is indirect.

Corticosteroids induce cell death, whereas increased ascorbate concentration (up to 100 μg/mL) clearly extended the life span of mice spleen immune-T cells.[85] T cells produce cytokines and become effector cells responding to antigens (viruses). Increased survival of these cells may be key to enhanced immune response and control of inflammatory reaction as a result of elevated levels of vitamin C. It remains debatable whether inhibition of ascorbic acid-induced lipid peroxidation[86] is the only mechanism involved in regulating gene expression by vitamin C. For instance, Farquaharson et al.[87] observed higher levels of 1,25 dihydroxyvitamin D3 concentrations and increased vitamin D3 receptor number. They suggested that the resulting amplification of vitamin D3 promotes chondrocyte differentiation and synthesis of cartilage matrix proteins.

In summary, evidence has been provided for the role of ascorbate in cell proliferation and differentiation, upregulation of cell membrane receptors, synthesis of peptide hormones and gene expression. *In vitro* and *in vivo* studies with teleost fish using the described approaches to understand growth factors, hormonal regulation, and nutrient absorption require urgent consideration.

17.4.3   *Transgenic fish with GLO gene*

We have demonstrated that gulonolactone oxidase (GLO) relatedness-matched actinopterygian fish phylogeny and modern teleosts may have lost the ability to synthesize ascorbic acid in the late Triassic period.[88] Therefore, absence of the GLO protein and GLO activity in teleosts suggests that the lack of transcription of the enzyme is highly specific. Furthermore, Krasnov et al.[89] claimed that small amounts of ascorbate were synthesized when rainbow trout kidney preparations were supplemented with rat GLO. If GLO is the only enzyme missing in teleosts incapable of ascorbic acid biosynthesis, then production of transgenic scurvy-resistant fish would require only the translation and stability of one enzyme. Rainbow trout embryos microinjected with rat GLO construct were found to transcribe the construct after hatching. Abnormalities of the trout hatchlings indicated expression of the foreign gene. However, both Western blotting and GLO activity failed to provide evidence of rat GLO presence in trout.

In major contrast to studies on rainbow trout, Toyohara et al.[90] claimed that rat GLO construct was successfully transferred to genomic DNA of teleost fish, *Oryzias latipes,* and was inherited in the second generation originating from the first transgenic males. Although the authors provide data on GLO activity in the trunk muscle of the transgenic fish, they reported that no differences were noted in ascorbate levels in bodies of controls and those expressing GLO activity.

The assay did not include an appropriate control (D-gulonolactone), and activity recorded (0.08 μM ascorbate/mg protein/hour at 37°C) was two orders of magnitude lower than that found in microsomal preparations of fish synthesizing ascorbate in kidneys (1.5 μM /mg /h at 25°C).[88] Therefore, without confirmation of ascorbate synthesizing ability in *Oryzias,* Toyohara's study with ascorbate-devoid feeds leaves some doubts as to the evidence presented.

Nishikimi and Yagi[91] demonstrated that overall homology in nucleotide sequences of human pseudogene (not expressing) and rat GLO (expressed) was ~80%. Taking into account that the average nucleotide substitution in rodents is 4 to 10 times higher than in primates, the estimated loss of GLO activity can be dated at ~70 million years ago. Given the monophyly of telosts and the presence of GLO activity in the Amiidae family, one can reason that the loss of GLO activity in the ancestors of modern teleosts occurred 200 million years ago.[88] Therefore, the number of substitutions in the GLO gene sequence have not been restricted by selective pressure during teleost evolution over a much longer time. The rat GLO cDNA probe used in the Southern blot analysis of human and guinea pigs genomic DNA gave positive signals,[91] so if the number of mutations accumulated in cold blooded animals is not excessive,[92] the probe may be used with extant actinopterigians. Consequently, GLO gene transfer from fish to fish may have a reasonably higher rate of success and will be ethically acceptable.

References

1. Carpenter, J. K., The history of scurvy and vitamin C, *Cambridge University Press,* Cambridge, U.K., 288, 1986.
2. Holst, A., and Frolich, T., Experimental studies relating to ship beri-beri and scurvy. II. On the etiology of scurvy, *J. Hygiene,* 7, 634, 1907.
3. Svirbely, J. L. and Szent-Gyorgyi, A., Hexuronic acid as the antiscorbutic factor, *Nature,* 129, 576, 1932.
4. King, C. G. and Waugh, W. A., The chemical nature of vitamin C, *Science,* 75, 1944: 357, 1932.
5. Haworth, W. N., Hirst, E. L., and Reynolds, R. J. W., Hexuronic acid as the antiscorbutic factor, *Nature* 129, 3259: 576, 1932.
6. Szent-Gyorgyi, A., Identification of vitamin C, *Nature,* 131, 225, 1933.
7. Stare, F. J. and Stare, I. M., Charles Glen King, 1896-1988, *J. Nutrition,* 118, 1272, 1988.
8. Reichstein, T., Grussner, A., and Oppenheimer, R., synthese der d-und 1-ascorbinsaure (C-vitamin), *Helvet. Chim. Acta,* 26, 1019, 1933.
9. McCay, C. M. and Tunison, A. V., The nutritional requirement of trout, *Annual Report,* Cortland Hatchery, New York Conservation Department, 3, 1933.

10. McLaren, B. A., Keller, E., O'Donnell, D. J., and Elvehjem, C. A., The nutrition of rainbow trout. I. Studies of vitamin requirements, *Arch. Bioch. Biophys.*, 15, 169, 1947.

11. Coates, J. A. and Halver, J. E., Water-soluble vitamin requirements of silver salmon; *Bureau of Sport Fisheries and Wildlife, Report 281*, Washington, DC, 1958.

12. Kitamura, S., Ohara, S., Suwa, T., and Nakagawa, K., Studies on vitamin requirements of rainbow trout, *Salmo gairdneir*-I. On the ascorbic acid, *Bull. Japan. Soc. Sci. Fish.*, 31, 818, 1965.

13. Poston, H. A., Effect of dietary L-ascorbic acid on immature brook trout, *Fish. Res. Bull.*, 35, 46, 1967.

14. Halver, J.E., Ashley, L. M., and Smith, R. M., Ascorbic acid requirements of coho salmon and rainbow trout, *Trans. Am. Fish. Soc.*, 90, 762, 1969.

15. Rucker, R. R., Yasutake, W. T., and Wedemeyer, G., An obscure disease of rainbow trout, *Prog Fish Culturist*, 3, 1970.

16. Primbs, E. R. J. and Sinnhuber, R. O., Evidence for the nonessentiality of ascorbic acid in the diet of rainbow trout, *Prog. Fish Culturist*, 33, 141, 1971.

17. Takeuchi, T., Dedi, J., Ebisawa, C., Watanabe, T., Seikai, T., Hosoya, K., and Nakazoe, J. I., The effect of beta-carotene and vitamin-A envriched artemia nauplii on the malformation and color abnormality of larval Japanese flounder, *Fish. Sci.*, 61, 141, 1995.

18. Dabrowski, K., Matusiewicz, M., and Blom, J. H., Hydrolysis, absorption and bioavailability of ascorbic acid esters in fish, *Aquaculture* 124, 169, 1994.

19. Kittakoop, P., Piyatiratitivorakul, S., and Menasveta, P., Detection of metabolic conversions of ascorbate-2-monophosphate and ascorbate-2-sulfate to ascorbic acid in tiger prawn (*Penaeus monodon*) using high-performance liquid chromatography and colorimetry, *Comp. Biochem. Physiol.*, 113B, 737, 1996.

20. Khaled, M. Y., Simultaneous HPLC analysis of L-ascorbic acid, L-ascorbyl-2-sulfate and L-ascorbyl-2-polyphosphate, *J. Liq. Chrom, Rel. Tech.*, 19, 3105, 1996.

21. Kim, H. R. and Seib, P. A., Assay of L-ascorbic 2-monophosphate in aquaculture feeds by high-performance anion-exchange chromatography with conductivity detection, *J. Chrom. A.*, 803, 141, 1998.

22. Huang, T. and Kissinger, P. T., Simultaneous LC determination of ascorbic acid and dehydroascorbic acid in foods and biological fluids, *BAS*, 29, 1996.

23. Esteve, M. J., Farre, R., Frigola, A., and Garcia-Cantabella, J. M., Determination of ascorbic and dehydroascorbic acids in blood plasma and serum by liquid chromatography, *J. Chromat. B.*, 688, 345, 1997.

24. Bates, C.J., Plasma Vitamin C assays: a European experience, *Int. J. Vit. Nutr. Res.*, 64, 283, 1994.

25. Moeslinger, T., Brunner, M., Volf, I., and Spieckermann, P. G., Spectrophotometric determination of ascorbic acid and dehydroascorbic acid, *Clin. Chem.*, 41, 1177, 1995

26. Koshiishi, I., Mamura, Y., Liu, J., and Imanari, T., Degradation of dehydroascorbate to 2,3-diketogulonate in blood circulation, *Bioch. Biophy. Acta*, 1425, 209, 1998.

27. Dabrowski, K., Administration of gulonolactone does not evoke ascorbic acid synthesis in teleost fish, *Fish Physiol. Biochem.*, 9, 5, 1991.

28. Tsujimura, M., Fukuda, T., and Kasai, T., Studies on the excretion of ascorbic acid 2- sulfate and total vitamin C into human urine after oral administration of ascorbic acid 2-sulfate, *J. Nutr. Sci. Vitamin*, 28, 467, 1982.

29. Machlin, L.J., Garcia, F., Kuenzig, W., and Brin, M., Antiscorbutic activity of ascorbic phosphate in the rhesus monkey and the guinea pig, *Am. J. Clin. Nutr.*, 32, 325, 1976.

30. Tsujimura, M., Thirty-year history of research about vitamin C activity in a naturally occurring ascorbic acid derivative; L-ascorbic acid 2-sulfate, *J. Kagawa Nutrit. Univ.*, 28, 31, 1997.

31. Dabrowski, K., Lackner, R., and Doblander, C., Effect of dietary ascorbate on concentration of tissue ascorbic acid, dehydroascorbic acid, ascorbic sulfate and ascorbic sulfate sulfohydrolase in rainbow trout, *Can. J. Fish. Aquat. Sci.*, 47, 1518, 1990.

32. Dabrowski, K., Comparative bioavailability of ascorbic acid and its stable forms in rainbow trout, in: C. Wenk, R. Fenster, and L. Volker, Eds., *Ascorbic Acid in Animal Nutrition*, Proceedings of the 2nd Symposium in Zurich Ittingen, Switzerland. 344, 1992.

33. Halver, J. E. and Hardy, R. W., L-ascorbyl-2-sulfate alleviates Atlantic salmon scurvy, *Proc. Soc. Exp. Biol. Med.*, 206, 421, 1994.

34. Cho, C. Y. and Cowey, C. B., Utilization of monophosphate ester of ascorbic acid by rainbow trout (*Oncorhynchus mykiss*), in S.J. Kaushik and P. Luquet, Eds., *Fish Nutrition in Practice*, INRA, Paris, 149, 1993.

35. Halver, J. E., Felton, S., and Pallmisano, A. N., Efficiacy of L-ascorbyl-2-sulfate as a vitamin C source for rainbow trout, in S.J. Kaushik and P. Luquet, Eds., *Fish Nutrition in Practice*, INRA, Paris, 137, 1993.

36. Felton, S. P., Dukelow, A., and Felton, H. M., Ascorbyl-2-sulfate compared with ascorbic acid in Atlantic salmon: uptake and distribution confirmed by mass spectroscopy, *Proc. Soc. Exp. Biol. Med.*, 215, 248, 1997.

37. Hilton, J. W., Cho, C. Y., and Slinger, S. J., Effect of graded levels of supplemental ascorbic acid in practical diets fed to rainbow trout (*Salmo gairdneri*), *J. Fish. Res. Board Can.*, 35, 431, 1978.

38. Matusiewicz, M., Dabrowski, K., Volker, L., and Matusiewicz, K., Ascorbate polyphosphate is a bioavailable vitamin C source in juvenile rainbow trout: tissue saturation and compartmentalization model, *J. Nutr.*, 125, 3055, 1995.

39. Shiau, S-Y. and Hsu, T-S., L-ascorbyl-2 sulfate has equal antiscorbutic activity as L- ascorbyl-2 monophosphate for tilapia *Oreochromis miloticus* x *O. aureus*, *Aquaculture*, 133, 147, 1995.

40. Tsukaguchi, H., Tokui, T., MacKenzie, B., Berger, U. V., Chen, X-Z., Wang, Y., Brubaker, R. F., and Hediger, M. A., A family of mammalian Na^+-dependent L-ascorbic acid transporters, *Nature*, 399, 70, 1999.

41. Nelis, H. J., Merchie, G., Lavens, P., Sorgeloos, P., and De Leenheer, A. P., Solid-phase extraction of ascorbic acid 2-sulfate from cysts of the brine shrimp *Artemia franciscana*, *Anal. Chem.*, 66, 1330, 1994.

42. Mohamram, M., Rucker, R. B., and Hodges, R. E., Formation *in vitro* of ascorbic acid 2-sulfate, *Bioch. Bioph. Acta*, 437, 305, 1976.

43. Baker, E. M., Hammer, D. C., March, S. C., Tolbert, B. M., and Canham, J. E., Ascorbate sulfate: a urinary metabolite of ascorbic acid in man, *Science*, 173, 826, 1971.

44. Lyle, S., Ozeran, J. D., Stanczak, J., Westley, J., and Schwartz, N. B., Intermediate channeling between ATP sulfurylase and adenosine 5'-phosphosulfate kinase from rat chondrosarcoma, *Biochemistry*, 33, 6822, 1994.

45. Gallice, P., Sarazin, F., Polverelli, M., Cadet, J., Berland, Y., and Crevat, A., Ascorbic acid-2-0-β-glucuronide, a new metabolite of vitamin C identified in human urine and uremic plasma, *Bioch. Bioph. Acta*, 1199, 305, 1994.

46. Crespy, V., Morand, C., Manach, C., Besson, C., Demigne, C., and Remesy, C., Part of quercetin absorbed in the small intestine is conjugated and further secreted in the intestinal lumen, *Am. J. Physiol.*, 277, G120, 1999.

47. Dabrowski, K., Gastrointestinal circulation of ascorbic acid, *Comp. Biochem. Physiol.*, 95A, 481, 1990.

48. Sobala, G. M., Schorah, C. J., Pignatelli, B., Crabtree, J. E., Martin, I. G., Scott, N., and Quirke, P., High gastric juice ascorbic acid concentrations in members of a gastric cancer family, *Carcinogenesis*, 14, 291, 1993.

49. Rathbone, B. J., Johnson, A. W., Wyatt J. I., Kellcher J., Heatley, R. V., and Losowsky, M. S., Asocrbic acid: a factor concentrated in human gastric juice, *Clin. Sci.*, 76, 237, 1989.

50. Naito, Y., Yoshikawa, T., Yoneta, T., Yagi, N., Matsuyama, K., Arai, M., Tanigawa, T., and Kondo, M., A new gastric ulcer model in rats produced by ferrous iron and ascorbic acid injection, *Digestion*, 56, 472, 1995.

51. Ekman, T., Risberg, B., and Bagge, U., Ascorbate reduces gastric bleeding after hemorrhagic shock and retransfusion in rats, *Europ. Sug. Res.*, 26, 187, 1994.

52. Rose, R., Intestinal absorption of water-soluble vitamins, *Proc. Soc. Esp. Biol. Med.*, 212, 191, 1996.

53. Rumsey, S. C., Kwon, O., Xu, G. W., Burant, C. F., Simpson, I., and Levine, M., Glucose transporter isoforms GLUT1 and GLUT3 transport dehydroascorbic acid, *J. Biol. Chem.*, 272, 18982, 1997.

54. Mambrini, M., Roem, A. J., Cravedi, J. P., Lalles, J. P., and Kaushik, S. J., Effects of replacing fish meal with soy protein concentrate and of DL-methionine supplementation in high-energy extruded diets on growth and nutrient utilization of rainbow trout, *Onchorhynchus mykiss*, *J. Animal Sci.*, 77, 2990, 1999.

55. Vera, J. C., Reyes, A. M., Carcamo, J. G., Velasquez, V., Rivas, C. I., Zhang, R. H., Strobel, P., Iribarren, R., Scher, H. I. Slebe, J. C., and Golde, D. W., Genistein is a natural inhibitor of hexose and dehydroascorbic acid transport through the glucose transporter, GLUT1, *J. Biol. Chem.*, 271, 8719, 1996.

56. Park, J. B. and Levine, M., Intracellular accumulation of ascorbic acid is inhibited by flavonoids via blocking of dehydroascorbic acid and ascorbic acid uptakes in HL-60, U937 and jurkat cells, *J. Nutr.*, 130, 1297, 2000.

57. Kuo, S-M., Morehouse, H. F., and Lin, C-P., Effect of antiproliferative flavonoids on ascorbic acid accumulation in human colon adenocarcinoma cells, *Cancer Letters*, 116, 131, 1997.

58. Young, V. R., Evidence for a recommended dietary allowance for vitamin C from pharmacokinetics: a comment and analysis, *Proc. Natl. Acad. Sci., U.S.A.*, 93, 14344, 1996.

59. Reddy, H. R. V. and Ramesh, T. J., Dietary essentiality of ascorbic acid for common carp *Cyprinus carpio* L., *Indian J. Exper. Biol.*, 34, 1144, 1996.

60. Phromkunthong, W., Boonyaratpalin, M., and Storch, V., Different concentrations of ascorbyl-2-monophosphate-magnesium as dietary sources of vitamin C for seabass, *Lates calcarifer*, *Aquaculture*, 151, 225, 1997.

61. Martins, M. L., Effect of ascorbic acid deficiency on the growth, gill filament lesions and behavior of pacu fry (*Piaractus mesopotamicus* Holmberg, 1887), *Braz. J. Med. Biol. Res.*, 28, 563, 1995.

62. Martins, M. L., Castagnolli, N., Zuim, S. M. F., and Urbinati, E. C., Influencia de diferentes niveis de vitamina C na racao sobre parametros hematologicos de alevinos de *Piaractus mesopotamicus* Holmberg (Osteichthyes, Characidae), *Revista Bras. Zool.*, 12, 609, 1995.

63. Araujo-Lima, C. and Goulding, M., *So Fruitful a Fish. Ecology, Conservation, and Aquaculture of the Amazon's Tambaqui*, Columbia University Press, New York, 191, 1997.

64. Gieseg, S. P., Cuddihy, S., Hill, J. V., and Davison, W., A comparison of plasma vitamin C and E levels in two Antarctic and two temperate water fish species, *Comp. Bioch. Physiol. B*, 125, 371, 2000.

65. Fracalossi, D. M., Allen, M. E., Nicholas, D. K., and Oftedal, O. T., Oscars, *Astronotus ocellatus*, have a dietary requirement for Vitamin C, *J. Nutr.* 128, 1745, 1998.

66. Saroglia, M.G. and Scarano, G., Fabbisogno di vitamina C nella dieta di spigola (*Dicentrarchus labras*) allevata in acqua di mare, *Rivista Italiana di Piscicoltura e Ittiopatologia*, 19, 1, 1984.

67. Verlhac, V. and Gabaudan, J., Influence of vitamin C on the immune system of salmonids, *Aqua. Fish. Mgmt.*, 25, 21, 1994.

68. Matusiewicz, M. and Dabrowski, K., Utilization of the bone/liver alkaline phosphatase activity ratio in blood plasma as an indicator of ascorbate deficiency in salmonid fish, *Proc. Soc. Exp. Biol. Med.*, 212, 44, 1996.

69. Sato, M., Hatano, Y., and Yoshinaka, R., L-ascorbyl 2-sulfate as a dietary vitamin C source for rainbow trout *Oncorhynchus mykiss*, *Nippon Suisan Gakkaishi*, 57, 717, 1991.

70. England, S. and Seifter, S., The biochemical functions of ascorbic acid, *Ann. Rev. Nutr.*, 6, 365, 1986.

71. Chojkier, M., Spanheimer, R., and Peterkofsky, B., Specifically decreased collagen biosynthesis in scurvy dissociated from an effect on proline hydroxylation and correlated with body weight loss, *J. Clin. Invest.*, 72, 826, 1983.

72. Peterkofsky, B., Palka, J., Wilson, S., Takeda, K. and Shah, V., Elevated activity of low molecular weight insulin-like growth factor-binding proteins in sera of vitamin C-deficient and fasted guinea pigs, *Endocrinology*, 128, 1769, 1991.

73. Kipp, D., Wilson, S., Gosiewska, A., and Peterkofsky, B., Differential regulation of collagen gene expression in granulation tissue and nonrepair conective tissues in vitamin C-deficient guinea pigs, *Wound Rep. Reg.*, 3, 192, 1995.

74. Kim, Y-R. and Peterkofsky, B., Differential effects of ascorbate depletion and α, α'-dipyridyl treatment on the stability, but not on the secretion, of type IV collagen in differentiated F9 cells, *J. Cell. Biochem,*. 67, 338, 1997.

75. Mahmoodian, F. and Peterkofsky, B., Vitamin C deficiency in guinea pigs differentially affects the expression of type IV collagen, laminin, and elastin in blood vessels, *J. Nutr.*, 129, 83, 1999.

76. Ragab, A. A., Lavish, S. A., Banks, M. A., Goldberg, V. M., and Greenfield, E. M., Osteoclast differentiation requires ascorbic acid, *J. Bone Miner. Res.*, 13, 970, 1998.

77. Nowak, G. and Schnellmann, R. G., Renal cell regeneration following oxidant exposure: inhibition by TGF-β and stimulation by ascorbic acid, *Toxicol. Appl. Pharmacol.*, 145, 175, 1997.
78. Kato, T. and Royce, P. M., Different responses of human and rat dermal fibroblasts to L-ascorbic acid 2-phosphate, *Biomed. Res.*, 16, 191, 1995.
79. Lee, L. E. J., Caldwell, S. J., and Gibbons, J., Development of a cell line from skin of goldfish, *Carassius auratus*, and effects of ascorbic acid on collagen deposition, *Histochem. J.*, 29, 31, 1997.
80. Xiao, G., Cui, Y., Ducy, P., Karsenty, G., and Franceschi, R. T., Ascorbic acid-dependent activation of the osteocalcin promoter in MC3T3-E1 preosteoblasts: requirement for collagen matrix synthesis and the presence of an intact OSE2 sequence, *Mol. Endocrin.*, 11, 1103, 1997.
81. Otsuka, E., Yamaguchi, A., Hirose, S., and Hagiwara, H., Characterization of osteoblastic differentiation of stromal cell line ST2 that is induced by ascorbic acid, *Am. J. Physiol.*, 277, C132, 1999.
82. Phillips, C. L., Tajima, S., and Pinnell, S. R., Ascorbic acid and transforming growth factor–β1 increase collagen biosynthesis via different mechanisms: coordinate regulation of proα1 (I) and proα1(III) collagens, *Arch. Biochem. Biophys.*, 295, 397, 1992.
83. Mahmoodian, F., Gosiewska, A., and Peterkofsky, B., Regulation and properties of bone alkaline phosphatase during vitamin C deficiency in guinea pigs, *Arch. Biochem. Biophys.*, 336, 86, 1996.
84. Ikeda, S., Horio, F., and Kakinuma, A., Ascorbic acid deficiency changes hepatic gene expression of acute phase proteins in scurvy-prone ODS rats, *J. Nutr.*, 128, 832, 1998.
85. Campbell, J. D., Cole, M., Bunditrutavorn, B., and Vella, A. T., Ascorbic acid is a potent inhibitor of various forms of T cell apoptosis, *Cellul. Immunol.*, 194, 1, 1999.
86. Chojkier, M., Houglum, K., Solis-Herruzo, J., and Brenner, D.A., Stimulation of collagen gene expression by ascorbic acid in cultured human fibroblasts, *J. Biolog. Chem.*, 264, 16957, 1989.
87. Farquharson, C., Berry, J. L., Mawer, E. B., Seawright, E., and Whitehead, C. C., Ascorbic acid-induced chondrocyte terminal differentiation: the role of the extracellular matrix and 1, 25-dihydroxyvitamin D, *Europ. J. Cell Biol.*, 76, 110, 1998.
88. Moreau, R. and Dabrowski, K., Biosynthesis of ascorbic acid by extant actinopterygians, *J. Fish Biol.*, 57, 733, 2000.
89. Krasnov, A., Reinisalo, M., Pitkanen, T. I., Nishikimi, M., and Molsa, H., Expression of rat gene for L-gulono-γ-lactone oxidase, the key enzyme of L-ascorbic acid biosynthesis, in guinea pig cells and in teleost fish rainbow trout (*Oncorhynchus mykiss*), *Bioch. Bioph. Res. Comm.*, 223, 650, 1996.
90. Toyohara, H., Nakata, T., Touhata, K., Hashimoto, H., Kinoshita, M., Sakaguchi, M., Hishikimi, M., Yagi, K., Wakamatsu, Y., and Ozato, K., Transgenic expression of L-gulono-γ-lactone oxidase in medaka (*Oryzias latipes*), a teleost fish that lacks this enzyme necessary for L-ascorbic acid biosynthesis, *Biochem. Biophys. Res. Comm.*, 223, 650, 1996.
91. Nishikimi, M. and Yagi, K., Molecular basis for the deficiency in humans of gulonaolactone exidase, a key enzyme for ascorbic acid biosyntheis, *Am. J. Clin. Nutr.*, 54, 1203S, 1991.

92. Martin, A. P., Naylor, G. J. P., and Palumbi, S. R., Rates of mitochondrial DNA evolution in sharks are slow compared with mammals, *Nature*, 357, 153, 1992.

Index